Epoxy-Based Biocomposites

Epoxy-Based Biocomposites highlights the influence of fibre type, nanofillers, and ageing conditions on the performance of epoxy-based biocomposites subjected to various loading conditions. This book serves as a useful reference for researchers, graduate students, and engineers in the field of polymer composites. In addition to investigating the behaviour of hybrid biocomposites and biocomposites reinforced with various nanofillers, this book discusses the response of epoxy-based biocomposites exposed to moisture absorption, accelerated weathering, and hygrothermal ageing. This book also considers the static and dynamic properties, such as creep, fatigue, and free vibration properties.

Epoxy-Based Biocomposites

Edited by
Chandrasekar Muthukumar, Senthil Muthu Kumar
Thiagamani, Senthilkumar Krishnasamy,
and Ahmad Ilyas Rushdan

CRC Press
Taylor & Francis Group
Boca Raton London New York

CRC Press is an imprint of the
Taylor & Francis Group, an **informa** business

Designed cover image: © Shutterstock

First edition published 2024
by CRC Press
4 Park Square, Milton Park, Abingdon, Oxon, OX14 4RN

and by CRC Press
6000 Broken Sound Parkway NW, Suite 300, Boca Raton, FL 33487-2742

CRC Press is an imprint of Taylor & Francis Group, LLC

© 2024 The right of Chandrasekar Muthukumar, Senthil Muthu Kumar Thiagamani, Senthilkumar Krishnasamy and Ahmad Ilyas Rushdan to be identified as the authors of the editorial material, and of the authors for their individual chapters, has been asserted in accordance with sections 77 and 78 of the Copyright, Designs and Patents Act 1988.

British Library Cataloguing-in-Publication Data
A catalogue record for this book is available from the British Library

Library of Congress Cataloging-in-Publication Data
Names: Muthukumar, Chandrasekar, editor. |
Thiagamani, Senthil Muthu Kumar,
editor. | Krishnasamy, Senthilkumar, editor. |
Bin Rushdan, Ahmad Ilyas, editor.
Title: Epoxy-based biocomposites / edited by Chandrasekar Muthukumar,
Senthil Muthu Kumar Thiagamani, Senthilkumar Krishnasamy, and
Ahmad Ilyas Bin Rushdan.
Description: First edition. | Boca Raton : CRC Press, [2023] |
Includes bibliographical references and index. |
Identifiers: LCCN 2022061064 | ISBN 9781032220512 (hbk) |
ISBN 9781032220529 (pbk) | ISBN 9781003271017 (ebk)
Subjects: LCSH: Polymeric composites. | Fibrous composites–Materials. |
Epoxy compounds. | Thermoplastic composites.
Classification: LCC TA455.P58 E665 2023 | DDC 620.1/92–dc23/eng/20230124
LC record available at https://lccn.loc.gov/2022061064

ISBN: 978-1-032-22051-2 (hbk)
ISBN: 978-1-032-22052-9 (pbk)
ISBN: 978-1-003-27101-7 (ebk)

DOI: 10.1201/9781003271017

Typeset in Times
by codeMantra

This book is dedicated to my father, mother, wife and sister,
Muthukumar, Thilagavathy, Aradhani and Alamelumangai.
It is their unconditional love and support
that drives my quest for research.

Contents

Preface

Fibre-reinforced polymer composites with thermoset matrices such as epoxy, polyester, vinyl ester, and bismaleimide resins have been used in widespread applications ranging from aerospace sector, automotive parts, construction and building materials, sports equipment, and household appliances. Natural fibre-based composites though having inferior thermal, physicochemical, and mechanical properties over conventional synthetic fibre-reinforced polymer composites are considered as a potential substitute in the applications requiring moderate strength and stiffness. Natural fibres have advantages such as better damping characteristics, low density, biodegradability, abundant availability at low cost, and non-abrasiveness to tooling, making them a cost-effective alternative reinforcement material for composites in certain applications.

This book specifically focuses on the epoxy-based biocomposites and is organized as follows: It contains six chapters on biocomposites reinforced with various natural fibres extracted from different parts of the plants; the seventh chapter focuses on the bio-epoxy-based composites which are completely biodegradable in nature, followed by the eighth chapter on the hybrid composite with various natural fibres in the stacking sequence or in prticulate form mixed together. Chapters 9 and 10 involve characterization of composites reinforced with the shell nanoparticles and other nanofillers. Chapters 11–13 cover ageing studies on the epoxy-based biocomposites. Chapters 14–17 cover up specific properties of the biocomposites such as creep, fatigue, fracture toughness, and free vibration properties. Chapters 18–20 highlight the suitability of the epoxy-based biocomposites in various applications.

Each chapter of this book is written by established researchers in the field of epoxy-based biocomposites. We would like to appreciate and thank all the authors of this book for their contribution. We are grateful to the publisher and their staff members associated with this book for their cooperation and support in successful completion of this book. The content in this book could be of help to undergraduate and postgraduate students, research scholars, academic researchers, professionals, and scientists looking for fundamental knowledge on the characterization of epoxy-based biocomposites and recent advancements in this area.

Editors

Dr. Chandrasekar Muthukumar is presently working as an Associate Professor at the Department of Aeronautical Engineering, Hindustan Institute of Technology & Science, Chennai, India. He graduated with a Bachelor's in Aeronautical Engineering from Kumaraguru College of Technology, Coimbatore, India. He obtained his Master's in Aerospace Engineering from Nanyang Technological University-TUM ASIA, Singapore. He earned his PhD in Aerospace Engineering from Universiti Putra Malaysia (UPM), Malaysia. His PhD was funded through a research grant from the Ministry of Education, Malaysia. During his association with the UPM, he obtained an internal research fund of 16,000 and 20,000 MYR from the University. He has five years of teaching and academic research experience. His field of expertise includes fibre metal laminate (FML), natural fibres, biocomposites, aging, and their characterization. He has authored and co-authored research articles in reputed SCI Journals, book chapters, and in conference proceedings. He has published edited books with CRC Press, Wiley & Sons, Springer, and Elsevier. He is a peer reviewer for the *Journal of Composite Materials*, *Polymer Composites*, *Materials Research Express*, and the *Journal of Natural Fibres*.

Dr. Senthil Muthu Kumar Thiagamani is working as an Associate Professor in the Department of Mechanical Engineering at Kalasalingam Academy of Research and Education (KARE), Tamil Nadu, India. He received his Diploma in Mechanical Engineering from the Directorate of Technical Education, Tamil Nadu in the year 2004, obtained his Bachelor of Engineering in Mechanical Engineering from Anna University Chennai in 2007 and Master of Technology in Automotive Engineering from Vellore Institute of Technology, Vellore in 2009. He received his Doctor of Philosophy in Mechanical Engineering (specialized in Biocomposites) from KARE in 2018. He has also completed his Postdoctoral Research at the Materials and Production Engineering Department at The Sirindhorn International Thai-German Graduate School of Engineering (TGGS), KMUTNB, Thailand. He started his academic career as an Assistant Professor in Mechanical Engineering at KARE in 2010. He has 13+ years of teaching and research experience. He is a member of international societies such as the Society of Automotive Engineers and the International Association of Advanced Materials. His research interests include biodegradable polymer composites and characterization. He has authored several articles in peer-reviewed international journals, book chapters, and conference proceedings. He has also published edited books on the theme of biocomposites. He is a member of the editorial board of the ARAI Journal of Mobility Technology. He is also serving as a reviewer for various journals from reputed publishers like Elsevier, Springer Nature, Wiley, Taylor and Francis, etc.

Dr. Senthilkumar Krishnasamy is an Associate Professor in the Department of Mechanical Engineering at PSG Institute of Technology and Applied Research, Coimbatore, Tamil Nadu, India. Dr. Krishnasamy graduated with a Bachelor's degree in Mechanical Engineering from Anna University, Chennai, India, in 2005. He then chose to continue his master's studies and graduated with a Master's degree in CAD/CAM from Anna University, Tirunelveli, in 2009. He obtained his PhD from the Department of Mechanical Engineering at Kalasalingam University (2016). He had been working in the Department of Mechanical Engineering, Kalasalingam Academy of Research and Education (KARE), India, from 2010 (January) to 2018 (October). He has completed his postdoctoral fellowship at Universiti Putra Malaysia, Serdang, Selangor, Malaysia and King Mongkut's University of Technology North Bangkok (KMUTNB) under the research topics of "Experimental investigations on mechanical, morphological, thermal and structural properties of kenaf fibre/mat epoxy composites" and "Sisal composites and Fabrication of Eco-friendly hybrid green composites on tribological properties in a medium-scale application," respectively. His areas of research interest include the modification and treatment of natural fibres, nanocomposites, 3D printing, and hybrid-reinforced polymer composites. He has published research papers in international journals, book chapters, and conferences in the field of natural fibre composites. He is also editing books from different publishers.

Ahmad Ilyas Rushdan is a senior lecturer in the Faculty of Chemical and Energy Engineering, Universiti Teknologi Malaysia, Malaysia. He is also a Fellow of the International Association of Advanced Materials (IAAM), Sweden; Fellow of the International Society for Development and Sustainability (ISDS), Japan; a member of the Royal Society of Chemistry, UK; and the Institute of Chemical Engineers (IChemE), UK. He received his Diploma in Forestry at Universiti Putra Malaysia, Bintulu Campus (UPMKB), Sarawak, Malaysia from Mei 2009 to April 2012. In 2012, he was awarded the Public Service Department (JPA) scholarship to pursue his Bachelor's Degree (BSc) in Chemical Engineering at Universiti Putra Malaysia (UPM). Upon completing his BSc. program in 2016, he was again awarded the Graduate Research Fellowship (GRF) by the Universiti Putra Malaysia (UPM) to undertake a Ph.D. degree in the field of Biocomposite Technology & Design at the Institute of Tropical Forestry and Forest Products (INTROP) UPM. Ahmad Ilyas Rushdan was the recipient of the MVP Doctor of Philosophy Gold Medal Award UPM 2019, for Best Ph.D. Thesis and Top Student Award, INTROP, UPM. He was awarded with Outstanding Reviewer by Carbohydrate Polymers, Elsevier United Kingdom, Top Cited Article 2020-2021 Journal Polymer Composite, Wiley, 2022, and Best Paper Award at various International Conferences. Ahmad Ilyas Rushdan also was listed and awarded Among World's Top 2% Scientist (Subject-Wise) Citation Impact during the Single Calendar Year 2019, 2020, and 2021 by Stanford University, US, PERINTIS Publication Award 2021 and 2022 by Persatuan Saintis Muslim Malaysia, Emerging Scholar Award by Automotive and Autonomous Systems 2021, Belgium, Young Scientists Network - Academy of Sciences Malaysia (YSN-ASM) 2021, UTM Young Research Award 2021, UTM Publication Award 2021, and UTM Highly Cited Researcher Award 2021. In 2021, he won the Gold Award and Special

Award (Kreso Glavac (The Republic of Croatia) at the Malaysia Technology Expo (MTE2022), a Gold Award dan Special Award at International Borneo Innovation, Exhibition & Competition 2022 (IBIEC2022), and, a Gold Award at New Academia Learning Innovation (NALI2022). His main research interests are (1) Polymer Engineering (Biodegradable Polymers, Biopolymers, Polymer composites, Polymer-gels) and (2) Material Engineering (Natural fibre reinforced polymer composites, Biocomposites, Cellulose materials, Nano-composites). To date, he has authored or co-authored more than 431 publications (published/accepted): 188 Journals Indexed in JCR/ Scopus, 3 non-index Journal, 17 books, 104 book chapters, 78 conference proceedings/seminars, 4 research bulletins, 10 conference papers (abstract published in the book of abstract), 17 Guest Editor of Journal special issues and 10 Editor/ Co-Editor of Conference/Seminar Proceedings on green materials related subjects.

Contributors

Tarkan Akderya
Department of Biomedical Engineering,
  Faculty of Engineering and
  Architecture
Izmir Bakırçay University
Izmir, Turkey

Anand Gobiraman
Department of Mechanical Engineering
Achariya College of Engineering
  Technology
Pondicherry, India

J. Arulmozhivarman
Natural Composites Research
  Laboratory, Department of Materials
  and Production Engineering, The
  Sirindhorn International Thai-
  German Graduate School of
  Engineering (TGGS)
King Mongkut's University of
  Technology North Bangkok
  (KMUTNB)
Bangkok, Thailand

Amritha Bemplassery
Department of Chemistry
National Institute of Technology
Calicut, India

Priscilla Quenia Muniz Bezerra
College of Chemistry and Food
  Engineering
Federal University of Rio Grande
Rio Grande, Brazil

Shashikant Chaturvedi
Department of Mechanical Engineering
Invertis University
Bareilly, India

Jorge Alberto Vieira Costa
College of Chemistry and Food
  Engineering
Federal University of Rio Grande
Rio Grande, Brazil

Michele Greque de Morais
College of Chemistry and Food
  Engineering
Federal University of Rio Grande
Rio Grande, Brazil

Shanmugam Dharmalingam
Department of Mechanical Engineering
Dr Mahalingam College of Engineering
  and Technology
Pollachi, India

Hossein Ebrahimnezhad-Khaljiri
Department of Materials Science and
  Engineering, Faculty of Engineering
University of Zanjan
Zanjan, Iran

A. Felix Sahayaraj
Department of Mechanical Engineering
KIT-Kalaignarkarunanidhi Institute of
  Technology
Coimbatore, India

Dheeraj Gunwant
Department of Mechanical Engineering
Invertis University
Bareilly, India
Department of Mechanical Engineering
Apex Institute of Technology
Rampur, India

C.S. Hassan
Faculty of Engineering, Technology and
    Built Environment
UCSI University
Kuala Lumpur, Malaysia

Aswathy Jayakumar
Materials and Production Engineering,
    The Sirindhorn International
    Thai-German Graduate School of
    Engineering (TGGS)
King Mongkut's University of
    Technology
Bangkok, Thailand
Department of Food and Nutrition,
    BioNanocomposite Research Center
Kyung Hee University
Seoul, Republic of Korea

Kishor Kalauni
Department of Mechanical Engineering
Invertis University
Bareilly, India

Jasila Karayil
Department of Applied Science
Government Engineering College
Kozhikode, India

Jun Tae Kim
Department of Food and Nutrition,
    BioNanocomposite Research Center
Kyung Hee University
Seoul, Republic of Korea

Suelen Goettems Kuntzler
College of Chemistry and Food
    Engineering
Federal University of Rio Grande
Rio Grande, Brazil

Emel Kuram
Department of Mechanical Engineering
Gebze Technical University
Gebze, Turkey

Jithun Lal
Department of Materials Engineering
Indian Institute of Science
Bangalore, India

Jaewoo Lee
Department of Polymer-Nano Science
    and Technology
Jeonbuk National University
Jeonju-si, Korea
Department of Bionanotechnology and
    Bioconvergence Engineering
Jeonbuk National University
Jeonju-si, Korea

M.F. Mohamed Nazer
Faculty of Engineering, Technology and
    Built Environment
UCSI University
Kuala Lumpur, Malaysia

Juliana Botelho Moreira
College of Chemistry and Food
    Engineering
Federal University of Rio Grande
Rio Grande, Brazil

A.S.B. Musamih
Faculty of Engineering, Technology and
    Built Environment
UCSI University
Kuala Lumpur, Malaysia

Deivanayagampillai Nagarajan
Department of Mathematics
Rajalakshmi Institute of Technology
Chennai, India

Harsha Negi
Department of Polymer and Process
    Engineering
IIT Roorkee
Roorkee, India

Kanakaraj Niranjana
Department of Aeronautical
 Engineering
KIT-Kalaignarkarunanidhi Institute of
 Technology
Coimbatore, India

Uğur Özmen
Department of Mechanical Engineering,
 Faculty of Engineering
Manisa Celal Bayar University
Manisa, Turkey

Sivasubramanian Palanisamy
Department of Mechanical Engineering
Dilkap Research Institute of Engineering
 and Management Studies
Mumbai, India

Jyotishkumar Parameswaranpillai
Department of Science, Faculty of
 Science & Technology
Alliance University
Bengaluru, India

Vellaichamy Parthasarathy
Rajalakshmi Institute of Technology
Chennai, India

Thirawudh Pongprayoon
Department of Chemical Engineering,
 Faculty of Engineering
King Mongkut's University of
 Technology
Bangkok, Thailand

Harikrishnan Pulikkalparambil
Materials and Production Engineering,
 The Sirindhorn International
 Thai-German Graduate School of
 Engineering (TGGS)
King Mongkut's University of
 Technology
Bangkok, Thailand

Sabarish Radoor
Department of Polymer-Nano Science
 and Technology
Jeonbuk National University
Jeonju-si, Korea
Department of Materials and Production
 Engineering, The Sirindhorn
 International Thai-German Graduate
 School of Engineering (TGGS)
King Mongkut's University of
 Technology
Bangkok, Thailand

L. Rajeshkumar
Natural Composites Research
 Laboratory, Department of Materials
 and Production Engineering, The
 Sirindhorn International Thai-
 German Graduate School of
 Engineering (TGGS)
King Mongkut's University of
 Technology North Bangkok
 (KMUTNB)
Bangkok, Thailand

Manickam Ramesh
Department of Mechanical Engineering
KIT-Kalaignarkarunanidhi Institute of
 Technology
Coimbatore, India

Sanjay Mavikere Rangappa
Materials and Production Engineering,
 The Sirindhorn International
 Thai-German Graduate School of
 Engineering (TGGS)
King Mongkut's University of
 Technology
Bangkok, Thailand

Jong Whan Rhim
Department of Food and Nutrition,
 BioNanocomposite Research Center
Kyung Hee University
Seoul, Republic of Korea

B.B. Sahari
Faculty of Engineering, Technology and
    Built Environment
UCSI University
Kuala Lumpur, Malaysia

M. R. Sanjay
Natural Composites Research
    Laboratory, Department of Materials
    and Production Engineering, The
    Sirindhorn International Thai-
    German Graduate School of
    Engineering (TGGS)
King Mongkut's University of
    Technology North Bangkok
    (KMUTNB)
Bangkok, Thailand

Santhosh Nagaraja
Department of Mechanical Engineering
MVJ College of Engineering
Bangalore, India

Carlo Santulli
School of Science and Technology
Università degli Studi di Camerino
Camerino, Italy

Saravana Kumar M
Department of production Engineering
National Institute of Technology
Tiruchirappalli, India

P. Sathish Kumar
Natural Composites Research
    Laboratory, Department of Materials
    and Production Engineering, The
    Sirindhorn International Thai-
    German Graduate School of
    Engineering (TGGS)
King Mongkut's University of
    Technology North Bangkok
    (KMUTNB)
Bangkok, Thailand

Arumugam Senthil
Department of Physics
SRM Institute of Science and
    Technology
Chennai, India

Senthilkumar Krishnasamy
Department of Mechanical Engineering
PSG Institute of Technology and
    Applied Research
Coimbatore, India

Jyothi Mannekote Shivanna
Department of Chemistry
AMC Engineering College
Bengaluru, India

Suchart Siengchin
Natural Composites Research
    Laboratory, Department of Materials
    and Production Engineering, The
    Sirindhorn International Thai-
    German Graduate School of
    Engineering (TGGS)
King Mongkut's University of
    Technology North Bangkok
    (KMUTNB)
Bangkok, Thailand

Reshma Soman
School of Biosciences
Mahatma Gandhi University
Kottayam, India

Balakrishnan Sundaresan
Department of Physics
Ayya Nadar Janaki Ammal College
Sivakasi, India

M. Tamil Selvan
Department of Mechanical Engineering
KIT-Kalaignarkarunanidhi Institute of
    Technology
Coimbatore, India

Larissa Herter Centeno Teixeira
College of Chemistry and Food
  Engineering
Federal University of Rio Grande
Rio Grande, Brazil

Ajitanshu Vedrtnam
Department of Mechanical Engineering
Invertis University
Bareilly, India
Departamento de Arquitectura
Escuela de Arquitectura -Universidad
  de Alcalá
Madrid, Spain

Vishvanathperumal Sathiyamoorthi
Department of Mechanical Engineering
S.A. Engineering College
Chennai, India

1 Epoxy Resin as Matrix for Polymer Composites

Factors Influencing the Properties of Polymers and Their Composites

Anand Gobiraman
Achariya College of Engineering Technology

Santhosh Nagaraja
MVJ College of Engineering

Vishvanathperumal Sathiyamoorthi
S.A. Engineering College

CONTENTS

DOI: 10.1201/9781003271017-1

1.1   INTRODUCTION

Development in composites has prompted the requirement for the improvement of materials as far as strength, stiffness and maintainability characteristics are concerned. Composite materials have become one of the materials with such enhanced qualities, enabling their potential in a variety of applications (Alothman et al., 2020; Chandrasekar et al., 2020; Chittimenu et al., 2021). Composite materials are a combination of at least two parts, one of which is present in the matrix stage and another may be in the fibre stage. The usage of normal or engineered filaments and the epoxy resin in the manufacture of composite materials has uncovered critical applications in an assortment of fields like construction, automobile, aviation, biomedical, and marine components (Nasimudeen et al., 2021; Senthil Muthu Kumar et al., 2021).

Epoxies are viewed as quite possibly the main class of thermosetting polymers. They are presently used as an essential class of thermosetting resins in a few modern applications. Because of their reach of superior mechanical and physical properties, they are utilized as defensive coatings and binding agents for polymer composites.

Epoxy resins cost more compared to many other thermosetting resins (like polyester or vinyl esters), while they also have superior mechanical properties and better resistance to fluid ingestion and unfavourable environmental factors. When compared to different thermoset polymers, these fantastic real qualities and their durability in administration help to provide an ideal expense execution ratio.

Epoxies also exhibit high electrical resistivity and excellent mechanical capabilities at higher temperatures, which can be attributed to their greater thermal stability when compared to polyester lattices and higher transition temperatures (T_g) (Brent Strong, 2008). Furthermore, they exhibit an outstanding connection to a few substrates (such as metal or plastic) as well as strands utilized as support in composite materials (for example, glass, carbon, and Kevlar). Epoxy resins' little shrinkage throughout the relieving method is another benefit. Polyester and vinyl esters recoil up to 12% volumetrically (specifically, the decrease in the volumetric shrinkage is in the scope of 5%–12% for polyester and 5%–10% for vinyl esters) and on the grounds that the resin keeps on restoring throughout significant stretches of time. This effect might not be immediately obvious. A polyester or vinyl ester cover is significantly less stable over time than an epoxy overlay, which has a rebound rate of less than 5% (Zarrelli et al., 2002). Because they have much fewer volatile outflows throughout the restoration system than polyester and vinyl ester ones due to the absence of styrene, epoxies can be exploited in the same way as "open-form" innovation creation (for example, hand lay-up or vacuum bagging). Epoxies work well with the majority of composite manufacturing techniques, including vacuum bagging, autoclave forming, pressure shaping, fibre winding, and hand lay-up (Senthilkumar et al., 2021, 2022; Senthilkumar et al., 2015). The versatility of these systems, which may be altered to fit one or more applications by changing a small number of their physical, mechanical, and processing characteristics, is another crucial aspect that makes them particularly intriguing. The characteristics of the composites with the epoxy resins can be enhanced by the selective mixture of resin systems or by adding reagents (such as flame retardants or toughening agent). For instance, pure resins can be modified by adding fillers with the better electrical qualities, like carbon nanotubes or metal powder, to provide better conductivity than is required. The best epoxy resins are those that have a variety of characteristics. Epoxy thermosetting polymers are appropriate for the great majority of advanced applications since no volatile or harmful compounds are emitted during the curing process. It is possible to choose monomers in a flexible manner to produce a range of products, from low T_g rubber to high T_g polymer (Shahroze et al., 2021; Thomas et al., 2021).

In this regard, there is vast scope for the use of the epoxy resins for vessels, submersibles, marine structures, and other marine structural components that are prone to significant environmental challenges. Therefore, the epoxy resins are generally considered to have higher durability and require little or no maintenance over long periods of time. Additionally, lightweight and corrosion resistance are necessary for epoxy resins to meet design requirements and operate at impeccable speed and reliability. The composites, with special reference to epoxy resins, have found extensive use in the automobile, aerospace, and marine industries. Weight reduction is a crucial factor to take into account when using epoxy resins. Composites made of epoxy resins have developed over time to satisfy the changing and expanding demands of the automotive, aerospace, and marine industries while continuing to be a useful tool for creating consolidated components for various applications. To date, epoxy composite materials are used in all areas right from sporting goods, aerospace, and automotive components and for a variety of components and structures. In recent years, epoxy composites are being extensively used owing to high performance and safety considerations.

1.2 CURING OF EPOXY RESINS

The term "curing" refers to an interaction that involves a thermosetting resin's characteristics changing as a result of synthetic responses.

Typically, epoxy structures have two parts: resin and hardener, as well as a relieving agent. Even at room temperature, these parts react to come together to form a three-dimensional structure even though they are stable when kept apart. Also accessible are one-section epoxy frameworks, particularly those utilized in the automotive and aviation sectors (Sharma and Luzinov, 2011). These mixes, which combine epoxy resin and an expert in inert restoration, can only repair at temperatures as high as 180°C to 200°C. The hardener is recommended under the idea that it is imperative to lower the relieving temperature to around 100°C–120°C. The idle relieving specialist provides the epoxy, with its high liquefying temperature, inertness in these constructions. The number of epoxy rings is connected to how responsive groups called epoxy rings or gatherings, with a higher hot hardness and a poorer durability of the strong state, are present in the epoxy resin. Due to the value of the bond points (all equivalent to about 60°C), the epoxy rings exhibit a considerable planar design that is illustrated by a high-pressure state. Instead of a bivalent oxygen point of 110°, a tetrahedral carbon molecule typically exhibits a bond point equivalent to 109.5°. As a result of this powerful connection, the epoxy ring's design is incredibly responsive. The active hydrogen iotas of the restoring expert may react with these receptive clusters, which are positioned terminally or inside the epoxy ring, to produce covalent intermolecular connections, which are the cornerstone of the alleviating process. By blending atoms of pitch and relieving structure, covalent connections between monomer structures are established to enhance the bond strength and fasten the curing of the composites. The curing of the composites occurs in two phases, namely gelation and vitrification.

The resin system transforms from a liquid or liquid state to a rubbery one during the crucial curing phase known as gelation. After this time, the cross-links that are formed will prevent the resin system from flowing as easily. Before the gelation time, the resin system is soluble in the appropriate solvent. Past this critical limit, the network cannot dissolve and instead develops as it is absorbed by the solvent. The fractions "Sol" and "Gel," in particular at this stage of the curing process, make the resin system. While the latter can still be extracted using a proper solvent, the former is still a liquid phase. With more covalent bonds being created, its quantity drops. Since the gel phase is still fragile, the curing process must be prolonged to produce the structural material until the fraction of sol is negligible or close to zero. The majority of the molecules are now attached to their surroundings and the 3D network of epoxy resins is forming at this point (i.e., vitrification). The curing temperature is reached at the beginning of the vitrification process after T_g rises throughout the curing process. When epoxy resin hardens at 90°C, Figure 5.3 illustrates the T_g changes over time. Beginning with the curing process, the reaction happens in a liquid state and at a controlled rate (before the gelation time). It is crucial to note that this difference exists even after the resin system has begun to gel. It has a high rate of cure due to its chemical and kinetic control. When the gel point is reached, this difference decreases, the curing process switches to diffusion control, slows down, and eventually stops (Ellis, 1993). By this time, however, not all epoxy groups had reacted, and the resin system had solidified. The resin must go through higher temperature postcuring procedures in order to guarantee a complete reaction and maintain the resin's ultimate mechanical and physical qualities.

1.3 CURING AGENTS

The reactive groups on the molecules of epoxy resin interact with a variety of curing agents, including amines, anhydrides, acids, mercaptans, imidazoles, phenols, and isocyanates, to generate covalent intermolecular bonds and a three-dimensional network. The most often used curing agents among these compounds are primary and secondary amines because the amine-curing epoxy resin is more environmentally friendly than other curing agent kinds (Dyakonov et al., 1996). Epoxies are used as adhesives and coatings for making matrices for aliphatic or alicyclic amines and for fibre-reinforced composites, especially with low-temperature epoxies, aromatic amines, etc. (Pascault and Williams, 2010). The structures of aliphatic and aromatic amines play a significant role in the curing of epoxies. Aliphatic amines are typically mixed with epoxies at ambient temperature, thereby providing higher cure rates and shorter cure times (i.e., gelling and vitrification times) than for alicyclic or aromatic polyamines. We provide a complete system. Aliphatic amines are highly reactive hardeners and when mixed with epoxy molecules, the short distance between active centres allows for a densely cross-linked network. As a result, it is exceedingly resistant to water and solvents (albeit less so than many organic solvents), highly resistant to alkali, and exceptionally resistant to several inorganic acids. It also has outstanding mechanical and adhesive qualities. It is conceivable to develop a curing system with weak mechanical characteristics. These epoxies, which typically cure at room temperature, have improved qualities after hardening at a high temperature. Aromatic amines react with epoxies more slowly than aliphatic or alicyclic amines when it comes to all the curing agents (Weinmann et al., 1996). The system created when this kind of curing agent and epoxy resin are mixed takes a long time to cure and requires extensive heating to achieve its perfect properties. For example, two heating cycles are required to cure aromatic amines. Between 150°C and 170°C is the temperature at which the initial heating is done. When aromatic amine is utilized as a curing agent, an epoxy system is produced, that exhibits high mechanical, electrical, thermal, and chemical resistance, especially against alkalis.

The second most popular form of hardener for epoxies, after curing agents, is carboxylic acids and anhydrides, particularly in thermosetting applications. The bulk of other thermosetting applications use acids as anhydrous agents, especially for electrically insulators, and as protective coating (thermosetting surface coatings) in epoxy resin systems. The epoxy system is appropriate for the manufacturing of large moulded parts because anhydrides give it a lengthy cure period and produce relatively little heat during the cure process. They also produce a cured resin and have balanced mechanical, chemical, and electrical properties. Low-temperature curing agents are made of polymer-based reactants. In other words, the tertiary amine used as an accelerator allows an epoxy-based curing process to occur between −20°C and 0°C. Conversely, epoxy/polymer systems have quick cure periods at room temperature (i.e., a curing life between 2 and 10 minutes) and reach their ideal characteristics in just 30 minutes. It is possible to create resin that is quickly cured at a high deflection temperature under load by heating materials between 80°C and 120°C and employing the curing ingredient imidazole. The electronics industry frequently uses imidazole-cured epoxies as moulding and sealing compounds because they generally have better physical qualities than those cured with tertiary amines (Ghaemy and Sadjady, 2006). Imidazoles are also frequently employed, much like tertiary amines, as co-curing agents for polyvalent phenols, aromatic amines, and organic acid anhydrides.

1.4 TYPES OF EPOXY RESINS

Epoxies come in a variety of forms with a range of molecular weights and properties. The major types of epoxy resins are given in Figure 1.1. Some of them are discussed in this section.

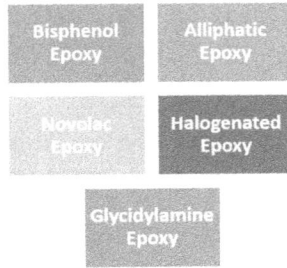

FIGURE 1.1   Schematic of the major types of epoxy resins.

1.4.1 POLYNUCLEAR PHENOL EPOXY

Epichlorohydrin and tetrakis(4-hydroxyphenyl)ethane can be combined to create these epoxies. The four epoxy rings found in the polynuclear phenolic epoxies, including tetraglycidyl methylene dianiline (TGMDA) epoxies, increase their wettability and strength when bonding reinforcements.

1.4.2 DGEBA EPOXY RESINS

The tried-and-true and most popular form of epoxy resin is the diglycidyl ether of bisphenol-A (DGEBA). Epichlorohydrin and bisphenol-A (BPA) react together with stoichiometric amounts of NaOH at a temperature of about 70°C to create this substance. For the generation of excess molecular-weight components to be controlled, epichlorohydrin is necessary. In reality, when DGEBA and BPA combine, epoxy resins with superior molecular weight and multiple repeating units of the molecules are produced. In the marketplace, DGEBA resins are offered with a range of 0–14 repeated molecular units. The refined variations (n = 0) of the resins have a molecular weight of 345 as opposed to the basic industrial model of the resins, which has a molecular weight of 385. Epoxy resins with such a better molecular weight (n = 14) could be produced in an alkaline environment with far less epichlorohydrin and reactants.

1.4.3 DGEBF EPOXY RESINS

Other than BPA, functional phenol groups may be utilized to react with epichlorohydrin in order to produce epoxy resins. The diglycidyl ether of bisphenol-F (DGEBF) epoxy is one such kind. The most often utilized substance is bisphenol-F. These epoxy resins are created through the process of the epichlorohydrin and bishpenol-F chemical reaction. Compared to DGEBA epoxy resins, the mechanical and chemical properties of DGEBF epoxy resins are superior and they have a reduced viscosity. Regular mixing of DGEBA and DGEBF prevents crystallization and lessens handling difficulties for the compounds at low temperatures by reducing their viscosity.

1.4.4 TGMDA Epoxy Resins

The interaction of epichlorohydrin and aromatic diamine results in the production of TGMDA epoxy resins. The functionality of TGMDA is higher than that of DGEBA and DGEBF resins. They provide composites with superior strength and stiffness and greater resilience to the effects of temperature (Campbell, 2010). To get superior flexibility and toughness with enhanced flexibility, the TGMDA is frequently combined with DGEBF and DGEBA.

1.4.5 Phthalonitrile/phenolic (PNP) Epoxy Resins

Tetrakis(4-hydroxyphenyl)ethane and epichlorohydrin are the reactants used to create the synthetic polynuclear phenol epoxy PNP epoxy resins. Four epoxy rings are also included in the PNP epoxy resins' molecules, which enhances wettability and stickiness.

1.4.6 Bio-Based Epoxy Resins

Since the beginning of the previous decade, intensive research has been devoted to the production of polymers from renewable resources. This is mostly because of rising petrochemical product costs associated with expanding environmental concerns. The components and characterization of bio-derived thermosetting resins, which include the replacement of BPA, are now being explored in relation to various plastic substances. A significant goal of the study is the production of epoxy resins using chemicals from natural sources. Cardanol, a phenol-based byproduct of the cashew nut industry, is one of the main precursors used to create several kinds of epoxy bio-based resins, for instance. The term "cashew nut shell liquid" (CNSL), which relates to the alkyl phenolic oil contained within the spongy mesocarp of the cashew nut shelled from the cashew tree *Anacardium occidentale* L., is used to describe the commercial-grade oil recovered during vacuum distillation. Over 25% of the entire weight of the nut is made up of CNSL, which is manufactured utilizing the most advanced roasting methods currently used in the cashew sector. It is anticipated that CNSL will be produced at a rate of roughly 300,000 piles per year globally (Calò et al., 2007). A thermosetting resin was created by mixing an epoxy monomer, an acid-based catalyst, and a resole chemical. This resin contains around 40% cardanol by weight (Maffezzoli et al., 2004). This closure was transformed into a synthetic molecule by polycondensing cardanol and formaldehyde with a simple catalyst. In order to create samples that could be tested in both tensile and bend configurations, the components were reinforced using the proper properties and curing temperatures. Natural fibres including shorter ramie, flax, hemp fibres, and a jute cloth supported the components. A DGEBA epoxy resin was cured using two different novolac resins, each of which included 35% and 20% via weight of unreacted cardanol (Campaner et al., 2009). The calorimetric analysis in this work showed that novolac/epoxy resin weight proportions lower than 60/40 resulted in the cross-linking of the resin, which is due to the rising presence of secondary hydroxyl groups for cross-linking activities. Mechanical components have advanced along with the development of epoxy resin thanks to bio-based epoxies. Instead, just a one-step mass loss simulating a single thermal degradation rate was seen at temperatures greater than 400°C. This is due to the fact that changing the amount of novolac did not noticeably change the cured resin's thermal resistance in a nitrogen environment. In order to compare the performance of an epoxy-cardanol resin-based paint (i.e., its physico-mechanical properties, chemical inertness, and corrosion safety efficiency) to paints made using epoxy resin that has not been modified, an epoxy-cardanol resin-based coat has been created and then detailed (Aggarwal et al., 2007). Since the new bio-based paints demonstrate stronger anticorrosive capabilities than the unaltered paints, epoxy resin with just a cardanol foundation creates an excellent bonding medium for the paint's constituent parts. A common herbal product called rosin is created by heating the liquid resin that pines and other conifers generate. Because of their chemical composition, rosin acids exhibit molecular stress similar to that of cycloaliphatic or aromatic chemicals. Their derivatives could make an epoxy resin with a bio-based basis by substituting modern, petroleum-based, fully rigid monomers. Researchers created a unique bio-based epoxy using rosin acid, and they demonstrated that it has the essential thermal stability, a high glass transition temperature ($T_g = 153.8°C$), and a high tensile modulus at ambient temperature. The results therefore lend credence to the idea that thermosetting resins with improved overall performance can be made using rosin. Wang et al. (2011) claim that an imide-shaped rosin-derived amine was developed as a curing agent for epoxy resins and tested against both commercially available aromatic amine-curing agents, as well as any other rosin-based completely anhydride curing products. The findings revealed that when compared to epoxy machines that were cured using the less-expensive amine-curing agent, those that were treated with this closure curing agent showed equal moduli, better T_g, and at best minimally decreased degradation temperature. However, because the imide form had developed, the breakdown temperature of the epoxy treated with the new curing agent was stronger than that of the epoxy cured with a rosin-based completely anhydride curing agent. Triglyceride-rich plant oils are a magnificent source of essential renewable resources. The four most significant vegetable oils in the world are palm oil, soybean oil (SBO), rapeseed oil, and sunflower oil, which together make up more than 80% of the market (Tan and Chow, 2010). Because most readily available vegetable oils are either unavailable or prohibitively expensive, linseed and soybean oils are found to be among the most widely used epoxidized vegetable oils at the moment. Fully polymeric materials based on soybean oil have been the subject of intensive research, especially in the last 10 years. Synthetic 3,4-epoxidized soybean oil (ESO) is produced by epoxidizing the double bonds in SBO triglycerides with hydrogen peroxide in either acetic or formic acid. It is currently commercially available for low cost and in great quantities. Epoxidized linseed oil and ESO are examples of functionalized vegetable oils that were treated with the help of an anhydride curing agent (Miyagawa et al., 2005). For each of the vegetable oils employed in this investigation, the authors modified the content of DGEBF by a specified proportion. A novel bio-based epoxy resin with a strong elastic modulus and high glass transition temperature was created by mixing DGEBF, vegetable oils, and the anhydride curing agent. This was a really important achievement. Based on the amount of vegetable oil, the Izod effect's energy and fracture toughness have significantly increased, resulting in a phase-separated morphology. Some writers have created epoxy resins by partially substituting a synthetic DGEBA epoxy prepolymer with increasing amounts of ESO, curing the combination with methyltetrahydrophthalic anhydride, and initiating the reaction with 1-methyl imidazole (Altuna et al., 2011). The authors changed the amount of DGEBF in each of the vegetable oils used in this experiment by a certain percentage. By combining DGEBF, vegetable oils, and the anhydride curing agent, a novel bio-based epoxy resin with a high glass transition temperature and strong elastic modulus was produced. This was an extremely significant accomplishment. The energy and fracture toughness of the Izod effect have dramatically increased based on the amount of vegetable oil, leading to a phase-separated morphology. According to some authors, epoxy resins can be produced by partially substituting a synthetic DGEBA epoxy prepolymer with increasing volumes of ESO, curing the mixture with methyltetrahydrophthalic anhydride, and starting the reaction using 1-methyl imidazole (Altuna et al., 2011). The scientists observed that when the machine was cured for 2 hours at 210°C with an epoxy/hydroxyl ratio of 1/1.4, the attributes were dispersed most uniformly. A few fundamental experiments on biocomposites have also been conducted by the researchers, starting with clean epoxy resin laminates and moving up to the utilization of microfibrillated cellulose reinforcements with weights ranging from 5% to 11%.

1.5 MECHANICAL PROPERTIES

Epoxy resins are commonly employed as the polymer matrix for composite materials because of their superior mechanical properties. Epoxy resins' propensity for brittleness is its most vulnerable trait when it comes to mechanical performances. Small quantities of liquid rubbers with highly reactive terminal carboxyl organizations are often added to epoxy resins to increase their elongation at wreck and, as a result, their impact characteristics. For instance, liquid rubber (carboxyl-terminated butadiene acrylonitrile, CTBN) contains copolymers with carboxyl-terminated butadiene and acrylonitrile (Valenza and Calabrese, 2003). Due to the possibility that amine-terminated poly(butadiene-co-acrylonitrile) (ATBN) utilized some of the exothermic energy created during epoxy cross-linking during its reaction with the epoxy resin, a drop in all reactivity metrics also happens when those chemicals are added. The gel time and treatment time both reduce as a result, in addition to the exothermic peak. As a result of adding rubber to epoxy resins, various properties, including the glass transition temperature, the Young's modulus, and the tensile strength, are diminished. Epoxy resins and thermoplastic polymers such polyethersulfones, polyetherimides, and polyphenylene oxides are mixed to improve the mechanical properties. In Table 1.1, some of the most significant mechanical characteristics of epoxy composites are mentioned.

In order to improve the mechanical properties of composites made of epoxy resin, Cha et al. (2018) investigated the reinforcement of the resin matrix utilizing functionalized carbon nanotubes (CNTs) and graphene nanoplatelets (GNP) with melamine.

TABLE 1.1

Some of the Important Mechanical Properties of the Epoxy Resin

Mechanical Properties	Metric	Comments
Hardness, Shore D	55.0–95.0	Average value: 80.2
Tensile strength, ultimate	5.17–97.0 MPa	Average value: 33.1 MPa
Tensile strength, yield	1.03–2900 MPa	Average value: 563 MPa
Elongation at break	0.000%–50.0%	Average value: 9.35%
Elongation at yield	1.00%–9.50%	Average value: 2.77%
Modulus of elasticity	0.0207–215 GPa	Average value: 35.2 GPa
Flexural yield strength	75.8–1890 MPa	Average value: 907 MPa
Flexural modulus	2.40–205 GPa	Average value: 58.5 GPa

Source: http://www.matweb.com/search/datasheet_print.aspx?matguid=956da5edc80f4c62a72c15ca2b923494.

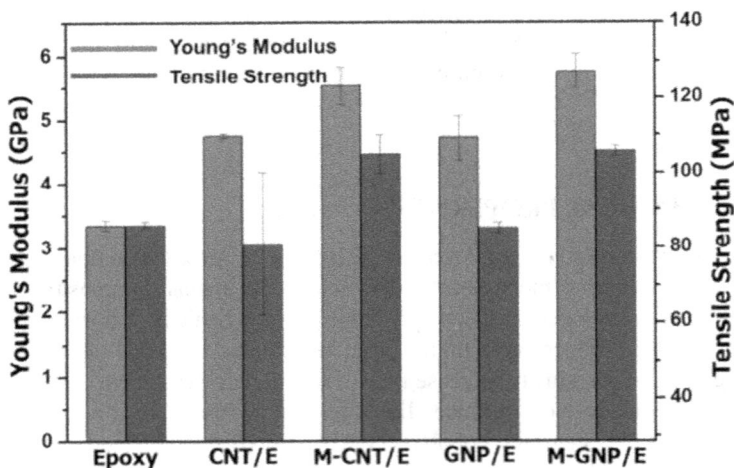

FIGURE 1.2 Improvement in the Young's modulus of the epoxy resin-based composites with functionalized reinforcements (Cha et al. 2018).

This is because melamine's amino groups make it easier for the epoxy system's strong covalent bonds to form. Figure 1.2 compares the performance of epoxy resins with CNT and GNP-reinforced composites as well as functionalized CNT and GNP-reinforced composites.

The inclusion of the melamine functionalization group inhibits the agglomeration and stacking of the CNT and GNP reinforcements and results in better interactions between the matrix and reinforcements and leads to better load transfer.

1.6 PHYSICOCHEMICAL PROPERTIES

In addition to having exceptional resistance to assault from seawater, epoxy resins are renowned to have the highest levels of chemical and corrosion resistance. These thermosetting resins are the best option for the matrix of composite systems since they keep their mechanical and chemical resistances and don't degrade even after being immersed in seawater or a harsh chemical environment for an extended period of time. Epoxy resins are frequently used in several shields, coatings, and paints to stop metal corrosion for the same reasons (Miskovic-Stankovic et al., 1999; Popovic et al., 2005; Armelin et al., 2008). Along with increasing these composites' resistance to seawater, the addition of polyester resins enhances their chemical qualities. In addition, hydrolysis events involving unsaturated groups in the resin may cause the polymer matrix and the fibre/matrix interface to deteriorate. The matrix swells and plasticizes as a result of this degradation, the fibre/matrix contact debonds, and the mechanical properties of the composites are significantly reduced. Epoxy-based composites, which frequently display advanced chemical balance in seawater, can be coated with vinyl ester coatings to address this problem. The ability of thermoset resins (orthophthalic polyester, isophthalic polyester, vinyl ester, and epoxy) to absorb moisture has been investigated by certain authors, as well as the composites made from these resins and reinforced with glass fibres. In particular, distilled water at three distinct immersion temperatures of 25°C, 55°C, and 55°C was used to mature composites and resins for 18 months. Experimental corrosion studies have shown that they have an impact on the ageing of the composites as well as the weight changes, particularly of the matrix resin (Davies et al., 2001). More study has been done on the effects of saltwater on climate change, salt and pollutant permeability, degradation in corrosive environments, and failure mechanisms of both glass/epoxy and glass/polyurethane composites (Mourad et al., 2010). To accomplish this, the laminates were submerged in seawater for periods of time varying from 3 months to a year, both at room temperature and 55°C. The results of the trials showed significant modifications in the internal corrosion resistance of both glass/epoxy and glass/polyurethane composites as well as the corrosion of glass/epoxy, indicating that high temperatures accelerated the breakdown of glass/polyurethane composites. The rate of corrosion of the glass/polyurethane composite, however, decreased as the composites grew older in seawater at room temperature due to passivation and also under high-temperature conditions in seawater at 65°C. Additionally, studies on the chemical ageing of epoxy and vinyl ester resins reinforced with various fibres (such as glass and carbon) have been conducted (Narasimha Murthy et al., 2010). Vinyl ester composites are less corrosion-resistant than epoxy ones, according to experimental results. Vinyl ester resins specifically deteriorate and lose mass faster than epoxy resins. As the matrix's glass fibre content rose to 35%, the rate of corrosion decreased. The authors also demonstrate that vinyl ester composites absorb more water than epoxy-based composites. Several physicochemical properties of epoxy resins are listed in Table 1.2.

TABLE 1.2

Some of the Important Physicochemical Properties of the Epoxy Resin

Physicochemical Properties	Metric	Comments
Density	0.860–2.60 g/cc	Average value: 1.35 g/cc
Water absorption	0.0300%–1.20%	Average value: 0.333%
Moisture absorption at equilibrium	2.80%–9.40%	Average value: 5.16%
Viscosity	1.00–1.20e+6 cP	Average value: 37100 cP
Molecular weight	471–1690 g/mol	Average value: 771 g/mol
Thickness	25.4–127 μm	Average value: 60.8 μm
Linear mould shrinkage	0.00100–0.00400 cm/cm	Average value: 0.00240 cm/cm

Source: http://www.matweb.com/search/datasheet_print.aspx?matguid=956da5edc80f4c62a7 2c15ca2b923494.

1.7 THERMAL AND ELECTRICAL PROPERTIES

Epoxy resins' thermal and electric characteristics are essential for many applications because these thermoplastic materials have a significantly reduced thermal and electric conductivity. To improve the thermal and electrical properties, numerous research projects and efforts were made to develop neat epoxy resins. Epoxy resins have grown significantly in popularity, especially during the past 10 years, as a result of the addition of fillers that improve their thermal and electrical qualities. Modern environments use a variety of microfillers and nanofillers as fillers, including carbon nanotubes, steel particles, and inorganic microfillers. One example is the use of inorganic microfillers to increase the heat conductivity of epoxy resins (Lee et al., 2008; Lee and Yu, 2005). To overcome the percolation threshold and achieve excessive thermal conductivity, more microfillers can be used, resulting in better thermal and electrical properties. Natural nanofillers enhance the specific characteristics of composites because of their excellent thermal conductivity. Multi-walled carbon nanotubes (MWCNTs) in particular have recently attracted a lot of attention outside of the biomedical domains due to their one-dimensional structure, high factor ratio, and advanced thermal conductivity (3000 W/mK for a single MWCNT and 200 W/mK for bulk MWCNTs at room temperature (Yang et al., 1991). Researchers have looked at how different carbon nanotube types, particle contents, interfacial contents, functionalization materials, and factor ratios affect epoxy resins' ability to conduct electricity and heat (Gojny et al., 2006). Due to their high aspect ratio and high factor ratio, MWCNTs have the highest potential for improving electric conductivity among all nanotube types. By boosting the percolation threshold, crucial components including functionalization, ultrasonication, and other processes enhance the specific properties. On the other hand, epoxy resins' thermal conductivity barely improves when carbon nanotubes are added. Although multi-walled nanotubes seem to have the highest ability to increase the thermal conductivity, epoxy resins can theoretically be predicted to have thermal conductivities that are higher than they actually do. According to Gojny et al. (2006), carbon nanotubes and maybe similar nanoparticles, which provide good surface surfaces, are not suitable to entirely increase the heat conductivity of polymer-based composites. The heat conductivity of the composites was demonstrated to be improved by adding 6.5% of MWCNTs or 71.7% of SiC microparticles to an epoxy resin, reaching levels that were 2.8 to 20 times higher than the pure epoxy (Zhou et al., 2010). The researchers also partially replaced microfillers with nanofillers in order to improve the composites' thermal conductivity. They were able to increase the thermal conductivity of normal epoxy by 24.5 times by mixing 5% MWCNTs by weight and 55% SiC microparticles by weight with it. To create polymers with enhanced electrical characteristics, carbon nanotubes with silver nanoparticle decoration (Ag-CNTs) were added to epoxy resin (Ma et al., 2008). The experiments' findings showed that composites with 0.1% of Ag-CNTs in weight had an increase in electric conductivity that was four orders higher than that of composites with the same amount of functionalized CNTs, and that this development was no longer at the expense of the composites' mechanical or thermal properties. Steel scraps have also been considered in various research investigations as fillers to improve the electrical and thermal conductivities of epoxy constructions. Researchers tested the electrical and thermal conductivities of epoxy structures loaded with powdered steel, copper, and nickel (Mamunya et al., 2002). Their research made it clear that the composites had qualities that enable the dispersion of steel debris in the polymer matrix at random, improving the electric and thermal properties. The percolation concept nevertheless holds true for structures with a random process of dispersed filler even though, unlike electric conductivity, the dependency of thermal conductivity on the rise of these factors suggests no modification in the percolation threshold region. Using the mean filler size, length distribution, particle morphology, loading, and epoxy cross-hyperlink density, the thermal and electrical properties of epoxy resins reinforced with alumina detritus were assessed (McGrath et al., 2008). The authors have demonstrated that the quantity of filler and the concentration of epoxy cross-hyperlink have been the most significant factors affecting the parameters of the composites, while other characteristics (such as particulate size, structure, and length distribution) have had little impact on the thermal and electrical properties. Due to its widespread availability, low density, and eternal conductivity, carbon black (CB) could be considered for various other essential filler components that could be employed to improve the electrical qualities of epoxy resins. Reduce the percolation threshold as much as you can for CB-filled polymer composites since too much CB in the volume concentration decreases effect resistance and improves electrical characteristics. Some writers have focused their research and development on developing novel composites with outstanding electrical conductivities by adding a small amount of plasticized CB to an epoxy resin (Abdel-Aal et al., 2008). According to the results of the experiment, adding material with CB content will enhance the conductance of the composites at room temperature. Modern-voltage conductivity also suggests a switching effect, allowing the proposed composites to switch voltage and current in contemporary digital devices, and thermal conductivity, in addition to the composites' poor temperature coefficient of conductivity and poor temperature dependence of electric conductivity.

The thermogravimetric study of the unfunctionalized CNT, GNP, and melanine-functionalized CNT, GNP-based composite composites has been extensively studied by Cha et al. (2018). It is evident from the thermogravimetric analysis (TGA) graphs (Figure 1.3) that the composites reinforced with melanine-functionalized CNT, GNP-reinforced epoxy composites exhibit more thermal stability than the pristine CNT, GNP-reinforced composites and the neat epoxy composites. This is the result of the functionalization group's creation of a solid link between the matrix and reinforcements, which enhances the thermal properties of composites made with epoxy resin.

1.8   APPLICATIONS

Epoxy resins have found multiple applications, especially when used as the matrix for polymer composites, and some of them are elucidated in this section of the chapter. The schematic in Figure 1.4 depicts the broad spectrum of application areas of epoxy resin.

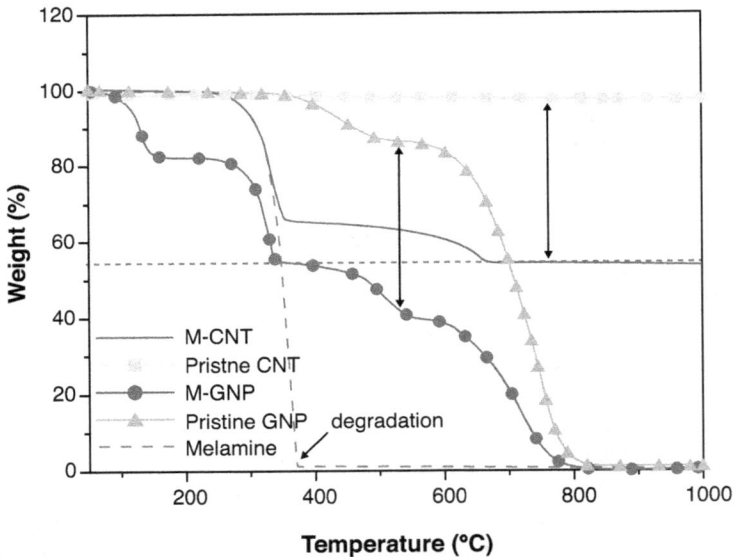

FIGURE 1.3   The thermogravimetric analysis of pristine CNT, GNP, and melamine-functionalized CNT and GNP (Cha et al. 2018).

FIGURE 1.4   Application areas of epoxy resin.

1.8.1 ADHESIVES

Adhesives are a vital component of industrial operations, helping to create a variety of unique product types. They provide a number of benefits over conventional joining methods, such as the ability of adhesive connections to advance in the presence of dynamic strain due to their flexibility. Furthermore, the breakdown of the material might be caused by the transmission of stress through rivets or bolts, either locally or regionally. Consequently, the adhesive bonds provide increased resistance to failure. Although the strain distribution in an adhesive bond is dispersed throughout the entire bond, the failure's propagation is more severely constrained. Adhesives are a crucial part of industrial processes since they enable the creation of numerous distinctive product types. Over traditional joining techniques, they offer a number of advantages, including the flexibility of adhesive connections' capacity to advance in the presence of dynamic strain. Additionally, local or regional stress transmission through rivets or bolts may contribute to the material's failure. As a result, the adhesive bonds offer higher failure resistance. Despite the fact that the tension is distributed evenly throughout an adhesive bond, the failure's spread is more severely restricted.

1.8.2 MARINE APPLICATIONS

Epoxy resins are extensively utilized in the marine industry as well as in paints, adhesives, and fibre-reinforced plastics. More than 90% of boats used for sporting activities that are less than 60 feet deep are currently manufactured with polyester resin, which is known to reduce their cost, as opposed to aerospace applications, where epoxy resins have long been used. In reality, epoxy resins when reinforced with the reinforcements and filler provide greater strength and better corrosion resistance characteristics in comparison with the vinyl ester resins and polyester ones. Since the epoxy resins can account for as much as 1/2 of the load of a composite component, the epoxy resins can enhance their strength and reduce the weight of the marine structures. To be useful in a maritime environment, resins probably need to have a high level of resistance to degradation brought on by water penetration. The moisture that is absorbed by each thermosetting resin increases the load on the composite structures. Additionally, because the covalent link between the network's molecules is broken, the water absorption has an impact on the resins' third-dimensional shape as well as their ability to adhere to reinforcements. The filler can also slow down deterioration and enhance the mechanical capabilities of the composites. Additionally, using epoxy resins for boat hulls can assist combat moisture absorption, quicker degradation, and delamination of laminates, which are common with other classes of resins like polyester, vinyl ester resins, etc.

1.8.3 AUTOMOTIVE APPLICATIONS

The automobile discipline has been historically characterized by their extensive use of metals; however, the metals are being replaced nowadays by composite materials. Since the 1970s, composite materials have developed. The best polyester resins reinforced with glass fibres were used in the beginning, and components were frequently created using the manual lay-up or spray-up process in open moulds or by utilizing sheet moulding compound. Since then, numerous new manufacturing techniques and materials have been developed, giving present-day producers of composite parts access to a wide variety of resources. In the modern automobile industry, composite materials make about 15%–20% of the typical weight of cutting-edge vehicles. Epoxy resins are frequently used to create primers in the automotive industry in order to cover metal, aluminium, and composite components before painting and serve as matrices for composite materials. These contemporary primers can be applied directly to a surface to give a solvent-based barrier coating and are more corrosion-resistant than older kinds. One of the key elements in corrosion prevention utilizing protective coatings is a lack of coating adhesion to prevent environmental damages. Numerous epoxy primers, or pigment-free coatings, have been investigated for their ability to adhere to metal (Bajat and Dedic, 2007). Each primer's dry and moist adhesion strengths have been measured in situ, and the corrosion imbalance of coated samples has been evaluated using electrochemical impedance spectroscopy. All of the samples have demonstrated exceptional adhesion in dry environments; however, after exposure to the corrosive agent for 25 days in a 3% NaCl environment, the epoxy primer showed the best corrosion resistance of all the samples that were tested. The greatest protective coatings of coloured epoxy primers on the metal surfaces of autos were shown by electrochemical impedance tests.

1.8.4 AEROSPACE APPLICATIONS

In aircraft structures, which have to possess strength, stiffness, sturdiness, and mild weight, the epoxy resins are most sought-after matrix materials, because they possess lower weight and higher strength. Epoxy resins have long been the standard for aviation parts because of these factors. Because material selection in this industry is solely focused on overall performance, material cost only plays a minor role. Aramid and glass fibres are frequently utilized to reinforce epoxy resins in non-structural components of aircraft, although carbon or boron fibres are favoured for structurally important additives like the plane tail and furniture for manipulating surfaces. The Boeing 767 boasts the most substantial usage of composite materials among the most prominent passenger aircraft, with a fuselage, skin, and rear wing shape built entirely of high-overall performance composites. Eighty percent of the load-bearing structure is made of graphite/epoxy. Currently, 10% of the weight of a spacecraft's structure is made up of epoxy-based composites. In terms of helicopter applications, the A139 model features a cockpit constructed of sandwich Kevlar/Nomex, while the tail boom and stabilizer components are composed of carbon-reinforced epoxy composites.

1.9 CONCLUSIONS

Epoxy resins exhibit better mechanical qualities than other thermosetting polymers, less shrinkage, increased resistance to moisture absorption, and improved corrosion resistance. In comparison to various thermoset polymers, these qualities and their durability in use give a favourable cost-effective performance ratio. Due to these reasons, epoxy resins are now used in a wide range of sectors, including construction, adhesives, electric and digital insulation, coatings, and matrix materials for the automotive, marine, and aerospace industries. Epoxy resins are used all over the world in items including coatings, electric/digital devices, adhesives, flooring and paving, composites, and moulding. Epoxy resin is also anticipated to be used in composite matrices for long-distance pipes and dental applications in the future. Last but not least, the creation of epoxy resins generated from biomaterials will be a crucial breakthrough for reaching sustainable objectives and reducing adverse environmental effects and CO_2 emissions.

REFERENCES

Abdel-Aal N, El-Tantawy F, Al-Hajry A and Bououdina M (2008), 'Epoxy resin/plasticized carbon black composites. Part I. Electrical and thermal properties and their applications', *Polym Compos*, 29, 511–517. doi: 10.1002/pc.20401.

Aggarwal L K, Thapliyal P C and Karade S R (2007), 'Anticorrosive properties of the epoxy-cardanol resin based paints', *Prog Org Coats*, 59, 76–80. doi: 10.1016/j.porgcoat.2007.01.010.

Alothman O Y, Jawaid M, Senthilkumar K, Chandrasekar M, Alshammari B A, Fouad H, Hashem M and Siengchin, S (2020). 'Thermal characterization of date palm/epoxy composites with fillers from different parts of the tree', *J Mater Res Technol*, 9, 15537–15546. doi: 10.1016/j.jmrt.2020.11.020.

Altuna F I, Espósito L H, Ruseckaite R A and Stefani P M (2011), 'Thermal and mechanical properties of anhydride-cured epoxy resins with different contents of bio-based epoxidized soybean oil', *J Appl Polym Sci*, 120, 789–798. doi: 10.1002/ app.33097.

Armelin E, Pla R, Liesa F, Ramis X, Iribarren J I and Aleman C (2008), 'Corrosion protection with polyaniline and polypyrrole as anticorrosive additives for epoxy paint', *Corros Sci*, 50, 721–728. doi: 10.1016/j.corsci.2007.10.006.

Bajat J B and Dedic O (2007), 'Adhesion and corrosion resistance of epoxy primers used in the automotive industry', *J Adhes Sci Technol*, 21, 819–831. doi: 10.1163/156856107781061512.

Brent Strong A (2008), *Fundamentals of Composite Manufacturing: Materials, Methods and Applications*, second edition, Dearborn, MI, Society of Manufacturing Engineers, vol. 2, p. 45.

Calò E, Maffezzoli A, Mele G, Martina F, Mazzetto S E, Tarzi A and Stifani C (2007), 'Synthesis of a novel cardanol-based benzoxazine monomer and environmentally sustainable production of polymers and bio-composites', *Green Chem*, 9, 754–759. doi: 10.1039/B617180J.

Campaner P, D'Amico D, Longo L, Stifani C and Tarzia A (2009), 'Cardanol-based novolac resins as curing agents of epoxy resins', *J Appl Polym Sci*, 114, 3585–3591. doi: 10.1002/ app.30979.

Campbell F C (2010), *Structural Composite Materials*, Materials Park, OH, ASM International, p. 68.

Cha J, Kim J, Ryu S and Hong S H, (2018), 'Comparison to mechanical properties of epoxy nanocomposites reinforced by functionalized carbon nanotubes and graphene nanoplatelets', *Composites Part B*. doi: 10.1016/j.compositesb.2018.11.011.

Chandrasekar M, Siva I, Kumar T S M, Senthilkumar K, Siengchin S and Rajini N. (2020). 'Influence of fibre inter-ply orientation on the mechanical and free vibration properties of banana fibre reinforced polyester composite laminates', *J Polym Environ*, 28, 2789–2800. doi: 10.1007/s10924-020-01814-8.

Chittimenu H, Pasupureddy M, Muthukumar C, Krishnasamy S, Muthu Kumar Thiagamani S and Siengchin S (2021), 'Fracture toughness of the natural fiber-reinforced composites: A review'. *Mech Dynam Prop Biocomposit*, 293–304.

Davies P, Mazéas F and Casari P (2001), 'Sea water ageing of glass reinforced composites: shear behaviour and damage modeling', *J Compos Mater*, 35, 1343–1372. doi: 10.1106/MNBC-81UB-NF5H-P3ML.

Dyakonov T, Chen Y, Holland K, Drbohlav J, Burns D, Velde D V, Seib L, Solosky E J, Kuhn J, Mann P J and Stevenson W T K (1996), 'Thermal analysis of some aromatic amine cured model epoxy resin systems – I: Materials synthesis and characterization, cure and post-cure', *Polym Degrad Stab*, 53, 217–242. doi: 10.1016/0141-3910(96)00085-7.

Ellis B (1993), 'The kinetics of cure and network formation', in Ellis B (ed.), *Chemistry and Technology of Epoxy Resins*, London, Blackie Academic and Professional.

Ghaemy M and Sadjady S (2006), 'Kinetic analysis of curing behavior of diglycidyl ether of bisphenol A with imidazoles using differential scanning calorimetry techniques', *J Appl Polym Sci*, 100, 2634–2641. doi: 10.1002/app.22716.

Gojny F H, Wichmann M H G, Fiedler B, Kinloch I A, Bauhofer W, Windle A H and Schulte K (2006), 'Evaluation and identification of electrical and thermal conduction mechanisms in carbon nanotube/epoxy composites', *Polymer*, 47, 2036–2045. doi: 10.1016/j.polymer.2006.01.029.

Lee E S, Lee S M, Shanefield D J and Cannon W R (2008), 'Enhanced thermal conductivity of polymer matrix composite via high solids loading of aluminum nitride in epoxy resin', *J Am Ceram Soc*, 91, 1169–1174. doi: 10.1111/j.1551–2916.2008.02247.x.

Lee W S and Yu J (2005), 'Comparative study of thermally conductive fillers in underfill for the electronic components', *Diam Relat Mater*, 14, 1647–1653. doi: 10.1016/j.diamond.2005.05.008.

Ma P C, Tang B Z and Kim J K (2008), 'Effect of CNT decoration with silver nanoparticles on electrical conductivity of CNT–polymer composites', *Carbon*, 46, 1497–1505. doi: 10.1016/j.carbon.2008.06.048.

Maffezzoli A, Calò E, Zurlo S, Mele G, Tarzia A and Stifani C (2004), 'Cardanol based matrix biocomposites reinforced with natural fibers', *Compos Sci Technol*, 64, 839–845. doi: 10.1016/j.compscitech.2003.09.010.

Mamunya Ye P, Davydenko V V, Pissis P and Lebedev E V (2002), 'Electrical and thermal conductivity of polymers filled with metal powders', *Eur Polym J*, 38, 1887–1897. doi: 10.1016/S0014-3057(02)00064-2.

McGrath L M, Parnas R S, King S H, Schroeder J L, Fischer D A and Lenhart J L (2008), 'Investigation of the thermal, mechanical, and fracture properties of alumina-epoxy composites', *Polymer* 49, 999–1014. doi: 10.1016/j.polymer.2007.12.014.

Miskovic-Stankovic V B, Zotovic J B, Kacarevic-Popović Z and Maksimovic M D (1999), 'Corrosion behaviour of epoxy coatings electrodeposited on steel electrochemically modified by Zn–Ni alloy', *Electrochim Acta*, 44, 4269–4277. doi: 10.1016/S0013-4686(99)00142-5.

Miyagawa H, Misra M, Drzal L T and Mohanty A K (2005), 'Fracture toughness and impact strength of anhydride-cured biobased epoxy', *Polym Eng Sci*, 45, 487–495. doi: 10.1002/pen.20290.

Mourad A H I, Beckry Mohamed A M and El-Maaddawy T (2010), 'Effect of seawater and warm environment on glass/epoxy and glass/polyurethane composites', *Appl Compos Mater*, 17, 557–573. doi: 10.1007/s10443-010-9143-1.

Narasimha Murthy H N, Sreejith M, Krishna M, Sharma S C and Sheshadri T S (2010), 'Seawater durability of epoxy/vinyl ester reinforced with glass/carbon composites', *J Reinf Plast Compos*, 29, 1491–1499. doi: 10.1177/0731684409335451.

Nasimudeen N A, Karounamourthy S, Selvarathinam J, Kumar Thiagamani S M, Pulikkalparambil H, Krishnasamy S and Muthukumar C (2021), 'Mechanical, absorption and swelling properties of vinyl ester based natural fibre hybrid composites', *Appl Sci Eng Prog*. doi: 10.14416/j.asep.2021.08.006.

Pascault J P and Williams R J J (2010), 'General concepts about epoxy polymers', in Pascault J P and Williams R J J (eds.), *Epoxy Polymers – New Materials and Innovations*, Weinheim, Germany, Wiley-VCH, p. 4.

Popovic M M, Grgur B N and Miskovic-Stankovic V B (2005), 'Corrosion studies on electrochemically deposited PANI and PANI/epoxy coatings on mild steel in acid sulfate solution', *Prog Org Coat*, 52, 359–365. doi: 10.1016/j.porgcoat.2004.05.009.

Senthil Muthu Kumar T, Senthilkumar K, Chandrasekar M, Karthikeyan S, Ayrilmis N, Rajini, N and Siengchin S (2021), 'Mechanical, thermal, tribological, and dielectric properties of biobased composites', in *Biobased Composites: Processing, Characterization, Properties, and Applications*, Wiley Online Library, pp. 53–73.

Senthilkumar K, Saba N, Chandrasekar M, Jawaid M, Rajini N, Siengchin S, Ayrilmis N, Mohammad F and Al-Lohedan H A (2021), 'Compressive, dynamic and thermomechanical properties of cellulosic pineapple leaf fibre/polyester composites: Influence of alkali treatment on adhesion', *Int J Adhes Adhes*, 106, 102823. doi: 10.1016/j.ijadhadh.2021.102823.

Senthilkumar K, Siva I, Rajini N and Jeyaraj P (2015). 'Effect of fibre length and weight percentage on mechanical properties of short sisal/polyester composite', *Int J Comput Aided Eng Technol*, 7(1), 60. doi: 10.1504/IJCAET.2015.066168.

Senthilkumar K, Subramaniam S, Ungtrakul T, Kumar T S M, Chandrasekar M, Rajini N, Siengchin S and Parameswaranpillai J (2022). 'Dual cantilever creep and recovery behavior of sisal/hemp fibre reinforced hybrid biocomposites: Effects of layering sequence, accelerated weathering and temperature', *J Ind Text*, 51(2_suppl), 2372S–2390S. doi: 10.1177/1528083720961416.

Shahroze R M, Chandrasekar M, Senthilkumar K, Senthil Muthu Kumar T, Ishak M R, Rajini N, Siengchin S and Ismail S O (2021), 'Mechanical, interfacial and thermal properties of silica aerogel-infused flax/epoxy composites', *Int Polym Proc*, 36(1), 53–59. doi: 10.1515/ipp-2020-3964.

Sharma S and Luzinov I (2011), 'Ultrasonic curing of one-part epoxy system', *J Compos Mater*, 45, 2217–2224. doi: 10.1177/0021998311401075.

Tan S G and Chow W S (2010), 'Biobased epoxidized vegetable oils and its greener epoxy blends: A review', *Polym Plast Technol Eng*, 49, 1581–1590. doi: 10.1080/03602559.2010.512338.

Thomas S K, Parameswaranpillai J, Krishnasamy S, Begum P M S, Nandi D, Siengchin S, George JJ, Hameed N, Salim NV and Sienkiewicz N (2021), 'A comprehensive review on cellulose, chitin, and starch as fillers in natural rubber biocomposites', *Carbohydr Polym Technol Appl*, 2, 100095. doi: 10.1016/j.carpta.2021.100095.

Valenza A and Calabrese L (2003), 'Effect of CTBN rubber inclusions on the curing kinetic of DGEBA–DGEBF epoxy resin', *Eur Polym J*, 39, 1355–1363. doi: 10.1016/S0014-3057(02)00390-7.

Wang H, Wang H and Zhou G (2011), 'Synthesis of rosin-based imidoamine-type curing agents and curing behavior with epoxy resin', *Polym Int*, 60, 557–563. doi: 10.1002/pi.2978.

Weinmann D J, Dangayach K and Smith C (1996), 'Amine-functional curatives for low temperature cure epoxy coatings', *J Coat Technol*, 68, 29–37.

Yang D J, Zhang Q, Chen G, Yoon S F, Ahn J, Wang S G, Zhou Q and Li J Q (1991), 'Thermal conductivity of multiwalled carbon nanotubes', *Phys Rev B*, 66, 165440. doi: 10.1103/PhysRevB.66.165440.

Zarrelli M, Skordos A A and Partridge I K (2002), 'Investigation of cure induced shrinkage in unreinforced epoxy resin', *Plast Rubber Compos Process Appl*, 31, 377–384. doi: 10.1179/146580102225006350.

Zhou T, Wang X, Liu X and Xiong D (2010), 'Improved thermal conductivity of epoxy composites using a hybrid multi-walled carbon nanotube/micro-SiC filler', *Carbon*, 48, 1171–1176. doi: 10.1016/j.carbon.2009.11.040.

2 Bast Fiber-Based Epoxy Composites

Manickam Ramesh and Kanakaraj Niranjana
KIT-Kalaignarkarunanidhi Institute of Technology

CONTENTS

2.1 INTRODUCTION

Numerous eco-sustainable cellulosic fiber products have emerged in the engineering and construction sectors. As a result of the global shift toward a bio-based economy, it is important to reduce carbon footprints as a supplement to fossil-based products.

DOI: 10.1201/9781003271017-2

The real-time sustainability model has developed with improvized policies in some of the developed countries [1,2]. However, even with large-scale implementation in other developing nations, more familiarity with the idea is still necessary. Since the nineteenth century, there has been a significant decrease in the market for petroleum-based fibers and polyester fibers. This loss may be attributed to the fact that scientific advancements in synthetic fibers were made over the twentieth century [3,4].

Primary and secondary fibers are the two main types of plant fibers that are assessed. Primary plant fiber is typically identified in the plant's more significant parts, such as its stem, seed, core, and roots. Meanwhile, secondary fibers are produced from the by-product of the predominant fiber. Seed, leaf, bast, core, grass or reed, and root fibers are the six key categories of plant fiber (collectively called as cellulosic fibers) [5,6]. The use of bio-fillers as a reinforcement in composite materials offers a number of advantages over synthetic fillers, including higher strength-to-weight ratios, lower environmental impact, lower production costs, and a sustainable and plentiful supply. Finally, the materials' limitations are to overcome. On the application of bio-fillers and fibers in composites, it is found that the materials that are damaged by moisture and unsuitable for use at elevated temperatures when exposed to bacterial and insect assault. It is more durable, flexible, resistant to deterioration in sea water, and dye-friendly. The bast fiber has the highest aspect ratio of any of the other fibers, which gives it the reputation of being the strongest fiber [7,8]. The natural cell structure of bast fiber is shown in Figure 2.1.

There are primarily two categories of polymers, thermoplastics, and thermosets. The most often used thermosetting resins are polyester, phenolic, and epoxy. The origin of the word "epoxy" can be traced back to two Greek words: epi, which means "over," and oxy, which means "acid" [11]. An early nineteenth century Russian chemist named N. A. Prilezhaev first discovered epoxide through the epoxidation reaction, which involves the interaction of unsaturated molecules with peroxybenzoic acid [12]. Epoxy-based composites are frequently utilized in load-bearing applications due to their inexpensive cost, high specific strength, outstanding adhesiveness, heat and solvent resistance, and excellent mechanical qualities [13,14].

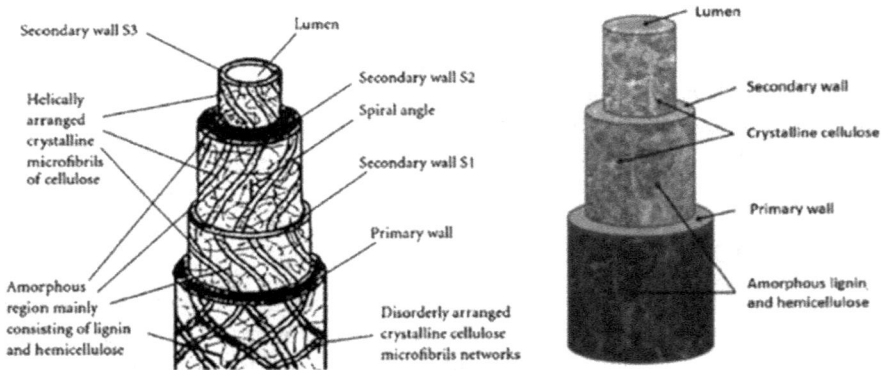

FIGURE 2.1   Natural cell walls of plant [9,10].

To meet engineering needs, it is crucial to have a firm grasp of the characteristics of natural materials. In the last two decades, epoxy composites based on natural fibers have gained a lot of popularity as structural materials. As the years progressed, scientists from all around the world conducted countless experiments. Later, in 1939, an American chemist Greenly produced a comparable resin under the name "epoxydiane," which might be called a binder for protective coats [11]. Two-thirds of all epoxy come from DOW Chemical and Shell, the two largest manufacturers in the world. Epoxies always have little resistance to the spread of fatigue cracks, poor fracture characteristics, and low impact strength. Epoxy has a wide range of applications across a variety of industries, including automotive, ship building, and petrochemical. Epoxy is used in the field of civil engineering for flooring, sealants, and polymer concretes. Epoxy is used in electro-technical business for potting and printed circuit boards [15].

The major classes of thermosetting polymers most frequently used as matrices for fiber-reinforced composite materials and structural adhesives are polymer matrices. Because they are made of amorphous, cross-linked polymers, they exhibit desired features such high tensile strength, elastics modulus, simple processing, improved thermal resistance, chemical resistance, and dimensional stability [16]. On the other hand, this results in a low toughness and poor fracture resistance, both of which need to be improved before they can be considered for many end-use applications. Improving these properties is one of the most successful strategies for improving the toughness of polymers [17].

Despite the possibility that the latter can be substituted for chemically inerted or treated substances. Epoxy resins are synthesized when the reactant epichlorohydrin is mixed with bisphenol-A. The monomers of polyamines, such as triethylenetetramine, make up the hardener. As a result of the chemical reaction between these molecules, a covalent link is formed between the amine and epoxide groups. The resultant polymer will have many cross-links since each OH group will react with an epoxide group [11]. Epoxy is utilized as a structural matrix material and a glue reinforced by fiber in the aerospace sector. Glass, carbon, kevlar, and boron are among the most often used reinforcing fibers. Therefore, researchers are consistently striving to make bio-composites with both robust and easily implemented nature in the industry. The most recent studies on bio-fillers and natural fibers-based epoxy and hybrid composites cover a wide range of factors impacting fiber performance [18,19].

2.2 BAST FIBERS

In the natural fiber, the cell walls are composed of lignocellulosic components like cellulose, hemicellulose, and lignin. Bast fibers are readily biodegradable by microbes, bacteria, and fungi. The cell walls of the fibers contain lignocellulosic components, and the next step is for the bacteria and fungi to develop the specialized enzymes needed to break these down into units small enough to be digested by the microorganisms and used for their own growth and sustenance [20,21]. Bacteria were shown to be less efficient than fungi at breaking down lignocellulosic materials. Naturally occurring fibers lose mechanical strength as they biodegrade, which has a knock-on effect on the mechanical strength and service life of composites made

Hemp   Jute   Flax   Kenaf   Ramie

FIGURE 2.2   Bast fiber plants.

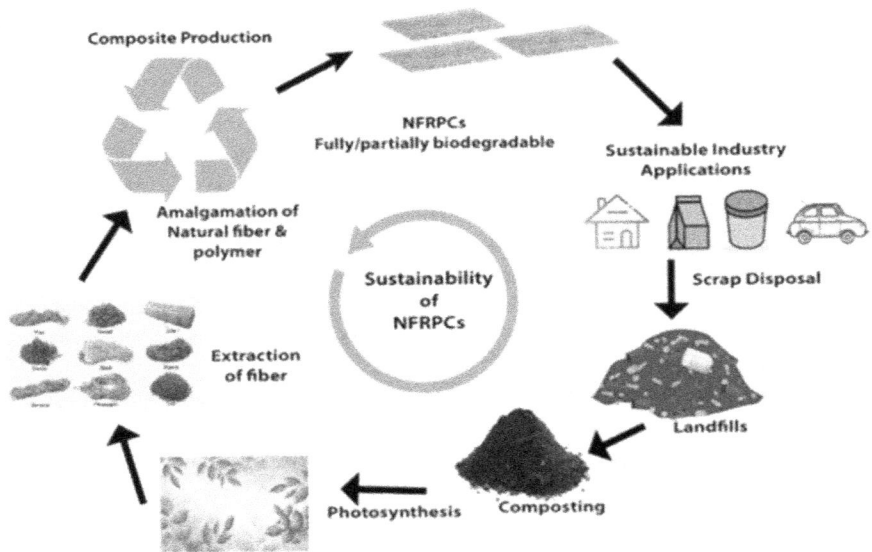

FIGURE 2.3   Life cycle and sustainability of bast fiber-reinforced composite [24].

from naturally occurring fibers. The biodegradability of polymers can be improved by adding biodegradable natural fibers to them; this is true for both biodegradable and nonbiodegradable polymers [22]. The lack of weather resistance means that these polymer composites can't be used in the outdoors. However, there are a number of steps that may be done to lessen the biodegradability of the fibers. These steps involve changing the chemistry of the cell wall through the use of chemical treatments [23]. Major bast fiber plants are depicted in Figure 2.2. Life cycle and sustainability of bast fiber-reinforced composite are presented in Figure 2.3 [24].

2.3   CLASSIFICATION OF BAST FIBER

Bast fibers are classified based on several factors and the major bast fibers are hemp, jute, kenaf, flax, and ramie [25]. The extracted bast fibers are presented in Figure 2.4. The longitudinal and cross-sectional views of these fibers are depicted in Figure 2.5. The structure of bast fiber is given in Figure 2.6. The prospects and

| Hemp | Jute | Flax | Kenaf | Ramie |

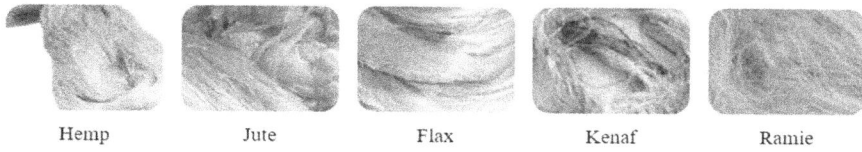

FIGURE 2.4 Extracted bast fibers.

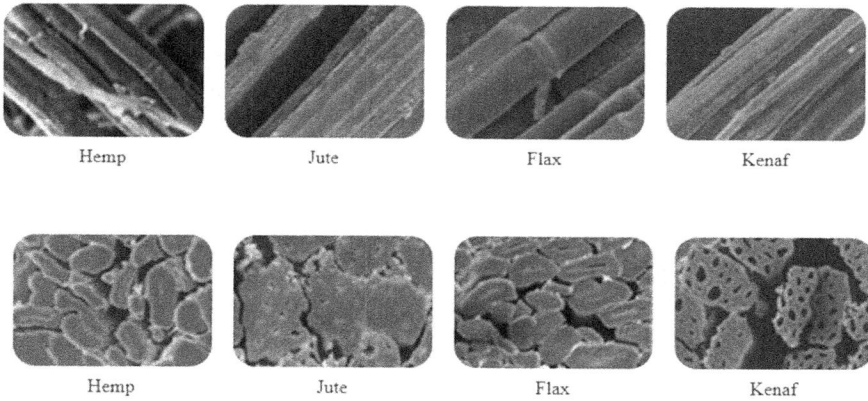

| Hemp | Jute | Flax | Kenaf |

| Hemp | Jute | Flax | Kenaf |

FIGURE 2.5 SEM images showing longitudinal and cross-sectional view of the bast fibers [25].

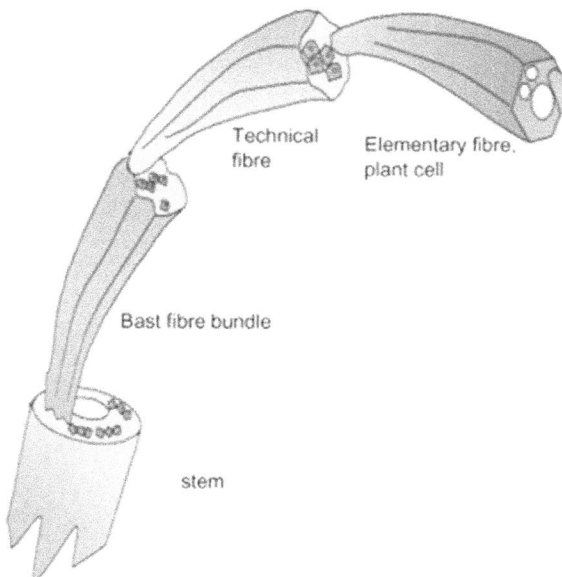

FIGURE 2.6 Structure of bast fiber [25].

chemical composition of different bast fibers are listed in Tables 2.1 and 2.2. The influence of chemical constituents is presented in Table 2.3.

TABLE 2.1
Prospect of the Bast Fibers

Fiber	Family Name	Also Known as	Botanical Name	Other Sphere	Primary Cultivator	Applications	Ref.
Jute	Malvaceae	Glorious fiber	*Corchorus capsularis* and *Colitorius*	Second largest vegetable fiber	India, China, Bangladesh	Yarn, string, carpet backing fabric, hessian	[26]
Flax	Linaceae	-	*Linum usitatissimum*	Crispier and more difficult to handle	Canada, France, Belgium, China, Ukraine and Belarus	Ropes, twines, cordage, sails, towels, canvas, and tents	[27]
Hemp	Cannabaceae	-	*Cannabis sativa*	-	Germany, China, France, Philippines	Household utensils and cosmetics	-
Kenaf	Malvaceae	Mesta/ Stockroot /Ambari	*Hibiscus cannabinus*	Inherent capacity for absorption as well as fire resistance	India, Bangladesh, USA	Cordage	[28]
Ramie	Urticaceae	China Grass	*Boehmeria nivea*	Superior elasticity and dyeability	China, Brazil, Philippines, India	Textile-based fibers	[29]

TABLE 2.2
Chemical Composition of Bast Fibers [12]

Fiber	Cellulose (wt.%)	Hemicellulose (wt.%)	Lignin (wt.%)	Pectin (wt.%)	Wax (wt.%)
Jute	61–71	13.6–20.4	11.8–13	0.2–0.4	0.5
Flax	70–73	18.6–20.6	2–5	2.3	1.5–1.7
Hemp	70–74	15–22.4	3.7–10	0.9	0.8
Kenaf	45–57	20.3–21.5	8–19	3–5	–
Ramie		13–16.7	0.5–0.7	1.9	0.3

TABLE 2.3
Influence of Chemical Constituents

Influencing Parameter	Effect
Hemicellulose	Hydrophilic tendency
	Absorbs the moisture from the atmosphere and forms a new hydrogen atom on the surface
	It protects the cellulose from the outer damage
	Thermal stability
Cellulose	Higher cellulose content
	High degree of polymerization
	Increased fibers Young's modulus and tensile strength
	Responsible for the mechanical strength
	Thermal stability
Micro-fibrillar angle	Smaller angle
	High degree of polymerization
	Stiffness
	Ductile
	Greater angle
	Non-ductile
	Higher tensile strength
Lignin, oils, and wax	Subsequent removal of this layer enhances mechanical interlocking by boosting interfacial shear strength and fiber wettability
	Its high hydrophilicity
	It alters the surface chemistry of fibers and is consequently incompatible with hydrophobic polymer matrices
Filler as an additive	Fillers are additives used to increase durability, temperature stability, and safety from spontaneous combustion
	Filler's size, aspect ratio, and chemical composition all have an impact on its properties
Pectin	Thermal stability

2.3.1 HEMP

Hemp is a kind of plant that belongs to the Cannabaceae family and is primarily cultivated in France and China to harvest its seeds and fiber [30]. *Cannabis* plants originated in Central Asia, and it wasn't until the iron age that they were brought to Europe. Today, it is grown in temperate regions such as North Korea, Chile, Japan, India, and Europe, among other places. The plant may grow to around 12 feet, with an average fiber bundle length of 10–50 mm and a diameter of 20–550 μm [31]. Several different products, including paper, textile fiber, seed food, oil, pulp, and biofuel, all make use of hemp in some capacity. It contributes less than 0.5% to the total output of natural fibers across the globe at this time. The tensile strength of the fiber may vary anywhere from 270 to 690 MPa, and its modulus can be anywhere from 23.5 to 90 GPa. Hemp fiber may have micro-fibrillar angles ranging anywhere from 2.2° to 6.2°. Hemp fiber is used in the production of automobile components by Bayerische Motoren Werke AG (BMW), which is a well-known automotive company [30,32].

2.3.2 JUTE

Corchorus olitorius or *Corchorus capsularis* is the scientific name for jute. Jute is the second least expensive bast fiber option available in primarily tropical areas [26]. It is a member of the Malvaceae family and has the most significant production volume compared to other types of fibers. The history of jute may be traced back to the dynasty of Akbar, and there is evidence of ancient documents of India's inhabitants using fibers [33,34].

2.3.3 FLAX

The Linaceae family comprises 13 genera and 300 different species of plant. Flax is a member of this family [35]. It is known to have been grown in Georgia and Egypt far before the year 5000 BCE and was the first plant to be woven into textiles. Because of the high demand for its fiber and linseed oil, it is primarily cultivated in France, Belgium, and Canada. Low-grade fibers are utilized for reinforcement in composites, while high-grade fibers are spun into yarns for use in the textile industry [31]. The fiber's impressive tensile strength is the result of this minimal angle. The tensile strength of the fiber ranges between 345 and 1100 MPa, while its modulus ranges between 27.6 and 103 GPa. The micro-fibrillar angle of flax fiber ranges from 5° to 10°, and its use is possible with thermoplastics and thermosets [36,37].

2.3.4 KENAF

Cultivation of the Malvaceae family plant known as kenaf dates back to 4000 BCE. It's a staple crop in modern-day Bangladesh and India, although it originated in southern Asia and central Africa. The people of Bengal and India use the name mesta, whereas those of Indonesia, South Africa, and Taiwan call it stockroot or ambari [38]. The herbaceous kenaf plant may grow to a height of 3 m, and its thick woody stem and base have a diameter of 3–5 cm. In tropical climates, the plant is mostly farmed for the seed oil and fiber it produces. It may reach a height of 10 cm/day when grown in natural settings, and it can mature in 3 months [39]. About 40% of the plant material that is extracted for its fibers is located inside the kenaf tree's bark, while the remaining 60% consists of the core wood. In comparison, the energy required to process 1 kg of glass fiber is just 15 MJ for the low-cost fiber made from kenaf. The tensile strength and modulus of kenaf range from 223 to 930 MPa and 14.5 to 53 GPa, respectively, giving it excellent potential as a raw material for thermoplastics and thermosets [40,41].

2.3.5 RAMIE

The Urticaceae plant family, from which ramie is derived, is largely cultivated in Brazil and China. It is widely known as "China Grass" due to the fact that it has been cultivated in China for many decades [42]. It is a perennial herb that lives for 7–20 years and reaches a height of 1–2.5 m. This bast fiber output is the smallest in the world because of its low production and the inclusion of impure components such as gum and pectin; using it as reinforcement in composites might be challenging. The use of ramie fibers is relatively underexplored as a result of the challenges that have been mentioned (impurity and availability) [43]. The fiber has a modulus that ranges from 24.5 to 128 GPa, and its tensile strength may be anywhere between 400 and 1000 MPa. According to the micro-fibrillar angle of the fiber is 7.5°. Ramie is a kind of fiber that may be utilized in thermoplastic and thermosetting resins [44,45].

2.4 EXTRACTION AND PROCESSING OF BAST FIBERS

The bast fibers are extracted from the respective plants in several ways. The common extraction process is depicted in Figure 2.7.

2.4.1 MECHANICAL EXTRACTION TECHNIQUES

The relationship between a core and its fiber can be broken by mechanical forces, and the process is known as mechanical extraction. The fundamental objective of mechanical processing of bast fibrous plants is to extract as much high-quality fiber as possible to facilitate further processing. The process of extracting fibers results in intertwined strands of fiber. The extraction of fiber-containing tissues from the plant's stem or other portions is required to obtain technical fiber. Decortication, or the removal of green fiber prior to retting, is an essential part of any discussion of fiber extraction technologies [46]. The quality of the bast fiber obtained in this manner is lower than that of fiber obtained via retting. Therefore, decorticated fiber is typically utilized in contexts outside of textile production. The fiber is often extracted from the woody components by means of mechanical processing, such as crushing and squeezing. Fibers may become damaged or broken as a result of the tension stress to which they are subjected during the processing of raw materials, which may be present in a greater or smaller amount depending on the circumstances. This may have a direct and adverse effect on the product's quality. Mechanical processing has the potential to significantly increase the purity of the fibers but also has the risk of significantly reducing their length. Instances in which the proportion of long bast fibers to impurities is reversed could make it possible for there to be a greater proportion of long bast fibers despite there being a greater concentration of impurities [27].

2.4.2 RETTING PROCESS

Bast fibers are typically extracted from plants using one of three methods: retting, mechanical, or a hybrid. The goal of the retting procedure in biology is to

FIGURE 2.7 Fiber extraction process.

remove fiber bundles from their native context with as little disruption as possible. The conditions under which the fibers are retted have a major impact on the quality of the extract. Due to over-retting, fibers can become brittle and even break at their weakest points. Because of the connection between the cores and the fiber bundles that occur from under retting, the purity of the fiber is compromised [47].

This method is far more efficient than retting in terms of tons per hour handled. However, it is challenging to exert precise control over a collection of mechanical pressures operating on the stalk, and the bond-breaking process does not respond well to external mechanical forces. This method also produces fiber with wildly varying lengths, which is another drawback of mechanical extraction [48]. In this scenario, relying solely on mechanical extraction for removal does not produce particularly satisfying results. During the mechanical extraction process, plant stalks that have been pre-retted before the process can have improved fiber separation. Consequently, the combined use of retting and mechanical extraction has increased the efficiency of fiber extraction [3].

The most suitable approach of retting is called water retting, and it entails submerging bundles of stalks in water for a period. When the stalk is exposed to water, it swells up from the inside, which enables more moisture and natural decay bacteria to seep in via the cracks in the outer covering. An excessive amount of retting causes the fiber to become brittle, whereas an insufficient amount of retting makes separation more challenging. In double retting, the stalks are removed from the water before the process is finished, dried for several months, and then retted again; this gentle process results in excellent fiber. The process of retting with water is the oldest technique for which there is documentary evidence. This method formerly enjoyed a great deal of notoriety due to its capacity to generate retted bast fibers of an exceptionally high quality. On the other hand, the production of a big quantity of sewage water is a significant problem that cannot be ignored [49].

2.4.3 DECORTICATION

The plant stalks are fed into the decorticator, where they experience the three types of force: impact, shear, and compression. As a consequence of the stalks being cut into several pieces, the bast fiber is subsequently separated to varying degrees to become the core. The fiber bundle is the primary product of the decortication process, although it is possible that in some circumstances it could also be considered a by-product. Processing fibers has historically involved the utilization of a wide variety of decorticators, including but not limited to crushing rollers, ball mills, hammer mills, drop weight systems, and many more [3].

2.4.4 SURFACE TREATMENT OF BAST FIBERS

The different physical and chemical surface modification of bast fibers is presented in Figure 2.8 and the effect of surface modification is listed in Table 2.4.

Physical method

Corona treatment
Plasma treatment
Gamma radiation

Chemical method

Alkali treatment
Silane treatment
Enzyme treatment
Peroxide treatment
Sodium chloride treatment
Stearic treatment
Etherification treatment
Acetylation
grafting
Benzoylation

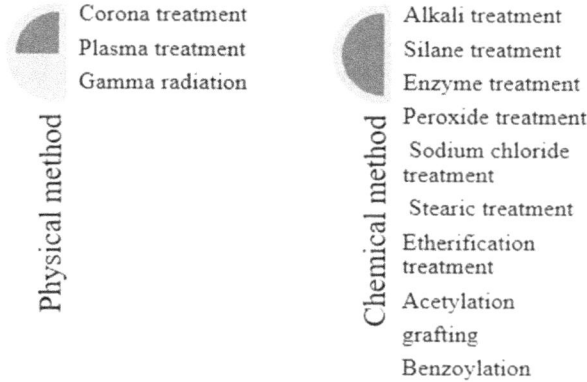

FIGURE 2.8 Physical and chemical modifications of natural fibers.

TABLE 2.4
Surface Modification of Bast Fibers and Their Effects

Treatment Method	Effects	Effect Identified	Ref.
Alkali treatment	• The elevated temperature will increase the fiber from cellulose I to cellulose II • The removal of hemicellulose, cellulose, lignin, and wax which create a crack in the fiber region • Treatment results in a rougher topography on the fiber surface, which promotes mechanical interaction with the polymer surface. The enhanced compatibility between matrix and reinforcement	It is defined in the crystallinity study using X-ray diffraction (XRD)	[13]
Sodium hydroxide treatment	Rough surface can be formed due to the removal of noncellulosic components	SEM images shown a micro-hole which results its absence of non-cellulosic	[50]

2.5 FABRICATION OF BAST FIBER EPOXY COMPOSITES

Bast fiber-based epoxy resin is a thermoset composite where the bonding ingredient gives the fiber strength. It has improved thermal qualities, in addition to higher strength and stiffness, as well as a high fatigue load. The epoxy resin and the hardener need to be thoroughly combined in order to obtain a homogenous solution. The epoxy is always mixed with hardener to initiate the polymerization process to cure itself. Usually, they are mixed in a ratio of 10:1. This is necessary in order to ensure that the matrix has consistent physical properties and that intermolecular bonds are formed [3]. Among the many types of polymer matrices, epoxy is one of the most common. Epoxy is a far more robust matrix system than its rival. In terms of physical, chemical, electrical, and thermal behavior, they have superior properties. Isophthalic and orthophthalic resins are the two types of polyester resin often utilized in industries for composite laminating. For applications that require high corrosion resistance and low temperature tolerance, orthophthalic resins with less styrene content are typically employed. Isophthalic polyester resin with styrene contents between half portions of the material also has a stellar reputation for resistance to corrosion and strength under stress. Through the utilization of this resin, the high chemical stability, high temperature resistance up to 350°C, great fire resistance, low toxic emissions, and good friction qualities were achieved [51].

2.5.1   HAND LAY-UP

Contact molding, also known as hand lay-up, is used for thermosetting resins. Bast fiber composites may be made with high fiber content by using either long or short fibers in a variety of shapes (such as the chopped mat, stitched or woven cloth). Having a low viscosity makes resin more workable. In this technique, a releasing agent (silicone spray or wax) is applied to the mold before liquid resin is brushed or rolled on top [52]. Overlaying resin-impregnated fibers and rollers are used to smooth down the air pockets. Curing takes place in a room with average room temperature and humidity. It is a method that requires a lot of labor but inexpensive and is used for fibers that are longer and have a higher fiber content [10].

2.5.2   MOLDING USING A RESIN TRANSFER SYSTEM

Vacuum-assisted resin transfer molding, co-injection resin transfer molding, and so on, are all types of resin transfer molding. Liquid resin is pumped into stationary preforms during this manufacturing procedure. After laying down preforms in the mold chamber, a second corresponding mold is fastened on top [53]. Resin is injected under pressure into the hot mold via a single or several ports. After it has had time to cool, the product is taken out. This method applies to any kind of fiber. The holes in stitched materials provide quick transit of the resin, which improves the material's performance. Utilizing this method, high fiber-volume composites with low-void contents may be produced [46].

2.5.3   COMPRESSION MOLDING

Compression molding is often utilized for rapid bulk manufacturing. Molding a composite into the proper form requires pre-heating and pressurization. The fibers are weighed before being poured into a hot mold. Pressurizing the mold before heating helps the material cure faster. Due to the lack of shear loads and the absence of high velocities, the fiber sustains little to no damage. These liquid chemicals cure into the form of a cavity, and the finished product is discharged. This method may be utilized with either long (unstitched form) or short fibers (mat, stitched form). The former method is used to lessen the contraction, whereas the latter creates a substantial fraction by volume [28].

2.6 PROPERTIES OF BAST FIBER-REINFORCED EPOXY COMPOSITES

The bast fibers that hold up the plant's stalk are very dense, highly crystalline, and cellulose-rich [3]. The physical parameters of bast fibers are given in Table 2.5 [10].

2.6.1 MECHANICAL PROPERTIES

Bast fibers or fiber bundles or unit fibers like other natural fibers exhibit good specific mechanical qualities, particularly when it comes to their strength and stiffness. Bast fibers are equivalent to synthetic fibers like glass fibers, but they have the added advantage of a lower density. Some of the bast fibers' key mechanical features include tensile strength, Young's modulus, elongation at break, fracture stress, specific modulus, and specific strength [28]. These mechanical characteristics are typically examined in order to determine whether bast fibers are suitable for a given application, which is one reason why they are of such considerable relevance. This is true even among fibers derived from plants of the same type, such as hemp bast fibers, because mechanical properties can vary from one fiber to another. These discrepancies are caused by the very same fibers that influence the physical characteristics of the bast fiber, which in turn has an effect on the mechanical properties of the fiber. Furthermore, the mechanical strengths of bast are affected by the fibers' physical characteristics [54]. The ASTM standards of mechanical property evaluation are presented in Figure 2.9 [46].

For instance, the mechanical properties, strength, and stiffness of a fiber are largely determined by its chemical composition, cellulose content, and also by its physical property, the micro-fibrillar angle, with higher cellulose content and a lower micro-fibrillar angle increasing these characteristics. The bast fibers can be broken down into primary and secondary cell walls. The secondary cell walls contain crystalline cellulose, and are responsible for the fibers' mechanical strength. Hemicellulose functions as a matrix material that surrounds the cellulose structure, while lignin offers additional strength by shielding the fiber from any harm that may

TABLE 2.5
Physical Parameters of Bast Fibers [10]

Fiber	Density (g/cm³)	Specific Gravity (g/cm³)	Diameter (μm)	Micro-Fibrillar Angle (°)
Jute	1.3–1.46	1.30	20–200	8
Flax	1.4–1.5	1.30	12–600	5–10
Hemp	1.4–1.5	1.30	25–500	2–6.2
Kenaf	1.4	1.30	–	–
Ramie	1.0–1.55	1.30	20–80	7.5

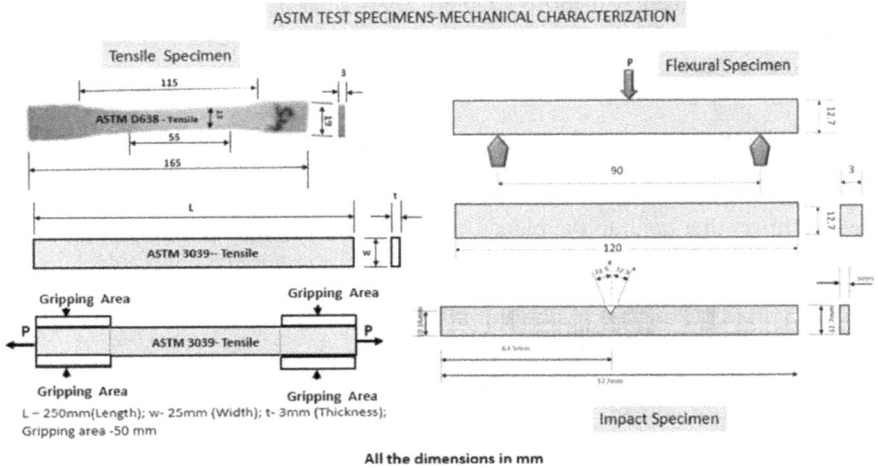

FIGURE 2.9   ASTM standards of mechanical property evaluation [46].

be caused by the environment. The result is a cellulose that is both robust and rigid. It is well recognized that the performance of composites, such as the mechanical properties, is reliant on the characteristics of the specific elements and the interfacial compatibility of such components [48]. One example of this is how composites perform mechanically. Cellulose, the primary constituent of flax and hemp fibers, is highly polar because of the presence of hydroxyl groups, acetate links (C–O–C), and ether linkages (C–O–C) in its structure. In comparison to non-polar polymers such as polypropylene, this makes cellulose more compatible with groups that are polar, acidic, or basic [31]. In order to ensure compatibility between the reinforcing fibers and the surrounding polymer matrix, it is necessary to eliminate many nanocellulose components for the composition of flax fibers. The major wall of the fiber includes practically all of the non-cellulose compounds, with the exception of proteins, inorganic salts, and coloring matter, which have been identified in the fiber lumen. It is this region of the fiber that causes issues, including poor absorbency, poor wettability, and other undesirable textile qualities [55]. To compensate for the alkali consumption required to neutralize various compounds during scouring of natural fibers, such as the acidic hydroxyl groups of cellulose, the carboxyl groups of pectin's, and also amino acids created by protein breakdown, a minimum concentration of 3% NaOH is required. Since the composite performance, i.e., the mechanical characteristics, depends on the properties of the separate components as well as on the interfacial compatibility between the materials, characterizing composite interfaces can be accomplished by examining the surfaces of the composite constituents prior to their combination [56].

Local surfaces have a distinct structure and composition in comparison to the bulk material, and this difference can provide valuable information that can be used to forecast the attributes and performance of the composite. In the composites interface, one or more of the interfacial bonding mechanisms, mechanical interlocking, chemical bonding, and inter-diffusion bonding may be at work simultaneously [57].

In composites, the shear strength is significantly improved by the rougher the fiber surface, but the transverse tensile strength is barely affected. In contrast to mechanical bonding, the strength of chemical bonding is not dependent on the roughness of the fiber surface; rather, it is dependent on the degree to which the chemical groups on the fiber surface react with the chemical groups present in the polymer matrix to produce chemical bonds per unit area, as well as the type of chemical bonds that are generated. Molecular inter-diffusion and inter-weaving between the two components (fiber and polymer matrix) results in inter-diffusion bonding, the strength of which depends on the number of molecules per unit area at the interface as well as the distance between the molecules. The incompatibility between the fiber (hydrophilic) and the polymer matrix (hydrophobic) results in weak interfacial bonding [58]. Extensive study has been carried out in an effort to improve this, and it was found that treating the fibers either mechanically or chemically helps to augment the weak interfacial connection. Corona, plasma, UV, heat treatments, electron radiation, and fiber beating are examples of mechanical treatments; alkali, acetyl, silane, acrylonitrile, and maleated anhydride-grafted coupling agent are examples of chemical treatment. The mechanical properties of different bast fibers are presented in Table 2.6 and their composites are given in Table 2.7.

TABLE 2.6
Mechanical Behavior of the Bast Fibers

Fiber	Tensile Strength (MPa)	Tensile Modulus (GPa)	Fracture Stress (MPa)	Elongation (%)
Jute	320–800	8–78	393–800	1.5–1.8
Flax	345–1100	27.6–103	345–1500	1.2–3.3
Hemp	270–690	23.5–90	270–900	1–3.5
Kenaf	223–930	14.5–53	-	1.5–2.7
Ramie	400–1000	24.5–128	-	1.2–4.0

TABLE 2.7
Mechanical Properties of Bast Fiber-Based Epoxy Composites

Matrix	Bast Fiber	Filler/ Synthetic Fiber	Fabrication Process	Combination	Tensile Stress	Flexural Strength	Impact Energy	Ref.
Epoxy	Jute	Calcium carbonate	Hand lay-up	Jute + epoxy Jute + epoxy + 3% of CaCO$_3$ Jute + epoxy + 6% of CaCO$_3$	55.15 N/mm^2 58.52 N/mm^2 63.54 N/mm^2	72.53 N/mm^2 105.51N/ mm^2 97.52 N/mm^2	0.52 J 0.72 J 0.91 J	[59]
Epoxy	Kenaf	Graphene	Compression molding	60 wt.% epoxy + variation in the fiber and filler	63 MPa	97 MPa	9.56 kJ/m^2	[50]
Epoxy	Flax Hemp	Basalt glass	Vacuum Infusion	Mixture and different stacking	153.16 MPa	137.95 MPa	-	-

2.6.2 Thermal Properties

The term "thermal stability" refers to the degree to which bast fibers withstand disintegration or degeneration when exposed to higher temperatures. The chemical components of the fibers, cellulose, hemicellulose, lignin, and pectin, all contribute to their thermal stability. This weight loss is the result of multiple stages of disintegration caused by the fibers' various chemical constituents, each of which is sensitive to a distinct temperature range [60]. At a temperature of 50°C, hemp fibers, for instance, begin to disintegrate and experience a loss of weight as a result of the evaporation of the moisture content of the hemp. When the temperature rises above 160°C, the lignin that is used to bond hemp fibers begins to weaken, which causes both physical and chemical changes within the fibers themselves. Hemp fibers lose weight due to the decomposition of hemicellulose or pectin at temperatures around 270°C and cellulose degradation at temperatures around 360°C. The decomposition of hemicellulose and pectin in hemp fibers has been correlated with temperatures between 320°C and 370°C, while the degradation of cellulose occurs between 390°C and 420°C [61]. To a greater or lesser extent, natural fibers begin to degrade at temperatures above 200°C. From this it is discovered that the fibers' physical and chemical properties alter when subjected to high temperatures. There was a negative impact on the fiber's mechanical qualities due to the changes in its physical and chemical properties, as demonstrated by the studies. This was confirmed by a study showing that 2 hours of vacuum heating at 300°C reduces the tensile strength of jute fibers by 60% [62]. The findings were similar, stating that ramie fibers' tensile strength dropped by approximately 10% after being heated to about 200°C for just 10 minutes. Weakening natural fiber-reinforced polymer composites are a direct result of degradation of natural fibers' mechanical characteristics. Since natural fibers within polymer composites can be degraded by heat, care must be taken during processing to maintain their integrity. In addition, grafting monomers onto the fibers is a proven method for increasing their thermal stability [42].

2.6.3 Acoustic Properties

It is possible to determine the sound absorption coefficient, abbreviated as SAC, by first measuring the total quantity of sound energy that is absorbed by the materials. The values 0 and 1 comprise the range of the SAC, with 1 indicating the highest possible absorption and 0 indicating that there is no absorption at all [29]. In comparison to the absorption of high-frequency sound waves, the assimilation of low-frequency sound waves, such as those with a frequency of 500 Hz, is an extremely challenging process. Transmission refers to the process in which sound waves travel across a medium without being absorbed and without suffering any loss in frequency. The transmission coefficient (t) measures the amount of incident energy that passes through a medium without being reflected or absorbed. The sound waves are partially absorbed and partially reflected when they hit objects. The reverberation chamber method and the impedance tube method are the two most common approaches to determining how well a material absorbs sound that has been documented. The sound absorption coefficient is typically measured using the reverberation chamber technique since it can be applied to larger samples. The normal incidence impedance

TABLE 2.8

Sound Absorption Characteristics of Bast Fiber Epoxy Composites [29]

Matrix	Bast Fiber	Fabrication Process	SAC
Epoxy	Ramie	Hot press compression machine	0.6 (2000 Hz)
	Flax		0.65 (2000 Hz)
	Jute		0.65 (2000 Hz)
Epoxy	Kenaf	Compression molding machine	0.085 (6000 Hz)
Epoxy	Flax	Hot press machine	0.96 (3200 Hz)
Epoxy	Flax	Compression (laminated)	0.11 (2000 Hz)

tube technique is favored when dealing with a limited sample size. The sound absorption characteristics of bast fiber-based composites are tabulated in Table 2.8 [29].

2.6.4 MOISTURE CONTENT

Similar to other forms of natural fibers, bast fibers are composed mostly of cellulose in their cell walls and typically include moisture concentrations in its voids and in its noncrystalline/amorphous regions. The chemical constituent of bast fibers, cellulose, has a large number of hydroxyl groups that are responsible for the formation of hydrogen bonds between the macro-molecules that make up the bast fiber cell walls. Upon interaction with ambient water molecules, the hydroxyl groups on the cellulose macro-molecules are released from their hydrogen bonds and are free to roam the cellular environment. Following this, the free hydroxyl groups on the cellulose macro-molecules will instead create new hydrogen bonds with the water molecules. As a direct consequence of this, the bast fibers have a hydrophilic characteristic. This results in a substantial moisture absorption in the presence of a hydrophobic polymer matrix in polymer composites reinforced with bast fibers when exposed to water [36].

Composites' micro-cracking, dimensional instability, and degraded mechanical qualities are all brought on by the absorption of moisture, which causes the bast fibers within it to swell. This reduces bonding strength at the interface. Jute fiber-reinforced epoxy composites lost mechanical characteristics, tensile and flexural strength, while submerged in water. In contrast, researchers found that adding jute fibers to glass fiber-reinforced thermoset composites slowed the composite's degradation when subjected to moisture for more than 70 hours. This was likely because the swelled jute fiber layers absorbed some of the strain caused by the resin's expansion, or because the swollen jute fibers provided insulation for the glass fibers in the core. Bast fibers must have their moisture removed before they may be employed as reinforcement in polymers. Chemically treating bast fibers to eliminate their hydrophilic hydroxyl groups would lessen their moisture absorption, as has been proposed by a few other studies [63]. Additionally, bast fiber chemical treatments improve the interfacial adhesion of bast fibers-reinforced polymer composites, which in turn reduces the composites' rate and amount of water absorption. Alternately, bast fiber-reinforced polymer composites might be coated with surface barriers, although this approach could prove too expensive to be practical [64].

2.7 MERITS AND DEMERITS OF BAST FIBERS

The merits of bast fiber-based composites over the synthetic fiber composites are presented in Figure 2.10 [24]. Merits and demerits of bast fiber-reinforced epoxy composites are listed in Table 2.9 [40,65]. Comparison of the benefits of natural fiber over synthetic fiber is given in Table 2.10.

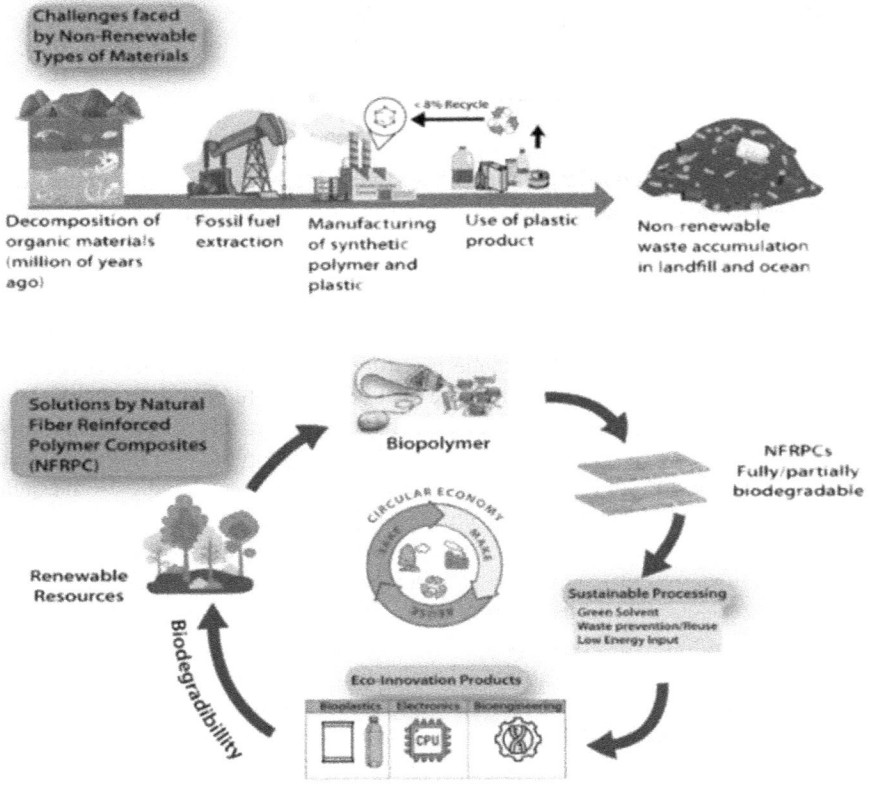

FIGURE 2.10 Demerits of synthetic fiber composites and merits of bast fiber composites [24].

TABLE 2.9
Merits and Demerits of Bast Fiber-Reinforced Composites

Merits	Demerits
• Sustainable and Biodegradable	• Superior ability to absorb moisture
• Manufacturing process consumes less energy	• Insufficient resistance to microbes
• Minimal waste	• Weak density
• Affordable, low cost to manufacture, possibly beneficial product in nations with low average wages	• Thermodynamically inefficient
• Elevated concentrations of thermal and acoustic insulation	• Quality changes based on location and time of year
• Less tool wears	• Variations in demand and availability
• Zero allergic reactions for workers while fabricating natural fiber composites	
• Non-abrasive	

TABLE 2.10
Comparing the Benefits of Natural Fiber Over Synthetic Fiber

Property	Natural Fiber	Synthetic Fiber
Raw Material	Low	High
Density	Low	High
Recyclability	Feasible	Not feasible
GHG during processing of either composite or fiber extraction	Minimal emission	Increased emission
Energy consumption during extraction or fabrication	Low	High

2.8 APPLICATIONS OF BAST FIBER-REINFORCED COMPOSITES

The applications of bast fiber-based epoxy composites are explained in Table 2.11 [9].

TABLE 2.11
Applications of Bast Fiber-Based Epoxy Composites [9]

Bast Fiber-Reinforced Composites	Applications
Flax	BMW 7 vehicle
Hemp	Lotus Eco Elise Sports Car: Seats and body panel
Kenaf	Toyota indoor trims
Jute fiber-reinforced polyester resins	Building Madras House and grain elevator buildings
	Automobile interior linings
	• Roof
	• Rear wall
	• Side panel lining
	Shipping pallets
	Construction products
	• Composite roof tiles
	Furniture and household products
	• Storage containers
	• Window
	• Picture frames
	• Food service
	• Trays
	• Toys
	• Flower pots
	• Fan blades

2.9 CONCLUSION

The purpose of this chapter was to investigate the works that are associated with bast fiber-reinforced epoxy polymer composites. Bast fiber-reinforced epoxy composites' tensile, flexural, and impact properties, as well as the effects of chemical treatments to increase the interfacial matrix–fiber bonding, leading to enhancements in the physical properties of the fiber polymer composites, were discussed. Scientists in the field of materials research have been focusing on bast fibers because of their lightweight, durability, efficiency, and excellent mechanical qualities as a potential replacement for more traditional materials. Lack of homogeneity of qualities due to environmental circumstances when cultivated, biological attack of fungi and mildew, and limited applications are some of the limitations of bast fiber-reinforced epoxy composites.

The high hydroxyl content of cellulose in bast fiber also renders it susceptible to increased water absorption, which in turn impacts the material's mechanical properties. Effective chemical treatment of the bast fiber can solve these issues. Therefore, bast fibers are superior to traditional materials as reinforcement in the composite world because of their superior physical strengths. In addition to providing a biodegradable resource, the polymer industry's reliance on native raw materials also offers the possibility for economic growth in rural areas. Overall, bast fiber-reinforced epoxy composites have a promising future due to their lightweight, low cost, decomposable nature, and demand as alternatives to glass fiber composites.

REFERENCES

[1] M. A. Madlan, "The use of natural fibers in making dissipative silencer," *Adv. Res. Nat. Fibers*, vol. 1, no. 1, pp. 1–3, 2019.

[2] A. Felix Sahayaraj, M. Muthukrishnan, R. Prem Kumar, M. Ramesh, and M. Kannan. "PLA based bio composite reinforced with natural fibres – Review," In *IOP Conference Series: Materials Science and Engineering*, vol. 1145, no. 1, p. 012069. IOP Publishing, 2021.

[3] G. Rajeshkumar. G. L. Devnani, S. Sinha, and M. R. S. Suchart, "Bast fibers and their composites," In *Bast Fibers and Their Composites Processing, Properties and Applications*, 2022, doi: 10.1007/978-981-19-4866-4.

[4] R. Bhoopathi, M. Ramesh, M. Naveen Kumar, P. Sanjay Balaji, and G. Sasikala. "Studies on mechanical strengths of hemp-glass fibre reinforced epoxy composites," In *IOP Conference Series: Materials Science and Engineering*, vol. 402, no. 1, p. 012083. IOP Publishing, 2018, doi:10.1088/1757-899X/402/1/012083.

[5] K. N. Keya, N. A. Kona, and F. A. Koly, "Natural fiber reinforced polymer composites: History, types, advantages, and applications," *Mater. Eng. Res.*, 2019, doi: 10.25082/ MER.2019.02.006.

[6] R. Bhoopathi, M. Ramesh, R. Rajaprasanna, G. Sasikala, and C. Deepa. "Physical properties of glass-hemp-banana hybrid fiber reinforced polymer composites", *Indian J. Chem. Technol.*, vol. 10, no. 7, 2017, doi: 10.17485/ijst/2017/v10i7/103310.

[7] T. K. Mulenga, A. U. Ude, and C. Vivekanandhan, "Techniques for modelling and optimizing the mechanical properties of natural fiber composites: A review," *Fibers*, vol. 9, no. 1, pp. 1–17, 2021, doi: 10.3390/fib9010006.

[8] M. Ramesh, K. Palanikumar, and K. Hemachandra Reddy. "Plant fibre based bio-composites: Renewable and sustainable green materials", *Renewable Sustainable Energy Rev.*, vol. 79, pp. 558–584, 2017.

[9] T. Mishra, P. Mandal, A. K. Rout, and D. Sahoo, "A state-of-the-art review on potential applications of natural fiber-reinforced polymer composite filled with inorganic nanoparticle," *Compos. Part C Open Access*, vol. 9, p. 100298, 2022, doi: 10.1016/j. jcomc.2022.100298.

[10] N. M. Nurazzi et al., "A review on natural fiber reinforced polymer composite for bullet proof and ballistic applications," *Polymers*, vol. 13, no. 4, pp. 1–42, 2021, doi: 10.3390/ polym13040646.

[11] A. D. Eselev and V. A. Bobylev, "History of creation and development of epoxy resins in Russia," *Polym. Sci. - Ser. D*, vol. 2, no. 4, pp. 265–269, 2009, doi: 10.1134/ S1995421209040169.

[12] N. Uppal, A. Pappu, V. K. S. Gowri, and V. K. Thakur, "Cellulosic fibers-based epoxy composites: From bioresources to a circular economy," *Ind. Crops Prod.*, vol. 182, p. 114895, 2022, doi: 10.1016/j.indcrop.2022.114895.

[13] F. M. Khan et al., "A comprehensive review on epoxy biocomposites based on natural fibers and bio-fillers: Challenges, recent developments and applications," *Adv. Fiber Mater.*, vol. 4, no. 4. pp. 683–704, 2022, doi: 10.1007/s42765-022-00143-w.

[14] M. Ramesh, K. Palanikumar, and K. Hemachandra Reddy. "Impact behaviour analysis of sisal/jute and glass fiber reinforced hybrid composites", *Adv. Mater. Res.*, vol. 984–985, pp. 266–272, 2014.

[15] K.-D. Bouzakis, I. Tsiafis, Nikolaos Michailidis, and A. Tsouknidas. "Determination of epoxy resin's mechanical properties by experimental-computational procedures in tension," In *Proceedings of 3rd International Conference on Manufacturing Engineering (ICMEN)*, UK, pp. 1–3. 2008.

[16] V. Mittal, R. Saini, and S. Sinha, "Natural fiber-mediated epoxy composites - A review," *Compos. Part B: Eng.*, vol. 99, pp. 425–435, 2016, doi: 10.1016/j. compositesb.2016.06.051.

[17] K. C. Shekar, B. A. Prasad, and N. E. Prasad, "Manufacturing and characterization of epoxy matrix hybrid manufacturing and characterization of epoxy matrix hybrid nano-composite," In *IOP Conference Series: Materials Science and Engineering*, 2021, doi: 10.1088/1757-899X/1057/1/012009.

[18] T. Liu, P. Butaud, V. Placet, and M. Ouisse, "Damping behavior of plant fiber composites: A review," *Compos. Struct.*, vol. 275, no. 14, 2021, doi: 10.1016/j.compstruct.2021.114392.

[19] R. Bhoopathi and M. Ramesh, "Influence of eggshell particles on mechanical and water absorption properties of hemp-glass fibres reinforced hybrid composites," In *IOP Conf. Series: Materials Science and Engineering*, vol. 923, p. 012042, 2020.

[20] O. Faruk, A. K. Bledzki, H. Fink, and M. Sain, "Progress report on natural fiber reinforced composites," *Macromol. Mater. Eng.*, vol. 299, no. 1, pp. 9–26, 2014, doi: 10.1002/mame.201300008.

[21] M. Ramesh, C. Deepa, L. Rajeshkumar, and M. R. Sanjay. "Life-cycle and environmental impact assessments of plant fibers and its bio-composites: A literature review", *J. Ind. Text.*, 2020, doi: 10.1177/1528083720924730.

[22] I. Garmendia, J. Garcı, and M. Garcı, "Influence of natural fiber type in eco-composites," *J. Appl. Polym. Sci.*, 2007, doi: 10.1002/app.27519.

[23] K. Rohit and S. Dixit, "A review - Future aspect of natural fiber reinforced composite," *Polym. Renew. Resour.*, vol. 7, no. 2, pp. 43–59, 2016, doi: 10.1177/204124791600700202.

[24] S. H. Kamarudin et al., "A review on natural fiber reinforced polymer composites (NFRPC) for sustainable industrial applications," *Polymers*, vol. 14, no. 17, pp. 1–36, 2022, doi: 10.3390/polym14173698.

[25] M. N. Fazita, H. A. Khalil, T. M. Wai, E. Rosamah, and N. S. Aprilia, "Hybrid bast fiber reinforced thermoset composites." In V. K. Thakur et al. (eds.) *Hybrid Polymer Composite Materials*, pp. 203–234, Woodhead Publishing, Duxford, 2017.

[26] M. Farzana, K. M. Maraz, S. N. Sonali, M. M. Hossain, M. Z. Alom, and R. A. Khan, "Properties and application of jute fiber reinforced polymer-based composites," *GSC Adv. Res. Rev.*, vol. 11, no. 1, pp. 084–094, 2022, doi: 10.30574/gscarr.2022.11.1.0095.

[27] M. S. Salit, M. Jawaid, N. Bin Yusoff, and M. E. Hoque, "Manufacturing of natural fiber reinforced polymer composites," In *Manufacturing of Natural Fibre Reinforced Polymer Composites*, pp. 1–383, 2015, doi: 10.1007/978-3-319-07944-8_16.

[28] A. G. N. Abbas, F. N. A. A. Aziz, K. Abdan, N. A. M. Nasir, and M. N. Norizan, "Kenaf fiber reinforced cementitious composites," *Fibers*, vol. 10, no. 1, pp. 1–24, 2022, doi: 10.3390/fib10010003.

[29] T. Hassan et al., "Factors affecting acoustic properties of natural-fiber-based materials and composites: A review," *Textiles*, vol. 1, no. 1, pp. 55–85, 2021, doi: 10.3390/textiles1010005.

[30] M. Pereira, L. De Mendonc, P. Henrique, and P. Mendonc, "Mechanical, thermal and ballistic performance of epoxy composites reinforced with *Cannabis sativa* hemp fabric," *J. Mater. Res. Technol.*, 2021, doi: 10.1016/j.jmrt.2021.02.064.

[31] T. U. Berlin, T. Lampke, B. Wielage, and F. U. Berlin, "Surface characterization of flax, hemp and water uptake behavior," *Polymer Composites*, vol. 23, no. 5, pp. 872–894, 2002.

[32] R. Bhoopathi and M. Ramesh (2020). "Influence of bio-based eggshell nano-particles on thermal and morphological properties of alkali-treated hemp fibre reinforced epoxy composites", *J. Polym. Environ.*, vol. 28, no. 8, pp. 2178–2190.

[33] A. Suresh, L. Jayakumar, and A. Devaraju, "Investigation of mechanical and wear characteristic of Banana/Jute fiber composite," *Mater. Today Proc.*, 2020, doi: 10.1016/j.matpr.2020.07.426.

[34] A. Felix Sahayaraj, M. Muthukrishnan, and M. Ramesh, "Experimental investigation on physical, mechanical and thermal properties of jute and hemp fibers reinforced hybrid polylactic acid composites", *Polym. Compos.*, 2022, doi: 10.1002/pc.26581.

[35] F. Abbasi, F. Samadi, S. M. Jafari, S. Ramezanpour, and M. Shams Shargh, "Ultrasound-assisted preparation of flaxseed oil nanoemulsions coated with alginate-whey protein for targeted delivery of omega-3 fatty acids into the lower sections of gastrointestinal tract to enrich broiler meat," *Ultrason. Sonochem.*, vol. 50, pp. 208–217, 2019, doi: 10.1016/j.ultsonch.2018.09.014.

[36] A. Moudood, A. Rahman, A. Öchsner, M. Islam, and G. Francucci, "Flax fiber and its composites: An overview of water and moisture absorption impact on their performance," *J. Reinf. Plast. Compos.*, vol. 38, no. 7, pp. 323–339, 2019, doi: 10.1177/0731684418818893.

[37] M. Ramesh. "Flax (*Linum usitatissimum* L.) fibre reinforced polymer composite materials: A review on preparation, properties and prospects", *Prog. Mater. Sci.*, vol. 102, pp. 107–166, 2019.

[38] Z. Y. Lim, A. Putra, M. J. M. Nor, and M. Y. Yaakob, "Sound absorption performance of natural kenaf fibers," *Appl. Acoust.*, vol. 130, pp. 107–114, 2018, doi: 10.1016/j.apacoust.2017.09.012.

[39] Z. N. Azwa and B. F. Yousif, "Characteristics of kenaf fiber/epoxy composites subjected to thermal degradation," *Polym. Degrad. Stab.*, vol. 98, no. 12, pp. 2752–2759, 2013, doi: 10.1016/j.polymdegradstab.2013.10.008.

[40] R. Mahjoub, J. M. Yatim, A. R. Mohd Sam, and M. Raftari, "Characteristics of continuous unidirectional kenaf fiber reinforced epoxy composites," *Mater. Des.*, vol. 64, pp. 640–649, 2014, doi: 10.1016/j.matdes.2014.08.010.

[41] M. Ramesh. Kenaf (*Hibiscus cannabinus* L.) fibre based bio-materials: A review on processing and properties", *Prog. Mater. Sci.*, vol. 78–79, pp. 1–92, 2016.

[42] F. M. Margem, S. N. Monteiro, J. B. Neto, R. J. S. Rodriguez, and B. G. Soares, "The dynamic-mechanical behavior of epoxy matrix composites reinforced with ramie fibers," *Rev. Mater.*, vol. 15, no. 2, pp. 167–175, 2010, doi: 10.1590/s1517-70762010000200012.

[43] Z. Djafar, I. Renreng, and M. Jannah, "Tensile and bending strength analysis of ramie fiber and woven ramie reinforced epoxy composite," *J. Nat. Fibers*, vol. 18, no. 12, pp. 2315–2326, 2021, doi: 10.1080/15440478.2020.1726242.

[44] A. P. Irawan, T. P. Soemardi, K. Widjajalaksmi, and A. H. S. Reksoprodjo, "Tensile and flexural strength of ramie fiber reinforced epoxy composites for socket prosthesis application," *Int. J. Mech. Mater. Eng.*, vol. 6, no. 1, pp. 46–50, 2011.

[45] M. Ramesh, L. Rajeshkumar, and D. Balaji, "Mechanical and dynamic properties of ramie fiber composites," In K. Senthil Kumar et al. (eds.) *Mechanical and Dynamic Properties of Biocomposites*, WILEY-VCH GmbH, Weinheim, pp. 275–293, 2021.

[46] P. Krishnasamy, G. Rajamurugan, B. Muralidharan, and R. Krishnaiah, "General practice to enhance bast fiber composite properties for state of art applications-A review," *Eng. Res. Express*, vol. 4, no. 1, p. 12002, 2022, doi: 10.1088/2631-8695/ac49d7.

[47] C. Obele and E. Ishidi, "Mechanical properties of coir fiber reinforced epoxy resin composites for helmet shell," *Ind. Eng. Lett.*, vol. 5, no. 7, pp. 67–75, 2015.

[48] N. Graupner, A. S. Herrmann, and J. Müssig, "Natural and man-made cellulose fiber-reinforced poly (lactic acid) (PLA) composites: An overview about mechanical characteristics and application areas," *Compos. - A: Appl. Sci. Manuf.*, vol. 40, no. 6–7, pp. 810–821, 2009, doi: 10.1016/j.compositesa.2009.04.003.

[49] T. Setyayunita, R. Widyorini, S. N. Marsoem, and D. Irawati, "Effect of different conditions of sodium chloride treatment on the characteristics of kenaf fiber-epoxy composite board," *J. Korean Wood Sci. Technol.*, vol. 50, no. 2, pp. 93–103, 2022, doi: 10.5658/WOOD.2022.50.2.93.

[50] R. R. Raj et al., "Effect of graphene fillers on the water absorption and mechanical properties of NaOH-treated kenaf fiber-reinforced epoxy composites," *J. Nanomater.*, vol. 2022, 2022, doi: 10.1155/2022/1748121.

[51] G. K. Sathishkumar et al., "Synthesis and mechanical properties of natural fiber reinforced epoxy/polyester/polypropylene composites: A review," *J. Nat. Fibers*, vol. 19, no. 10, pp. 3718–3741, 2022, doi: 10.1080/15440478.2020.1848723.

[52] G. B. Nyior, S. A. Aye, and S. E. Tile, "Study of mechanical properties of raffia palm fiber/groundnut shell reinforced epoxy hybrid composites," *J. Miner. Mater. Charact. Eng.* pp. 179–192, 2018, doi: 10.4236/jmmce.2018.62013.

[53] K. Agarwal, S. K. Kuchipudi, B. Girard, and M. Houser, "Mechanical properties of fiber reinforced polymer composites: A comparative study of conventional and additive manufacturing methods," *J. Compos. Mater.*, vol. 52, no. 23, pp. 3173–3181, 2018, doi: 10.1177/0021998318762297.

[54] M. A. Al Faruque, M. Salauddin, M. M. Raihan, I. Z. Chowdhury, F. Ahmed, and S. S. Shimo, "Bast fiber reinforced green polymer composites: A review on their classification, properties, and applications," *J. Nat. Fibers*, 2021, doi: 10.1080/15440478.2021.1958431.

[55] K. Charlet, S. Eve, J. P. Jernot, M. Gomina, and J. Breard, "Tensile deformation of a flax fiber," *Procedia Eng.*, vol. 1, no. 1, pp. 233–236, 2009, doi: 10.1016/j.proeng.2009.06.055.

[56] M. Asim, M. Jawaid, K. Abdan, and M. R. Ishak, "Effect of alkali and silane treatments on mechanical and fiber-matrix bond strength of kenaf and pineapple leaf fibers," *J. Bionic Eng.*, vol. 13, no. 3, pp. 426–435, 2016, doi: 10.1016/S1672-6529(16)60315-3.

[57] C. Das and S. Kumar, "Evaluation of surface response of *Ficus benghalensis* fiber — epoxy composites under dry sliding wear conditions," *J. Inst. Eng. Ser. E*, 2020, doi: 10.1007/s40034-020-00182-1.

[58] K. Malik, F. Ahmad, E. Gunister, K. Malik, and F. Ahmad, "Drilling performance of natural fiber reinforced polymer composites: A review," *J. Nat. Fibers*, 2021, doi: 10.1080/15440478.2020.1870624.

[59] K. Senthilnathan, R. Ravi, J. Stephen Leon, G. Suresh, K. Manikandan, and R. Lavanya, "Analysing the effect of mechanical properties of various proportions of filler material on jute fiber/epoxy reinforced composites," *J. Phys. Conf. Ser.*, vol. 1921, no. 1, 2021, doi: 10.1088/1742-6596/1921/1/012089.

[60] B. P. Nanda and A. Satapathy, "Processing and thermal characteristics of human hair fiber-reinforced polymer composites," *Polym. Polym. Compos.*, pp. 1–13, 2019, doi: 10.1177/0967391119872399.

[61] R. Rahman and S. Z. F. S. Putra, "Tensile properties of natural and synthetic fiber-reinforced polymer composites," In M. Jawaid et al. (eds.) *Mechanical and Physical Testing of Biocomposites, Fibre-Reinforced Composites and Hybrid Composites*, Woodhead Publishing, Duxford, pp. 81–102, Elsevier Ltd, 2019.

[62] S. Ozkur, H. Sezgin, and I. Yalcin-Enis, "The effect of curing and post-curing processes on physical and mechanical behaviors of jute fabric reinforced AESO/epoxy based bio-composites," *Fibers Polym.*, vol. 23, no. 5, pp. 1410–1421, 2022, doi: 10.1007/s12221-022-4447-y.

[63] A. Ashori and S. Sheshmani, "Hybrid composites made from recycled materials: Moisture absorption and thickness swelling behavior," *Bioresour. Technol.*, vol. 101, no. 12, pp. 4717–4720, 2010, doi: 10.1016/j.biortech.2010.01.060.

[64] A. Aditya, P. Pratim, and V. Chaudhary, "Effect of moisture absorption on the properties of natural fiber reinforced polymer composites: A review," *Mater. Today Proc.*, 2021, doi: 10.1016/j.matpr.2021.02.812.

[65] J. Wang, H. Xie, Z. Weng, T. Senthil, and L. Wu, "A novel approach to improve mechanical properties of parts fabricated by fused deposition modeling," *Mater. Des.*, vol. 105, pp. 152–159, 2016, doi: 10.1016/j.matdes.2016.05.078.

3 Leaf Fiber-Based Epoxy Composites

Thermal and Mechanical Properties

Hossein Ebrahimnezhad-Khaljiri
University of Zanjan

CONTENTS

3.1   INTRODUCTION

It is no secret that the composite materials have shown extremely progressive trend in the last decades; as a consequence, many metallic parts have been replaced with the composite parts. This can be due to having special properties of composite materials like lightweight, higher mechanical properties, higher chemical stability, etc. [1]. Today, new kinds of composite materials are being developed by various scientific groups for using them in new applications or solving primary composite issues [2]. Smart composites [3], multi-scale composites [4], hybrid structures, etc. are the examples of these new composite materials. Among them, bio-based composites are introduced to reduce the environmental issues of composites. Like other composites, bio-composites are fabricated from two distinct materials; either one or both materials are produced from the natural sources. One of the new bio-composites, which have potential for using as secondary or tertiary or in some cases primary applications, is

DOI: 10.1201/9781003271017-3

epoxy-based bio-composites. To fabricate these composites, epoxy polymers should be composed with the natural-based reinforcements like bio-fibers [5].

These bio-fibers can be extracted from seed, bast, leaf, stalk, fruit, etc., which have been developed in the past years. The production costs of some of them are negligible because they are considered agro-waste. So, it can be helpful for using them instead of synthetic fibers in the applications with lower mechanical or thermal performances. Among the known families of fibers, leaf-based fibers have the proper mechanical properties, so they can be a good choice for reinforcing the epoxy polymers. The sisal, abaca, henequen, banana, and pineapple are examples of famous leaf-based natural fibers. Besides, *Moringa oleifera*, *P. tenax*, *A. donax* L. and screw-pine are the unknown leaf fibers, which have proper potential for reinforcing the epoxy polymers. To develop epoxy bio-composites, knowing the mechanical and thermal performances of them is so necessary. Therefore, the aim of this chapter is to review the leaf-based epoxy bio-composites for their mechanical and thermal properties. Some famous and unknown bio-fibers are described in the chapter, which can be seen in Figure 3.1. It should be noted that the palm-based epoxy bio-composites

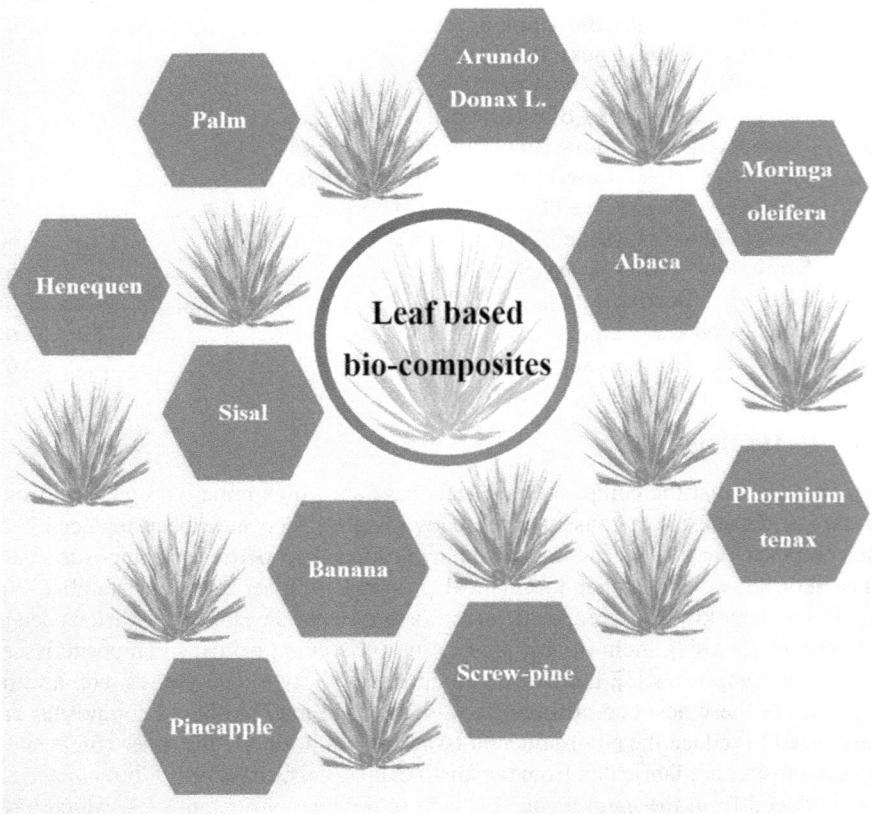

FIGURE 3.1   Various leaf fibers as natural reinforcements of epoxy bio-composites.

are one of the major families of leaf-based bio-composites with the highest variation in performances. Hence, the mechanical and thermal properties of these bio-composites would demand a separate chapter.

3.2 SISAL-BASED EPOXY BIO-COMPOSITES

Sisal or Agave Sisalana fibers are one of the leaf-based fibers, which have high cellulose content and low microfibrillar angle. These fibers have proper mechanical properties for incorporating them into the composites. Rapid growing in most of the climates and low production costs are other benefits of these fibers. Having these features make sisal fibers a potential candidate for developing them in order to use into the epoxy composites [5]. Various surface treatments, modifications, using bio-epoxy resins, adding nanomaterials, especially nano-cellulose, and characterizing the features of sisal/epoxy composites under the static and dynamic conditions are the processes/efforts involved in this bio-composite development. To have a comparative overview about the features of sisal/epoxy composites, it is better to investigate the effect of weight fraction or orientation of sisal fibers on the properties of these composites.

Given this viewpoint, Gupta et al. [6] added various weight fractions (15%, 20%, 25% and 30%) of sisal fibers into the epoxy resin to characterize the static and dynamic features of sisal/epoxy composites. The tensile strength, elastic modulus, flexural strength, flexural modulus, impact strength and impact energy had the highest values, when the weight fraction of sisal was 30 wt.%, which were 83.96 MPa, 1.58 GPa, 251 MPa, 11.32 GPa, 22.03 kJ/mm^2, and 1.09 J, respectively. By using the dynamic mechanical analysis (DMA), it has been found that the storage modulus was highest, when 25 wt.% sisal fibers were added into the epoxy polymer. Also, the maximum glass transition temperature (T_g) belonged to the composite with the 20 wt.% sisal fibers.

The effect of sisal fiber orientations into the epoxy matrix on the features of sisal/epoxy composites was investigated by Kumaresan et al. [7]. In this work, the stacking sequences of sisal fibers into the epoxy matrix were 0/90°, 90°, and +45°. The obtained results showed that the composite with layering of 90° had the highest tensile, flexural and impact strength, which were 38.8 MPa, 151 MPa, and 7.3 J, respectively. The sisal fibers have sub-fibrillar structures like aramid fibers, which help to exhibit the high impact strength. Therefore, researchers like Militello et al. [8] studied on this capability in the sisal/epoxy composites through a low velocity impact test. The reported variations of this work were in the weight fraction and fiber orientations (unidirectional, cross-ply, quasi-isotropic, and mat fibers). The highest impact strength achieved in the cross-ply composite with the 22.8 J impact energy, followed by quasi-isotropic (19.2 J), mat fibers (13.0 J) and unidirectional fibers (7.8 J). According to these results, it can be said that the angle-ply (quasi-isotropic and cross-ply) laminates had the specific impact resistance, so it can be introduced as the first candidate for using instead of the synthetic fibers into the epoxy polymers for energy absorption applications like helmets, protection systems, etc. This superior impact resistance can be attributed to the damage mechanisms, which were the longitudinal tensile failure of sisal fibers, splitting of those, delamination between layers and pulling out the fibers from the epoxy matrix.

Webo et al. [9] characterized the impact toughness and hardness of epoxy composites containing various weight fractions of sisal fibers. Also, they compared the untreated and treated sisal/epoxy composites. Their treatment involved the combination of alkali and silane treatments. The highest impact energy for epoxy containing 50 wt.% untreated sisal fibers was 10.34 J. This further enhanced up to 12.65 J by performing the surface treatment. In this weight fraction, the Barcol unit (BU) of untreated and treated composites was 45 and 49 BU, respectively. The wear properties of sisal/epoxy composites with various fiber orientations (longitudinal, transverse and normal fibers alignments) were assessed by Chand and Dwivedi [10]. The minimum wear rate was seen in the composite with the normal fibers alignment, followed by transverse and longitudinal fibers alignments. In another work, Behera et al. [11] investigated the effect of alkali, glutamic acid and combination of them on the dry wear performance of sisal/epoxy composites. The highest wear resistance was obtained, when the sisal fibers were modified with the alkali treatment, whereas the TGA analysis showed the maximum thermal resistance achieved by performing the glutamic acid treatment, which can be seen in Figure 3.2.

From various researches, it can be found that the effective way for enhancing the mechanical properties of sisal/epoxy composites can be surface modifications. For example, Rong et al. [12] treated the sisal fibers with the alkali, acetylated, cyanoethylated, organo-silanation, heat treatment, combinations of alkali/silanation, and alkali/heat treatment methods. These treatments changed the roughness of sisal fibers, which resulted in better adhesion between sisal fibers and epoxy matrix (seen Figure 3.3). It seems that each treatment can be useful for distinct mechanical

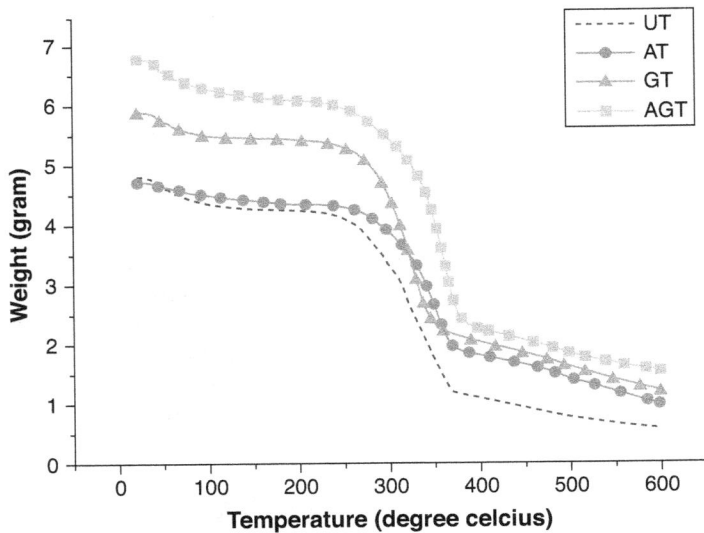

FIGURE 3.2 Effect of chemical treatments of sisal fibers on the TGA curves of sisal/epoxy bio-composites. AGT, alkali-glutamic acid treated; AT, alkali-treated; GT, glutamic acid-treated; UT, untreated. (Reproduced from [11] with the permission of Taylor & Francis (Free of charge).)

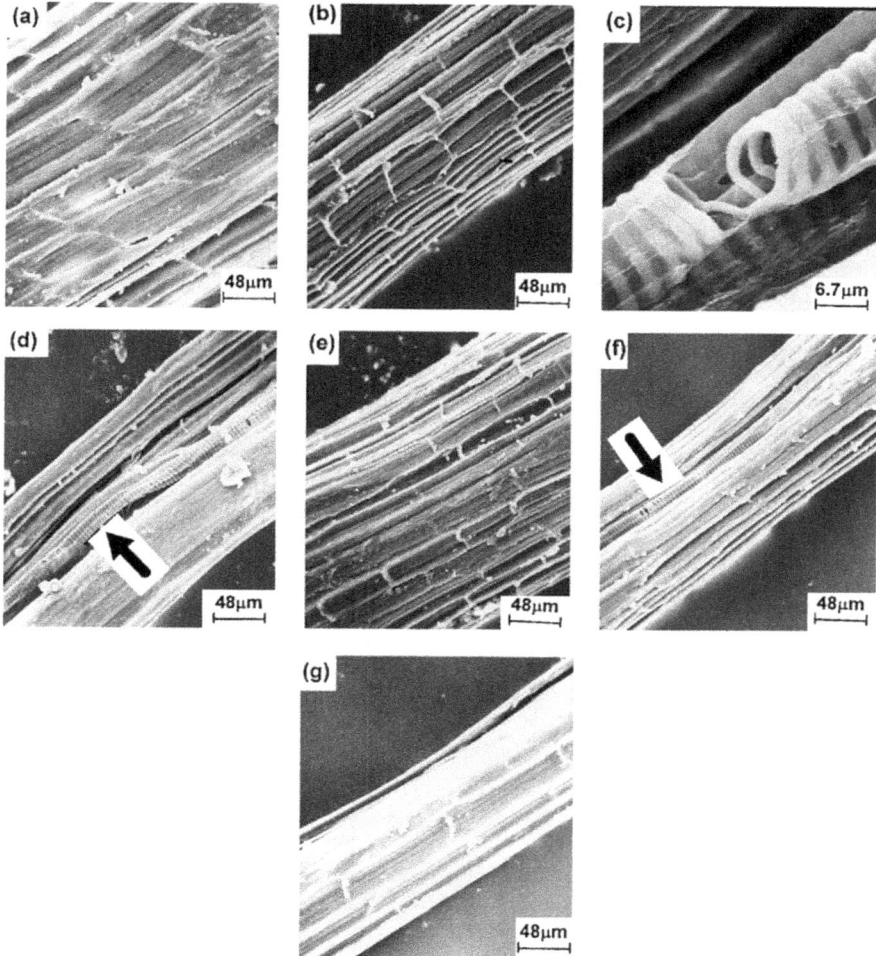

FIGURE 3.3 The effect of treatments on the surface morphology of sisal fibers for reinforcing the epoxy polymer: (a) untreated sisal fibers, (b and c) alkali-treated sisal fibers, (d) acetylated sisal fibers, (e) cyanoethylated sisal fibers, (f) alkali- and heat-treated sisal fibers, and (g) heat-treated sisal fibers. The arrows indicate the exposed helical fibrils. (Reprinted from [12] with the permission of Elsevier (License number: 5207151372361).)

properties. Therefore, based on the applied load conditions, the treatment method can be selected. Using potassium permanganate [13] and bezoxazine [14] are other modification methods, which have been performed by other researchers on the sisal fibers to improve the mechanical performances of sisal/epoxy composites. Using the sodium bicarbonate and coating fibers with the polylactic acid (PLA) is known as eco-friendly surface modification method. The effect of this treatment on the flexural and thermal properties of sisal/epoxy composites can be seen in Figure 3.4 [15].

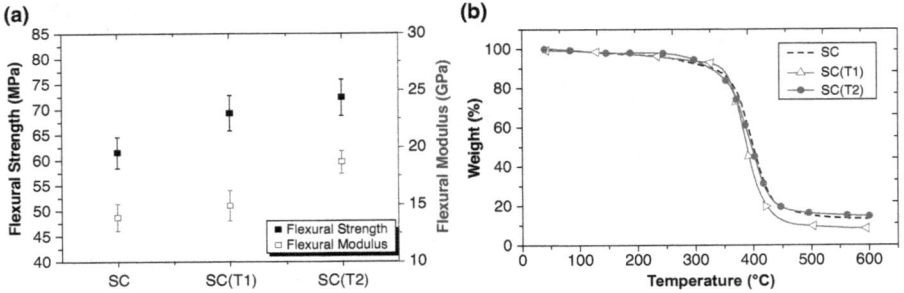

FIGURE 3.4   The effect of eco-friendly treatment on the properties of sisal/epoxy bio-composites: (a) flexural properties and (b) thermal properties. SC, sisal fibers; SC-T1, sodium bicarbonate-treated sisal fibers; SC-T2, sodium bicarbonate-treated and PLA-coated sisal fibers. (Adapted from [15] with the permission of Elsevier (License number: 5207161305961).)

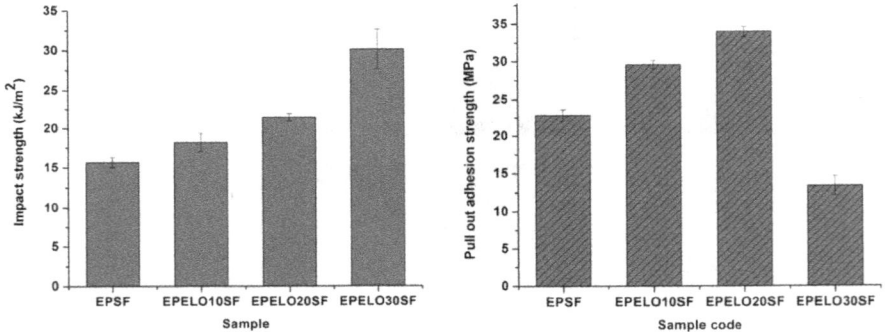

FIGURE 3.5   The effect of epoxidized linseed oil content on the mechanical properties of sisal/epoxy bio-composites. The EPSF, EPELO10SF, EPELO20SF and EPELO30SF are the epoxy composites with 0, 10, 20 and 30 wt.% of epoxidized linseed oil, respectively. (Adapted from [16] with the permission of John Wiley and Sons (License number: 5207161100938).)

The other type of modification method for improving the mechanical properties of this bio-composite is using epoxidized plant oils, which can increase the toughness, impact resistance, and pull-out strength, which can be observed in Figure 3.5. However, using these oils reduces the thermal stability of this bio-composite. Linseed [16], castor [17], and soybean oils [18] are the examples of modifying agents.

3.3   HENEQUEN-BASED EPOXY BIO-COMPOSITES

One of the leaf sources is the agave plants, which is used in Mexico in the textiles manufacture, due to having proper stiffness and simple knit process. Two kinds of agave plants are grown in this country; one of them is Ixtle (*Agave vivipara*) and the other is henequen (*Agave fourcroydes*). The evidence about using these fibers into the polymeric materials, especially epoxy families, is so limited [19]. Soto et al. [20] characterized the thermal properties of henequen/epoxy bio-composites. The obtained results showed that the composite containing 9 wt.% fibers had the highest thermal degradation temperature, which was about 398.89°C. Also, the fly ash content after TGA test in this composite was 5.37 wt.%. Sampieri-Bulbarela et al. [21] assessed the tensile and flexural properties of this bio-composite. According to the reported results, the tensile strength, elastic modulus, flexural strength and flexural modulus of composite containing 47 vol.% henequen fibers were about 175 MPa, 8 GPa, 270 MPa and 9 GPa, respectively. The impact features of composite with 64 vol.% henequen fibers were characterized by Gonzalez-Murillo and Ansell [22]. For un-notched and notched samples, the impact strengths were 104.4 and 116.4 kJ/m², respectively. Applying the NaOH treatment had the negative effect of these strengths, i.e., strengths reduced down to 99.61 and 90.81 kJ/m², respectively.

3.4 ABACA-BASED EPOXY BIO-COMPOSITES

Abaca, Musa textilis Nee or Manila hemp is a plant similar to the banana tree, which can provide high-quality fibers for fabricating nets, automobile parts and natural fibers. From this plant, fibers up to 30 m can be extracted [23,24]. Also, these fibers into the polymeric composites, due to having the rough surface, have stronger bonding, as compared with other natural fibers [25]. With these features, it can be expected that using these fibers into the epoxy polymers can be interesting for scientists. Like other natural fibers, the abaca fibers need the surface modifications for improving the mechanical properties of abaca/epoxy bio-composites. Punyamurthy et al. [26,27] investigated the simultaneous effect of abaca fibers loading and chemical treatments on the tensile, flexural and impact properties of these bio-composites. The obtained results from these composites are listed in Table 3.1. It should be noted that the maximum mechanical properties in these works belonged to the composite with the 40 wt.% abaca fibers.

These researchers explained the mechanisms of each treatment for improving the mechanical properties. The alkali treatment could remove the lignin and hemicellulose from the abaca fibers. Also, this treatment could expose more hydroxyl groups on the surface of fibers, which resulted in improved bonding with the epoxy matrix. The permanganate treatment caused the better interlocking at the surface

TABLE 3.1

The Mechanical Properties of 40 wt.% Abaca Fibers/Epoxy Bio-Composite [26,27]

Chemical Modification	Tensile Strength (MPa)	Flexural Strength (MPa)	Impact Energy (mJ/mm²)
Untreated	36.48	52.2	6.18
Alkali treatment	43.78	56.46	6.5
Acrylic acid treatment	50.28	60.24	7.12
Permanganate treatment	48.84	58.56	6.8
Benzenediazonium treatment	58.62	67.86	7.68

of fibers by creating the rough surface. The acrylic acid treatment could enhance the stress transfer between the abaca fibers and epoxy matrix, which could improve the mechanical properties of the bio-composites. Finally, benzenediazonium as coupling agent could improve the adhesion between abaca fibers and epoxy by chemical bonds [26,27].

Cai et al. [28] investigated the effect of alkali treatment on the interfacial bonding of abaca/epoxy composites. This treatment was performed by using 5, 10 and 15 wt.% NaOH solution. Figure 3.6 shows the digital images of the surface of untreated and treated fibers. The untreated fiber is included of the bundles of elementary fibers with the binding components (pectin, lignin and hemi-cellulose). The NaOH treatment caused that the bundles be free from these components, which resulted to create the grooves on the surface of fibers in the 5 wt.% NaOH solution and to fibrillate the bundle in the 15 wt.% NaOH solution. Therefore, it can be said that the maximum shear strength (42.2 MPa) can be achieved by using 5 wt.% NaOH solution, as confirmed by SEM images shown in Figure 3.7.

Kurien et al. [29] characterized the wear properties of these bio-composites by incorporating 5, 10, 15, 20, and 25 wt.% abaca fibers into the epoxy matrix. The highest wear resistance was seen in the composite with 25 wt.% abaca fibers. The characterized wear mechanisms in these bio-composites were micro-cracking, debonding, pit formation, matrix breakage, and fiber fractures. Liu et al. [30] investigated the effect of mercerization and silane treatments on the transverse thermal conductivity and transverse tensile strength of abaca/epoxy bio-composites. They mentioned that by applying this treatment, the weakest linkage changed from the abaca bundle/epoxy interface to the elementary fibers. Also, they found that the transverse thermal conductivity had the enhancing trend, when the transverse tensile strength increased or the void content of bio-composites had the reducing trend.

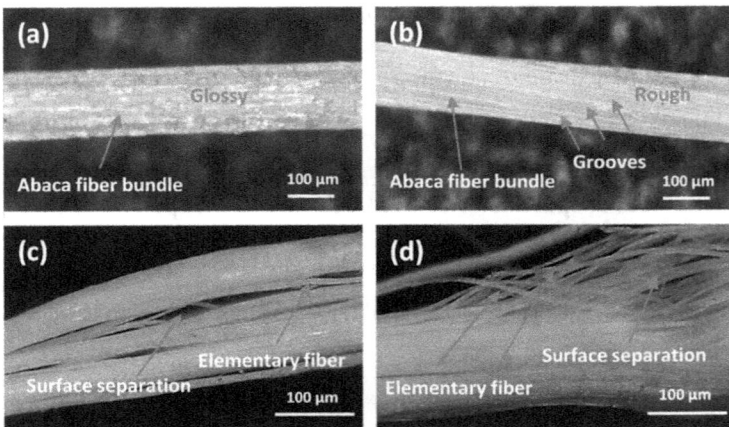

FIGURE 3.6   Digital images of the surface of abaca fibers: (a) untreated fibers, (b) 5 wt.% NaOH-treated, (c) 10 wt.% NaOH-treated and (d) 15 wt.% NaOH-treated. (Reprinted from [28] with the permission of Elsevier (License number: 5207161472293).)

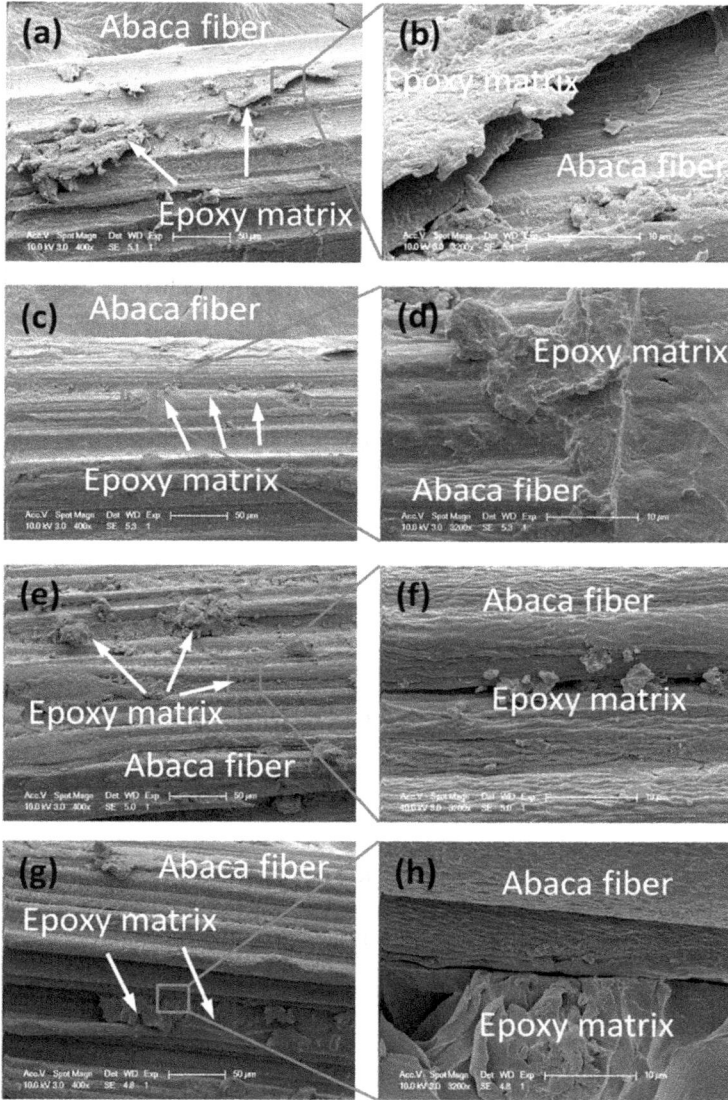

FIGURE 3.7 SEM observations from the surface of untreated and treated abaca fibers/epoxy bio-composites: (a and b) untreated, (c and d) 5 wt.% NaOH-treated, (e and f) 10 wt.% NaOH-treated and (g and h) 15 wt.% NaOH-treated. (Reprinted from [28] with the permission of Elsevier (License number: 5207161472293).)

3.5 BANANA-BASED EPOXY BIO-COMPOSITES

The stem section of banana plant can be used for extracting the natural fibers, which is done by mechanical or manual processes. Having 35% cellulose, 15% hemi-cellulose and 20% lignin makes banana fibers suitable for incorporating into the composite parts [31,32]. Balaji et al. [33] added the banana fibers with the length of 10 and 20 mm into the epoxy matrix. The weight contents of these fibers were 5%, 10%, 15%, and 20%. The obtained results showed that the composite with the 15 wt.% banana fibers had the maximum tensile, flexural and impact strengths, which were 30 MPa, 56 MPa, and 2.7 J/mm^2, respectively. Also, the maximum hardness belonged to the composite with 20 wt.% banana fibers, which was 68 HRB. The minimum thermal stability temperature for those bio-composites was 220°C. In the similar work, Subramanya et al. [34] added the 30 wt.% chopped banana fibers (5–12 mm) into the epoxy resin. The tensile strength, elastic modulus and impact strength of this composite were 15.74 MPa, 2.3 GPa, and 0.11 J/m, respectively. Also, the calculated fracture toughness and energy release rate were 201.70 MPa.m$^{1/2}$ and 15.77 kJ/m^2, respectively. In another study, this research group investigated the thermal conductivity of this composite and found a maximum thermal conductivity of 0.342 w/m.k [35].

Gairola et al. [36] added 10, 20, 30, 40 and 50 wt.% non-woven banana fibers into the epoxy resin to survey the mechanical properties of those. The maximum flexural and tensile strength were 38.1 and 65.6 MPa (at 30 wt.% filler addition), whereas the highest value of hardness and impact strength were seen at 40 wt.% filler addition, which were 45.6 HV and 2.8 J, respectively. It seems that in the higher content, the void fraction had the increment trend, which resulted in deterioration of mechanical properties of bio-composites. Venkateshwaran et al. [37] performed the alkali treatment on the banana fibers with various concentrations of NaOH solution (0.5, 1, 2, 5, 10, 15, and 20 wt.%). Then, they composed these treated fibers with the epoxy resin. Finally, they assessed the effect of these treatments on the mechanical properties of fabricated bio-composites. Based on this research, it seems that the 1 wt.% NaOH-treated fibers-reinforced bio-composite behaved superiorly than other bio-composites under mechanical loads. The tensile strength, elastic modulus, flexural strength, flexural modulus, and impact strength of this composite were 33.6 MPa, 1.68 GPa, 69.03 MPa, 13.22 GPa, and 12.25 J/m, respectively. Nguyen et al. [38] fabricated the NaOH-treated banana/epoxy bio-composites. The length and diameter of added banana fibers were 30 mm and 30 μm, respectively. The mechanical and thermal features of these composites are summarized in Figures 3.8 and 3.9, respectively.

3.6 PINEAPPLE-BASED EPOXY BIO-COMPOSITES

The pineapple (*Ananas comosus*) is one of the tropical fruits in the world. The leaves of this fruit are known as agro-waste, which have potential for producing natural fibers, due to their constituents. Its fibers are famous, because of having high content of cellulose, which can be reached up to 80%. This high content has both positive and negative effect on the performances of their polymeric composites. From the research studies, it is evident that the high content of cellulose can provide high specific strength and stiffness, but it can enhance the hydrophilic nature of extracted fibers. The other major component is lignin, the high content of which reduces the mechanical properties of pineapple-based bio-composites. The existence of waxy layer on the surface of these fibers is another challenge for using them as reinforcement into the polymers, because this layer reduces the adhesion bonding between the fibers and polymers [39]. To solve these challenges, many attempts have been done,

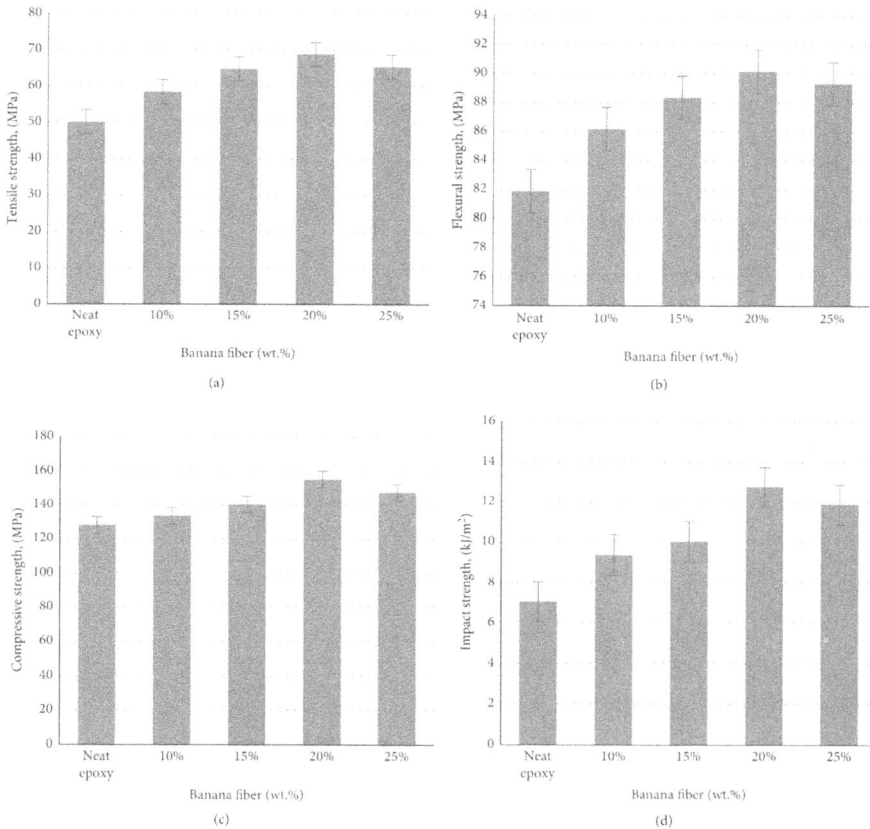

FIGURE 3.8 The mechanical properties of banana/epoxy bio-composites: (a) tensile strength, (b) flexural strength, (c) compressive strength and (d) impact strength. (Reprinted from [38] as open access source of Hindawi.)

which will be discussed in the following. Before that, the thermal and mechanical features of pineapple fibers/epoxy composites need to be introduced, which will be discussed in the following.

Glória et al. [40] added the continuous and aligned pineapple fibers with various volume fractions (10%, 20%, and 30%) into the epoxy matrix to characterize the flexural strength. In this work, the flexural strength of neat epoxy was about 65 MPa. The highest flexural strength was about 120 MPa, which belonged to the composite with the 30 vol.% pineapple fibers. Jagadish et al. [41] added various contents of pineapple fibers (1, 5, 10, 15, and 20 wt.%) with the thickness of 3 and 5 mm into the epoxy resin for assessing the tensile, flexural, impact, and hardness features of these bio-composites. The obtained results in all mechanical tests showed that the composite with the 10 wt.% short fibers (both thickness) had the highest performance compared to other bio-composites. This trend can be related to the weak bonding between the pineapple fibers/epoxy resin. The summary of these data is listed in Table 3.2.

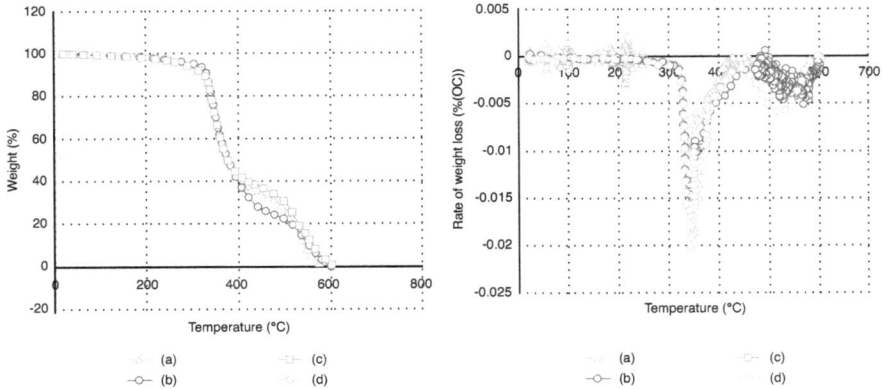

FIGURE 3.9 The thermal properties of epoxy bio-composite containing various contents of banana fibers: (a) 20 wt.%, (b) 25 wt.%, (c) 15 wt.% and (d) 10 wt.%. (Reprinted from [38] as open access source of Hindawi.)

TABLE 3.2
The Mechanical Properties of 10 wt.% Pineapple Fibers/Epoxy Bio-Composites [41]

Thickness of Fiber (mm)	Tensile Strength (MPa)	Elastic Modulus (GPa)	Flexural Strength (MPa)	Flexural Modulus (GPa)	Impact Energy (kJ/m^2)	Hardness (HRM)
3	79.98	3.45	138.45	7.54	97.51	79.58
5	83.75	3.75	140.65	7.92	98.28	80.21

Jain et al. [42] incorporated pineapple fibers (5, 10, 15, 20, 25 wt.%) into the epoxy resin to investigate the mechanical and thermal properties of these bio-composites. Like previous work, the maximum mechanical performance was seen in the composite with 10 wt.% reinforcement. But, the thermal properties had different trend and the highest thermal stability was observed in the 20 wt.% pineapple fibers composite. According to the TGA results, they mentioned that these composites have four steps for decomposition during the TGA test. The first of those was the dehydration or removal of the moisture at the temperature of 25°C–200°C. The second degradation occurred at the temperature of 250°C–477°C, which was due to the debonding of glycosidic linkage or decomposition of hemi-cellulose of pineapple fibers. The third step can be due to the breakage of cellulose linkages and bonding. The degradation of lignin as thermally stable compound was the fourth step, which happened at the temperature range of 500°C–800°C.

As previously mentioned, pineapple fibers have weak interface bonding with the polymers, especially epoxy families. So, several ways have been introduced by researchers to solve this issue. The simplest way is the alkali treatment, which can reduce or remove the lignin and waxes. For investigating this, Nagarajan et al. [43]

first treated the mat pineapple fibers with the 5 wt.% NaOH solution and then assessed the mechanical properties of treated fibers/epoxy composites. By this treatment, the elastic modulus of 20 wt.% pineapple fibers/epoxy composite reached from 3090 to 3440 MPa. This shows that the treated fibers have higher interaction with the epoxy polymer, as compared with the untreated fibers, thereby resulting in an increase in the stiffness of this bio-composite.

One of the interesting subjects about the pineapple is that pineapple can be used in the form of pineapple particles, which can be produced by using the ball milling of chopped fibers. The fabrication of pineapple fillers was done by Kumar et al. [44] for adding in the epoxy polymer. Figure 3.10 shows the flexural and fracture properties of epoxy composites containing these fillers. Incorporating this filler into the pineapple fibers/epoxy composite can influence on their mechanical properties, which was studied by Saha et al. [45]. These researchers added 2.5, 5, 7.5, and 10 wt.% pineapple fillers into 30 wt.% pineapple fibers/epoxy composite; the effect of fillers on the tensile and compressive properties can be seen in Figure 3.11. From this figure, it can be said

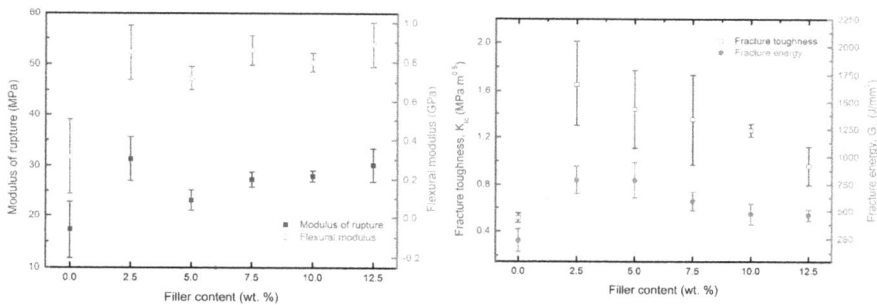

FIGURE 3.10 The flexural and fracture properties of pineapple filler/epoxy bio-composites. (Adapted from [44] with the permission of Springer Nature (License number: 5207170270705).)

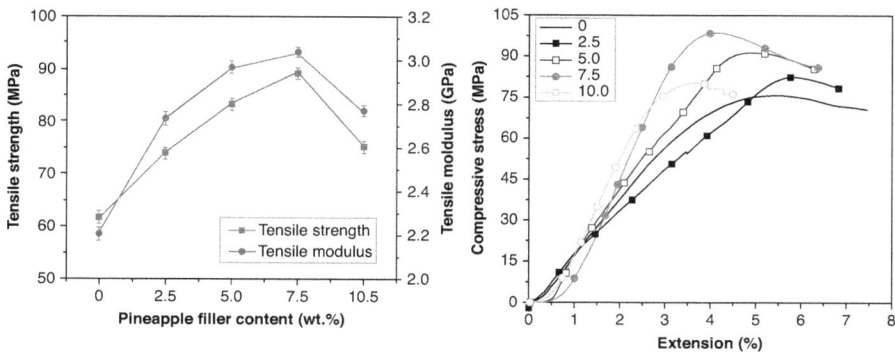

FIGURE 3.11 The effect of various pineapple contents on the tensile and comprehensive properties of 30wt.% pineal leaf fibers/epoxy bio-composite. (Adapted from [45] with the permission of Springer Nature (License number: 5207170514159).)

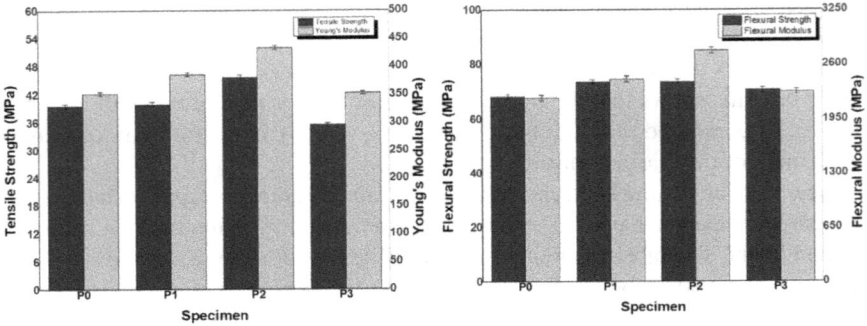

FIGURE 3.12 The tensile and flexural properties of pineapple fiber/epoxy composites containing various weight fractions of nano-calcium carbonate particles (P0: 0 wt.%, P1: 1 wt.%, P2: 3 wt.%, P3: 5 wt.%). (Adapted from [48] with the permission of Elsevier (License number: 5207170731013).)

that adding this filler can improve the mechanical performance of pineapple fibers/epoxy bio-composite by promoting the adhesion between the fiber reinforcement and epoxy matrix. The similar trend has also been reported by Cheirmakani et al. [46].

According to the literature, it can be said that by adding various fillers, especially bio-filler, into the pineapple/epoxy composites can tune the properties for multiple applications. With this view point, Dinesh et al. [47] included the nano-silica into this composite and used the Mahua oil as toughening agent for assessing thermomechanical and wear properties of that the composite. The composite with the 30 vol.% pineapple fibers, 15 vol.% Mahua oil and 1 vol.% nano-silica had the best mechanical performance, i.e., the tensile strength, elastic modulus, flexural strength, flexural modulus, impact energy and hardness of the composite were 160 MPa, 5.3 GPa, 225 MPa, 6.21 GPa, 5.5 J and 87 shore-D, respectively. Also, the wear properties of that composite showed the improvement trend. However, the Mahua oil caused the thermal stability of composite to deteriorate.

The nano-size bio-filler can be more effective than micro-size bio-filler on the performance of pineapple/epoxy composites. For example, Mahadevaswamy and Suresha [48] dispersed the 1, 3 and 5 wt.% nano-CaCO$_3$ into this composite to improve its mechanical properties, and results of this work can be seen in Figure 3.12.

3.7 *MORINGA OLEIFERA*-BASED EPOXY BIO-COMPOSITES

M. oleifera (Sahajana) fibers are one of the unknown leaf fibers, which grow in the regions of Bangladesh, Pakistan, India and Afghanistan. All parts of this plant like leaf, flour and seed can be utilized as food. The results showed that the leaves and seed of this plant can be used as filler and nano-cellulose into the epoxy matrix [49]. Mishra and Sinha [50] added the *M. oleifera* leaf powder into the epoxy to investigate the mechanical properties of this bio-composite, which can be seen in Figure 3.13. From this figure, it can be found that the optimum content of these powders into the epoxy polymer is about 20 wt.%. Ayrilmis et al. [51] extracted the nano-cellulose (whiskers) from *M. oleifera* fibers for using as bio-reinforcement into the epoxy resin. The content of these nano-whiskers into the epoxy matrix were 0.06, 0.12, and

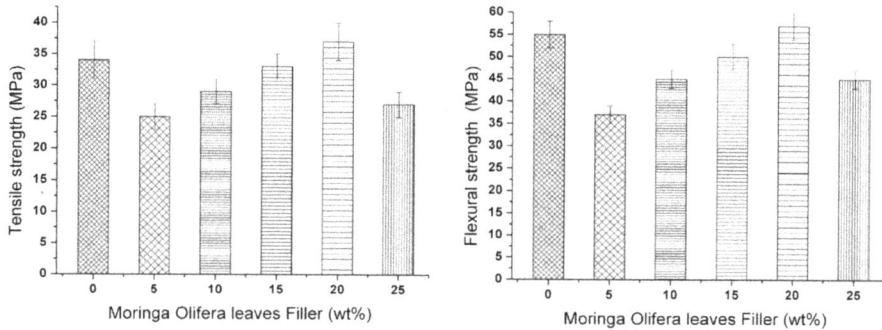

FIGURE 3.13 The tensile and flexural strength of *Moringa oleifera* leaves filler/epoxy bio-composites. (Adapted from [50] with the permission of John Wiley and Sons (License number: 5207170954571).)

0.18 wt.%. The improvement in the tensile and flexural modulus of filled composite with the 0.18 wt.% nano-whiskers were 31.4% and 38.2%, respectively. The thermal stability of these nanocomposites was slightly improved. However, the tensile and flexural strength had the decrement behaviors, which were 48.7% and 41.6%, respectively. The weak adhesion between these nano-whiskers and the epoxy matrix was the main reason of this reduction trend. According this research work and by applying the surface modifications on these nano-whiskers, it can be said that these nano-cellulose can be added as second bio-reinforcement into the bio-composites.

3.8 *PHORMIUM TENAX* BASED EPOXY BIO-COMPOSITES

The other unfamiliar leaf-based fiber is *P. tenax* with the common name of harakeke. New Zealand flax is another name for this plant, which is a kind of agave plant. This plant is cultivated in New Zealand and Norfolk Island. The leaves of this plant are so stiff and can grow up to 3 m long and 125 mm wide. A comparative study between phormium leaf fibers/epoxy and glass mat fibers/epoxy showed that this bio-composite had 30% lower flexural strength than glass fibers/epoxy composite. These acceptable mechanical properties were related to the rarity of kink bands in each cell of fibers and wrinkled cell-wall surface [52]. So, it can be said that these natural fibers can be a good choice for using as reinforcement into the epoxy composite.

De Rosa et al. [52] added 20 wt.% *P. tenax* unidirectional fibers into the epoxy matrix for surveying the thermal and mechanical properties of this bio-composite. The TGA results showed two degradation stages. The first of stages was at the temperature range of 200°C–305°C, which resulted in the depolymerization of hemicellulose, pectin, and cleavage of glycosidic linkage of cellulose. In this stage, the reported weight loss was about 20%. The second degradation occurred at the temperature of 305°C–370°C, which can be attributed to the decomposition of the α-cellulose. The weight loss in this stage was about 45.2%. The char ratio of this composite at the temperature of 700°C was 15.7%, which showed the proper thermal stability in this bio-composite. Adding these bio-fibers into the epoxy resin caused the tensile and flexural strength to improve to about 25% and 32%, respectively. Also, the obtained tensile and flexural moduli of this bio-composite were about two times those of neat epoxy. The impact properties of bio-composite containing 16 wt.% phormium fibers were investigated by Santulli et al. [53]. The maximum reported impact energy was 3.1 kJ/m^2.

Newman et al. [54] deacetylated phormium fibers by using the NaOH treatment for reinforcing the epoxy polymer. The obtained results showed that unlike other natural fibers, the alkali treatment is not useful for improving the mechanical properties of this bio-composite. In another study, Newman et al. [55] characterized the failure mechanisms of epoxy composite reinforced with the unidirectional phormium fibers. The crack growing through resin-filled thin-walled cell, crack pinning by coarse bundles of thick-wall cells, and debonding the fibers from the epoxy matrix were the characterized failure mechanisms in this bio-composite.

3.9 ARUNDO DONAX L.-BASED EPOXY BIO-COMPOSITES

A. donax L. or giant reed is another plant, which can grow up to 8 m. This plant has fast growing rate and is widely distributed in the Mediterranean countries. The contents of cellulose, hemi-cellulose and lignin of fibers of this plant are 31%–49%, 22%–35% and 9%–22%, respectively. With these constituents in the above proportion, this plant fiber can be a good candidate for reinforcing the epoxy polymers [56]. Fiore et al. [57] added these fibers into the grinding machine to fabricate the microfiller for using as bio-reinforcement into the epoxy resin. The elastic modulus of neat epoxy was 2.17 GPa. The maximum elastic modulus was obtained by adding 15 wt.% giant reed filler with the size of 500 μm to 2 mm, which was 3.04 GPa. The obtained flexural modulus of neat epoxy and this composite were 3 and 3.45 GPa, respectively. The tensile and flexural strength had the reducing trend by loading these microfillers, due to having weak adhesion between these hydrophilic fillers and hydrophobic matrix. So, it can be said that the surface modification is so necessary for enhancing the strength of these bio-composites. To do so, Scalici et al. [58] performed the plasma treatment on these fibers to enhance the mechanical strength of this bio-composite. The flexural test confirmed improving trend in the treated fibers/epoxy composite, which can be seen in Figure 3.14. These researchers also performed the DMA test on this bio-composite. The highest T_g belonged to the composite containing 2.5 wt.% treated fibers with the length of 3 cm, which was 90.5°C.

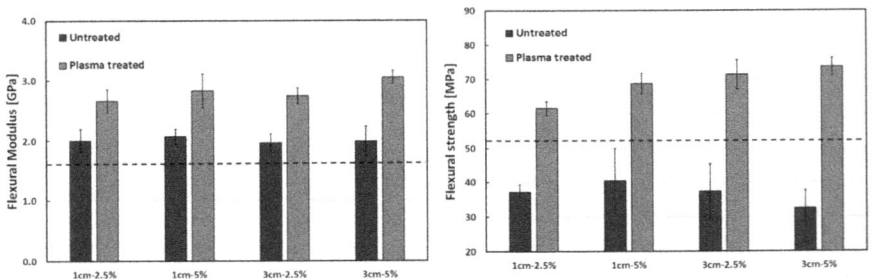

FIGURE 3.14   The untreated and plasma-treated *Arundo donax* L. leaf fibers/epoxy biocomposites with various lengths and fiber contents. (Adapted from [58] with the permission of Elsevier (License number: 5207171142648).)

3.10 SCREW-PINE-BASED EPOXY BIO-COMPOSITES

Screw-pine or *Pandanus odoratissimus* is one of the Pandanaceae family plants. The leaves of this plant have similar mechanical properties like pineapple leaf. The cellulose content of extracted fibers from screw-pine leaf is about 52%. Use of these leaf fibers into the polymeric composites is still limited [59]. According to the study of Hossain et al. [60], the maximum tensile and flexural strength of epoxy containing these fibers were obtained by incorporating 10 wt.% screw-pine fibers, which were 53 and 61 MPa, respectively. In the study of Deesoruth et al. [61], the highest compressive strength was achieved with this fiber fraction. Naik et al. [62] by performing alkali treatment improved the mechanical strength of screw-pine/epoxy composite. Also, the maximum impact energy was obtained for the composite reinforced with 30 wt.% screw-pine fibers.

3.11 FUTURE PERSPECTIVE

To date, many efforts have been made to characterize various mechanical and thermal features of bio-composites, especially leaf-based bio-composites. The progress in the subject of leaf-based epoxy composites shows various trends. In the future, some researchers will try to develop or introduce new modification methods, especially the environmental friendly methods, for reducing the hydrophilic nature of leaf fibers, to fabricate epoxy bio-composites with higher mechanical properties. The other trend can be characterizing the mechanical or thermal features of unknown leaf fibers like screw-pine added into the epoxy polymers. Synthesizing nano-fillers from leaf fibers like nano-cellulose and using them into the leaf-based epoxy composites can be another progressive trend in the near future.

3.12 CONCLUSION

To substitute natural fibers with synthetic fibers for reducing the production costs and environmental issues, many efforts have been made by different scientific groups. The leaf-based epoxy composites are one of the major groups of bio-composites, which have proper potential for using in the secondary, tertiary and in some cases primary applications. Among them, sisal, abaca, banana and pineapple leaf fibers had the highest developments for reinforcing the epoxy polymers. Reducing the hydrophilic nature of these fibers, modifying the surface of them with the surface treatments, creating the chemical bonds with the epoxy polymers through various methods, using another polymer as tuning agent of mechanical properties are the efforts which have been done to improve the mechanical and thermal performances of leaf-based epoxy bio-composites.

REFERENCES

1. Ebrahimnezhad-Khaljiri, H., and R. Eslami-Farsani. 2017. Thermal and Mechanical Properties of Hybrid Carbon/Oxidized Polyacrylonitrile Fibers-Epoxy Composites. *Polymer Composites* 38 (7): 1412–17.

2. Eslami-Farsani, R., and H. Ebrahimnezhad-khaljiri. 2021. Smart Epoxy Composites. In *Epoxy Composites: Preparation, Characterization and Applications*, ed. J. Parameswaranpillai, H. Pulikkalparambil, S. M. Rangappa, and S. Siengchin, pp. 349–94. Weinheim: Wiley-VCH Verlag GmbH & Co.
3. Mohammadi, M. A., R. Eslami-Farsani, and H. Ebrahimnezhad-Khaljiri. 2020. Experimental Investigation of the Healing Properties of the Microvascular Channels-Based Self-Healing Glass Fibers/Epoxy Composites Containing the Three-Part Healant. *Polymer Testing* 91: 106862.
4. Bigdilou, M. B., R. Eslami-Farsani, H. Ebrahimnezhad-Khaljiri, and M. A. Mohammadi. 2020. Experimental Assessment of Adding Carbon Nanotubes on the Impact Properties of Kevlar-Ultrahigh Molecular Weight Polyethylene Fibers Hybrid Composites. *Journal of Industrial Textiles*. doi:10.1177/1528083720921483.
5. Senthilkumar, K., N. Saba, N. Rajini, M. Chandrasekar, M. Jawaid, S. Siengchin, et al. 2018. Mechanical Properties Evaluation of Sisal Fibre reinforced Polymer Composites: A Review. *Construction and Building Materials* 174: 713–29.
6. Gupta, M. K., and R. K. Srivastava. 2016. Properties of Sisal Fibre Reinforced Epoxy Composite. *Indian Journal of Fibre and Textile Research* 41 (3): 235–41.
7. Kumaresan, M., S. Sathish, and N. Karthi. 2015. Effect of Fiber Orientation on Mechanical Properties of Sisal Fiber Reinforced Epoxy Composites. *Journal of Applied Science and Engineering* 18 (3): 289–94.
8. Militello, C., F. Bongiorno, G. Epasto, and B. Zuccarello. 2020. Low-Velocity Impact Behaviour of Green Epoxy Biocomposite Laminates Reinforced by Sisal Fibers. *Composite Structures* 253: 112744.
9. Webo, W., L. Masu, and M. Maringa. 2018. The Impact Toughness and Hardness of Treated and Untreated Sisal Fibre-Epoxy Resin Composites. *Advances in Materials Science and Engineering* 2018: 8234106.
10. Chand, N., and U.K. Dwivedi. 2007. Influence of Fiber Orientation on High Stress Wear Behavior of Sisal Fiber-Reinforced Epoxy Composites. *Polymer Composites* 28 (4): 437–41.
11. Behera, S., R. K. Gautam, S. Mohan, and A. Chattopadhyay. 2021. Dry Sliding Wear Behavior of Chemically Treated Sisal Fiber Reinforced Epoxy Composites. *Journal of Natural Fibers*. doi:10.1080/15440478.2021.1904483.
12. Rong, M. Z., M. Q. Zhang, Y. Liu, G. C. Yang, and H. M. Zeng. 2001. The Effect of Fiber Treatment on the Mechanical Properties of Unidirectional Sisal-Reinforced Epoxy Composites. *Composites Science and Technology* 61 (10): 1437–47.
13. Patra, A., D. K. Bisoyi, P. K. Manda, and A. K. Singh. 2013. Electrical and Mechanical Properties of the Potassium Permanganate Treated Short Sisal Fiber Reinforced Epoxy Composite in Correlation to the Macromolecular Structure of the Reinforced Fiber. *Journal of Applied Polymer Science* 128 (2): 1011–19.
14. Tragoonwichian, S., N. Yanumet, and H. Ishida. 2007. Effect of Fiber Surface Modification on the Mechanical Properties of Sisal Fiber-Reinforced Benzoxazine/Epoxy Composites Based on Aliphatic Diamine Benzoxazine. *Journal of Applied Polymer Science* 106: 2925–35.
15. Sahu, P., and M. K. Gupta. 2021. Eco-Friendly Treatment and Coating for Improving the Performance of Sisal Composites. *Polymer Testing* 93: 106923.
16. Sahoo, S. K., V. Khandelwal, and G. Manik. 2018. Influence of Epoxidized Linseed Oil and Sisal Fibers on Structure–Property Relationship of Epoxy Biocomposite. *Polymer Composites* 39: E2595–2605.
17. Reddy Paluvai, N., S. Mohanty, and S. Nayak. 2015. Mechanical and Thermal Properties of Sisal Fiber Reinforced Acrylated Epoxidized Castor Oil Toughened Diglycidyl Ether of Bisphenol A Epoxy Nanocomposites. *Journal of Reinforced Plastics and Composites* 34 (18): 1476–90.

18. Sahoo, S. K., S. Mohanty, and S. K. Nayak. 2017. Mechanical, Thermal, and Interfacial Characterization of Randomly Oriented Short Sisal Fibers Reinforced Epoxy Composite Modified with Epoxidized Soybean Oil. *Journal of Natural Fibers* 14 (3): 357–67.
19. Torres, M., V. R. Rodriguez, P. I. Alcantara, and E. Franco-Urquiza. 2020. Mechanical Properties and Fracture Behaviour of Agave Fibers Bio-Based Epoxy Laminates Reinforced with Zinc Oxide. *Journal of Industrial Textiles*. doi:10.1177/1528083720965689.
20. Rodríguez Soto, A. A., J. L. Valín Rivera, L. M.S. Alves Borges, and J. E. Palomares Ruiz. 2018. Tensile, Impact, and Thermal Properties of an Epoxynovolac Matrix Composites with Cuban Henequen Fibers. *Mechanics of Composite Materials* 54 (3): 341–48.
21. Sampieri-Bulbarela, S., A. Manzano-Ramírez, J. L. Reyes-Araiza, M. S. Muñiz Villareal, J. R. Gasca-Tirado, L. M. Apátiga, H. Savastano, and A. Marroquín de Jesús. 2011. Influences of a Novel Henequen Fabric Structure on the Mechanical Properties of a Polymeric Composite. *Scientific Research and Essays* 6(25): 5324–30.
22. Gonzalez-Murillo, C., M. P. Ansell. 2009. Mechanical Properties of Henequen Fibre/ Epoxy Resin Composites. *Mechanics of Composite Materials* 45: 435–42.
23. Liu, K., H. Takagi, R. Osugi, and Z. Yang. 2012. Effect of Physicochemical Structure of Natural Fiber on Transverse Thermal Conductivity of Unidirectional Abaca/Bamboo Fiber Composites. *Composites Part A: Applied Science and Manufacturing* 43 (8): 1234–41.
24. Singh, S., H. Khatri, S. Deshlahra, and T. N. Babu. 2021. Mechanical Properties of Graphene Reinforced Abaca Fiber Composites. *IOP Conference Series: Materials Science and Engineering* 1123 (1): 012023.
25. Vijaya R., B., V. M. Manickavasagam, C. Elanchezhian, C. V. Krishna, S. Karthik, and K. Saravanan. 2014. Determination of Mechanical Properties of Intra-Layer Abaca-Jute-Glass Fiber Reinforced Composite. *Materials and Design* 60: 643–52.
26. Punyamurthy, R., D. Sampathkumar, B. Bennehalli, R. Patel, and S. C. Venkateshappa. 2014. Abaca Fiber Reinforced Epoxy Composites : Evaluation of Impact Strength International Journal of Sciences : Abaca Fiber Reinforced Epoxy Composites : Evaluation of Impact Strength. *International Journal of Sciences: Basic and Applied Research* 18 (2): 305–17.
27. Punyamurthy, R., D. Sampathkumar, Ba. Bennehalli, R. Patel, G. R. Gouda, and C. V. Srinivasa. 2015. Influence of Fiber Content and Effect of Chemical Pre-Treatments on Mechanical Characterization of Natural Abaca Epoxy Composites. *Indian Journal of Science and Technology* 8 (11): 53236.
28. Cai, M., H. Takagi, A. N. Nakagaito, Y. Li, and G. I.N. Waterhouse. 2016. Effect of Alkali Treatment on Interfacial Bonding in Abaca Fiber-Reinforced Composites. *Composites Part A: Applied Science and Manufacturing* 90: 589–97.
29. Kurien, R. A., D. P. Selvaraj, and C. P. Koshy. 2021. Worn Surface Morphological Characterization of NaOH-Treated Chopped Abaca Fiber Reinforced Epoxy Composites. *Journal of Bio- and Tribo-Corrosion* 7: 31.
30. Liu, K., X. Zhang, H. Takagi, Z. Yang, and D. Wang. 2014. Effect of Chemical Treatments on Transverse Thermal Conductivity of Unidirectional Abaca Fiber/Epoxy Composite. *Composites Part A: Applied Science and Manufacturing* 66: 227–36.
31. Srinivasan, T., G. Suresh, K. Santhoshpriya, C. T. Chidambaram, K. R. Vijayakumar, and A. Abdul Munaf. 2021. Experimental Analysis on Mechanical Properties of Banana Fibre/Epoxy (Particulate) Reinforced Composite. *Materials Today: Proceedings* 45: 1285–89.
32. Mohan, T. P., and K. Kanny. 2016. Nanoclay Infused Banana Fiber and Its Effects on Mechanical and Thermal Properties of Composites. *Journal of Composite Materials* 50 (9): 1261–76.

33. Balaji, A., R. Purushothaman, R. Udhayasankar, S. Vijayaraj, and B. Karthikeyan. 2020. Study on Mechanical, Thermal and Morphological Properties of Banana Fiber-Reinforced Epoxy Composites. *Journal of Bio- and Tribo-Corrosion* 6: 60.
34. Subramanya, R., D. N. S. Reddy, and P. S. Sathyanarayana. 2020. Tensile, Impact and Fracture Toughness Properties of Banana Fibre-Reinforced Polymer Composites. *Advances in Materials and Processing Technologies* 6 (4): 661–68.
35. Subramanya, R., P. Sathyanarayana, M. Kn, and S. Naik. 2020. The Manufacture and Characterisation of Short Banana Fibre-Reinforced Polymer Composites. *Advances in Materials and Processing Technologies*. doi:10.1080/2374068X.2020.1833403.
36. Gairola, S. P., Y. K. Tyagi, B. Gangil, and A. Sharma. 2020. Fabrication and Mechanical Property Evaluation of Non-Woven Banana Fibre Epoxy-Based Polymer Composite. *Materials Today: Proceedings* 44: 3990–96.
37. Venkateshwaran, N., A. E. Perumal, and D. Arunsundaranayagam. 2013. Fiber Surface Treatment and Its Effect on Mechanical and Visco-Elastic Behaviour of Banana/Epoxy Composite. *Materials and Design* 47: 151–59.
38. Nguyen, T. A., and T. H. Nguyen. 2021. Banana Fiber-Reinforced Epoxy Composites: Mechanical Properties and Fire Retardancy. *International Journal of Chemical Engineering* 2021: 1973644.
39. Asim, M., K. Abdan, M. Jawaid, M. Nasir, Z. Dashtizadeh, M. R. Ishak, M. E. Hoque, and Y. Deng. 2015. A Review on Pineapple Leaves Fibre and Its Composites. *International Journal of Polymer Science* 2015: 950567.
40. Glória, G. O., M. C. A. Teles, A. C. Ce. Neves, C. M. F. Vieira, F. P. D. Lopes, Ma. A. Gomes, F. M. Margem, and S. N. Monteiro. 2017. Bending Test in Epoxy Composites Reinforced with Continuous and Aligned PALF Fibers. *Journal of Materials Research and Technology* 6(4): 411–16.
41. Jagadish, M. R., and A. Ray. 2020. Investigation on Mechanical Properties of Pineapple Leaf–Based Short Fiber–Reinforced Polymer Composite from Selected Indian (Northeastern Part) Cultivars. *Journal of Thermoplastic Composite Materials* 33(3): 324–42.
42. Jain, J., S. Jain, and S. Sinha. 2019. Characterization and Thermal Kinetic Analysis of Pineapple Leaf Fibers and Their Reinforcement in Epoxy. *Journal of Elastomers and Plastics* 51 (3): 224–43.
43. Nagarajan, T. T., A. S. Babu, K. Palanivelu, and S. K. Nayak. 2016. Mechanical and Thermal Properties of PALF Reinforced Epoxy Composites. *Macromolecular Symposia* 361 (1): 57–63.
44. Kumar, R., S. Bhowmik, K. Kumar, and J. P. Davim. 2020. Perspective on the Mechanical Response of Pineapple Leaf Filler/Toughened Epoxy Composites under Diverse Constraints. *Polymer Bulletin* 77 (8): 4105–29.
45. Saha, A., S. Kumar, and A. Kumar. 2021. Influence of Pineapple Leaf Particulate on Mechanical, Thermal and Biodegradation Characteristics of Pineapple Leaf Fiber Reinforced Polymer Composite. *Journal of Polymer Research* 28: 66.
46. Cheirmakani, B. M., B. Subburaj, and V. Balasubramanian. 2020. Exploring the Properties of Pineapple Leaf Fiber and *Prosopis julifora* Powder Reinforced Epoxy Composite. *Journal of Natural Fibers*. doi:10.1080/15440478.2020.1798844.
47. Dinesh, T., A. Kadirvel, and P. Hariharan. 2020. Thermo-Mechanical and Wear Behaviour of Surface-Treated Pineapple Woven Fibre and Nano-Silica Dispersed Mahua Oil Toughened Epoxy Composite. *Silicon* 12 (12): 2911–20.
48. Mahadevaswamy, H. S., and B. Suresha. 2020. Role of Nano-CaCO$_3$ on Mechanical and Thermal Characteristics of Pineapple Fibre Reinforced Epoxy Composites. *Materials Today: Proceedings* 22: 572–79.

49. Bharath, K. N., P. Madhu, T. G. Y. Gowda, M. R. Sanjay, V. Kushvaha, and S. Siengchin. 2020. Alkaline Effect on Characterization of Discarded Waste of *Moringa Oleifera* Fiber as a Potential Eco-Friendly Reinforcement for Biocomposites. *Journal of Polymers and the Environment* 28 (11): 2823–36.
50. Mishra, K., and S. Sinha. 2020. Development and Assessment of *Moringa Oleifera* (Sahajana) Leaves Filler/Epoxy Composites: Characterization, Barrier Properties and in Situ Determination of Activation Energy. *Polymer Composites* 41 (12): 5016–29.
51. Ayrilmis, N., F. Ozdemir, O. B. Nazarenko, and P. M. Visakh. 2019. Mechanical and Thermal Properties of *Moringa Oleifera* Cellulose-Based Epoxy Nanocomposites. *Journal of Composite Materials* 53 (5): 669–75.
52. Rosa, I. M. D., C. Santulli, and F. Sarasini. 2010. Mechanical and Thermal Characterization of Epoxy Composites Reinforced with Random and Quasi-Unidirectional Untreated *Phormium Tenax* Leaf Fibers. *Materials and Design* 31 (5): 2397–2405.
53. Santulli, C., G. Jeronimidis, I. M. De Rosa, and F. Sarasini. 2009. Mechanical and Falling Weight Impact Properties of Unidirectional Phormium Fibre/Epoxy Laminates. *Express Polymer Letters* 3 (10): 650–56.
54. Newman, R. H., E. C. Clauss, J. E. P. Carpenter, and A. Thumm. 2007. Epoxy Composites Reinforced with Deacetylated *Phormium Tenax* Leaf Fibres. *Composites Part A: Applied Science and Manufacturing* 38 (10): 2164–70.
55. Newman, R. H., M. J. L. Guen, M. A. Battley, and J. E. P. Carpenter. 2010. Failure Mechanisms in Composites Reinforced with Unidirectional Phormium Leaf Fibre. *Composites Part A: Applied Science and Manufacturing* 41 (3): 353–59.
56. Bessa, W., D. Trache, M. Derradji, H. Ambar, A. F. Tarchoun, M. Benziane, and B. Guedouar. 2020. Characterization of Raw and Treated *Arundo Donax* L. Cellulosic Fibers and Their Effect on the Curing Kinetics of Bisphenol A-Based Benzoxazine. *International Journal of Biological Macromolecules* 164: 2931–43.
57. Fiore, V., T. Scalici, G. Vitale, and A. Valenza. 2014. Static and Dynamic Mechanical Properties of *Arundo Donax* Fillers-Epoxy Composites. *Materials and Design* 57: 456–64.
58. Scalici, T., V. Fiore, and A. Valenza. 2016. Effect of Plasma Treatment on the Properties of *Arundo Donax* L. Leaf Fibres and Its Bio-Based Epoxy Composites: A Preliminary Study. *Composites Part B: Engineering* 94: 167–75.
59. Abral, H., H. Andriyanto, R. Samera, S. M. Sapuan, and M. R. Ishak. 2012. Mechanical Properties of Screw Pine (*Pandanus Odoratissimus*) Fibers-Unsaturated Polyester Composites. *Polymer - Plastics Technology and Engineering* 51 (5): 500–506.
60. Hossain, S., M. M. Rahman, A. Jamwal, P. Gupta, S. Thakur, and S. Gupta. 2019. Processing and Characterization of Pine Epoxy Based Composites. *AIP Conference Proceedings* 2148: 030017.
61. Deesoruth, A., H. Ramasawmy, and J. Chummun. 2014. Investigation into the Use of Alkali Treated Screwpine (*Pandanus Utilis*) Fibres as Reinforcement in Epoxy Matrix. *International Journal of Plastics Technology* 18 (2): 263–79.
62. Naik, V., M. Kumar, and V. Kaup. 2021. Study on the Mechanical Properties of Alkali Treated Screw Pine Root Fiber Reinforced in Epoxy Matrix Composite Material. *AIP Conference Proceedings* 2317: 020023.

4 Agro Waste-Based Epoxy Composites
Thermal and Mechanical Properties

Emel Kuram
Gebze Technical University

CONTENTS

4.1 INTRODUCTION

Epoxy is a significant class of commercial-grade thermosetting polymers. Epoxy has a large range of applications from packaging to electronics, thanks to different desirable features such as high stiffness, low shrinkage on curing, excellent chemical resistance and superior thermal stability (Mittal et al., 2018). Also, owing to its good environmental and thermal resistance, it could be employed for outdoor applications (Mittal et al., 2016; Saba et al., 2016a). However, fragility and weak mechanical properties of epoxy matrix have limited its performance (Rizal et al., 2020). Also, epoxy is easily flammable, and its combustion generates significantly toxic gases (Chen et al., 2019; Alothman et al., 2020; Chandrasekar et al., 2020).

Agricultural (agro) waste (residue) usage as a reinforcement in composites may induce several socioeconomic and environmental advantages (Chaudhary et al., 2015; Nasimudeen et al., 2021; Senthil Muthu Kumar et al., 2021). The incorporation of plant and waste fibers into polymeric materials will not only improve mechanical and thermal properties but also diminish materials cost, and create a green material. Also, this approach provides re-use of the waste or decreases the waste effectively (Shih et al., 2012), reduces waste disposal cost and enhances sustainability

DOI: 10.1201/9781003271017-4

(Rizal et al., 2020). Countries grow fruits and plants not only for agricultural purposes but also to produce raw materials for industry (Ashour, 2017).

The demand for raw materials to forest-based industries continues to increase. On the one hand, about 95% of lignocellulosic material for particleboard fabrication is wood (Nemli et al., 2008), and on the other hand, forests, which are primary resources of wood supply, are diminishing at an increasing rate (Ashori, 2006). This problem, together with necessity of protecting natural sources, has caused huge efforts to employ agro wastes in particleboard production (Chaudhary et al., 2015). The wide and large availability of natural fiber could decrease pressure on agriculture and forest (Ashour, 2017). Usage of agro waste can reduce deforestation which leads to global warming. Recently, measures have been taken to reduce deforestation and thus save the environment. In this initiative, agro waste products are used instead of wood for structural applications (Nagarajan et al., 2020).

Burning agro wastes induces environmental problems such as air pollution and soil erosion and reduces the biological activity of the soil. The use of agro wastes in composites not only protects the environment but can also help farmers generate an additional income from cultivation (Çöpür et al., 2007). Recycling of agro waste offers clear economic benefits: reducing costs associated with the usage of alternative raw materials and reducing raw material consumption (Barbieri et al., 2013; Zhang, 2013).

Natural fibers possess a great ability to absorb moisture; this is because of the hemicellulose content, which causes fiber degradation owing to processing temperatures (Nirmal et al., 2015; Senthilkumar et al., 2015, 2018, 2021). Natural fibers have hydrophilic characteristics and polymer matrix shows hydrophobic characteristics (Lazim et al., 2014). Strength of natural fiber-based composite is decreased owing to hydrophobic characteristics of epoxy matrix, causing weak bonding between matrix and fibers consisting of moisture (Jayamani et al., 2014). Therefore, natural fiber should be properly dried before utilizing in composites manufacturing. Interfacial bonding could be enhanced by treating fibers with alkaline (Nirmal et al., 2011). Surface of natural fibers is roughened with alkaline treatment by removing hemicellulose and lignin. Alkali treatment on fiber improved surface roughness because of the elimination of hemicellulose and other highly volatile materials from fiber surface (Ibrahim et al., 2011). Sodium hydroxide (NaOH) and hydrogen peroxide (H_2O_2) treatments have eliminated non-cellulose content (lignin and hemicellulose). Treatments enhanced interfacial adhesion between fiber and matrix. NaOH- and H_2O_2-treated composites displayed higher tensile strength than untreated one because of enhancement in fiber/matrix interaction adhesion in composite. Also, NaOH- and H_2O_2-treated composites were stiffer (Chun et al., 2020). Silane-treated soy stem fibers in composite displayed better mechanical result in comparison to untreated and other chemical-treated fibers in composites (oxalic acid and alkali treatment) (Vinod et al., 2021a). NaOH and silane modification of the fiber increased thermal stability and tensile strength. Silane treatment was the best for *Muntingia calabura* fiber in comparison to raw fiber and NaOH treatment (Vinod et al., 2021b). Green composites produced with 30 wt% *Punica granatum* agro waste and a 5-day chemical treatment employing aqueous solution of sodium bicarbonate have the best economic, environmental and technical indicators (Zindani et al., 2021).

There is a growing concern about sustainable usage of agro waste to decrease environmental burden and protect the ecosystem (Rizal et al., 2020; Thomas et al., 2021). The development of natural fiber-reinforced composite using highly environmentally friendly agro waste fibers as reinforcements has received more attraction and is increasing nowadays. However, improper adhesion between incompatible fiber and matrix has been counterproductive to the proper use of this abundant, naturally presence source. Various techniques of surface modification of fibers such as the use of compatibilizers, electron beam irradiation, mercerization (alkali treatment) (Van de Weyenberg et al., 2003; Edeerozey et al., 2007; Li et al., 2007), plasma treatment (Scalici et al., 2016) and treatment with other chemicals (Kushawa and Kumar, 2010; Lu et al., 2013) have been adopted till now to enhance adhesion between matrix and fiber. Recently, environmentally friendly treatment of natural fiber with sodium bicarbonate (Fiore et al., 2016) or sea water (Ishak et al., 2009) has been reported. The adhesiveness between the epoxy matrix and rice husk flour fiber was improved by treatment with NaOH (Bisht and Gope, 2018).

Green buildings have attracted attention as a tool for sustainable development within the current energy crisis and the degradation of the natural environment. Most countries around the world face three problems: the amount of climate change and carbon dioxide emissions, waste generation and energy consumption. The use of agro wastes requires knowledge of physical, mechanical, and thermal features of these materials. This information assists the decision maker and designer in evaluating the best ways to utilize these wastes, as well as avoiding hazards from burning or leaving the waste to decompose on site. Composites reinforced with natural fibers provide a highly comfortable construction and enable people to decrease energy consumption for cooling and heating (Ashour, 2017; Senthilkumar et al., 2022; Shahroze et al., 2021).

Agro wastes contribute to the development of composite materials having low environmental effect. Agro waste-based epoxy composites would be environmentally friendly materials. The current chapter discusses the possibility of employing agro wastes as reinforcing fibers in epoxy matrix. First, this chapter highlights the advantages of agro waste as reinforcing fibers. Second, this chapter draws attention to the thermal and mechanical features of some agro wastes-based epoxy composites.

4.2 AGRO WASTE FIBERS

The usability of agro wastes for the fabrication of green composites can aid to solve problems associated with the burning of agro residues, such as the emission of air pollutants (ammonia, carbon monoxide, elementary carbon, organic carbon, sulfur dioxide, etc.) (Jain et al., 2014; Ravindra et al., 2016; Oanh et al., 2018). Air quality deteriorates not only in region where burning occurs but also in surrounding regions, thus causing negative influences on human health (Awasthi et al., 2017). Emission of air pollutants might influence atmospheric climate (McNeill, 2017) and thus the danger of global warming is imminent (Zindani et al., 2021).

Renewable resources have been utilized as raw materials to diminish the adverse effect on the environment of existing composites. One of the most promising approaches is to replace reinforcing material with natural fibers, often from more sustainable sources (agro waste). But these materials possess various drawbacks and limitations for their usage in high-performance applications, one of the reasons being the weak interaction between matrix and fiber. New works display the possibility of enhancing their mechanical properties through the application of a third material to improve the interaction between matrix and fiber. Humins (a byproduct of furandicarboxylic acid production) was found to be an effective bio-binder for epoxy and natural fiber (Vidal et al., 2021).

Air, soil and water pollution have been the most serious problem. Soil pollution often influences lands and renders them unproductive. The primary reason of this is agro waste. We can reduce agro waste pollution by reusing agro wastes, namely natural fibers, by mixing them with polymeric material to reinforce composite. But this approach has its pros and cons. Polymer composites produced employing natural fibers are inferior in mechanical properties to polymer composites reinforced, for example, with carbon or glass fibers. Disadvantages of using natural fibers in the polymeric matrix are their high moisture sorption (because of the hydrophilicity of the natural fiber), causing high degradation and swelling rate, low resistance to chemicals and fire and low mechanical property. Natural fiber polymer composites display low levels of interfacial adhesion between matrix and fiber, which, if enhanced, ultimately overcomes all drawbacks and improves mechanical properties of produced composite. Optimal chemical treatments such as alkalization/mercerization of fibers have been investigated to improve interfacial adhesion causing improved mechanical properties. Chemical treatments could remove wax, impurities from fiber surface, creating a rough and hydrophobic surface (Verma and Goh, 2021). Alkali treatment is employed to separate fiber bundles into individual fibers. Processing causes an increase in aspect ratio of smaller fiber particles and creates rough fiber surface, which aids to increase interfacial bond between fiber and matrix (Jones et al., 2017).

Agro wastes are recognized as cheaper and valuable biological resources for the development of renewable high-performance materials (Mamat Razali et al., 2021). Agro wastes can be cobs, fiber, husk or straw. The internal structure of a single straw is tubular, efficient and tough. It comprises hemicellulose, cellulose, lignin and silica with great tensile and bending strength (Ashour, 2017). Mechanical properties of fibers derived from agro wastes depend on their chemical constituents. Cellulose helps to improve mechanical properties with hydrogen bonding, whereas hemicellulose helps biodegradation and hindrance in thermal stability, moisture sensitivity (Asim et al., 2020). Fiber rigidity and moisture resistance are dependent on lignin, which affects fiber structure, morphology, etc. (Verma et al., 2020).

It was concluded that the nano oil palm empty fruit bunch filler can be utilized as innovative and promising alternative to available expensive nanofillers, with relatively little effect on the environment and a significant effect on the aerospace, automotive, construction, electronics and semiconductor industries. The use of agro waste-based oil palm empty fruit bunch filler was effective and affordable for advanced applications, including highly thermally stable, superconducting and interfacial materials to replace expensive nanofillers (Saba et al., 2016b). Soy fiber (Vinod et al., 2021a) and *M. Calabura* (Vinod et al., 2021b) fiber were found as potential sources of sustainable and environmentally friendly raw materials for reinforcement that could be employed to manufacture green composites for lightweight structural applications (Vinod et al., 2021a, b).

4.3 EPOXY

Epoxies are important commercial grades of thermosetting polymer with a three-membered ring named oxirane, epoxide, glycidyl or ethoxylin group, which is highly strained and very reactive (Azeez et al., 2013; Saba et al., 2016a). Resins have low molecular weight (Vijayan et al., 2012). Properties of epoxy do it appropriate for reactivity to a large variety of curing agents. Various epoxy resins are present (cyclo-aliphatic epoxy resin, novolac, diglycidyl ether of bisphenol-A, triglycidyl *p*-amino phenol and tetraglycidyl diamino diphenyl methane) (Kausar, 2017). Additionally, modified epoxy systems are present, such as bio-based epoxy, and fluorine-, silicon- and phosphorus-containing epoxy. Curing agents are chiefly hardeners or catalysts, while catalysts are Lewis acids or tertiary amines, and their function is to start the polymerization of epoxy to create polyether (Hodd, 1989). The role of the epoxy in composites is to uniformly transfer force to filler and maintain integrity of the whole system (Abdellaoui et al., 2019).

Epoxy is extensively employed in industrial applications such as adhesives, aerospace structures, electronics, coatings (Shih, 2007; Shih et al., 2012), insulation materials and wind turbine blades. Epoxy has a wide range of applications from packaging to electronics, owing to different desirable features such as excellent chemical resistance, high stiffness, low shrinkage on curing and superior thermal stability (Mittal et al., 2018). Also, owing to its good environmental and thermal resistance, it could be employed for outdoor applications (Mittal et al., 2016; Saba et al., 2016a). However, fragility and weak mechanical properties of epoxy have limited its performance (Rizal et al., 2020). Also, epoxy is easily flammable; moreover, its combustion generates significantly toxic gases. Thus, the fire risk of epoxy is great, reducing usage of epoxy in aerospace, electrical and electronics areas. To enhance flame retardancy of polymers, halogen flame retardants are traditionally incorporated to create material less flammable. Flame retardancy mechanism is to stop chain reaction and content of incorporation is small. But when halogens are released into atmosphere after burning, they will destroy ozone layer and increase the amount of ultraviolet ray that is harmful to plants and animals. In addition, environmental and potential health hazards of halogen flame retardants have restricted their use. Thus, the production of an environmentally friendly, halogen-free flame retardant for epoxy is crucial (Chen et al., 2019). Agro wastes such as bagasse (Shen et al., 2021), rice straw (Jiang et al., 2018) and rice husk (Guna et al., 2020; Kavitha et al., 2021)-based flame retardants have been recommended as an alternative to halogen flame retardants. Adding modified sugarcane bagasse can improve mechanical properties, reduce flammability and decrease the amount of other chemical flame retardant required. Addition of sugarcane bagasse into epoxy could be applicable for safer epoxies for coating materials, in transportation area, developing composites and making building materials enabling for a variety of applications of polymeric materials (Liu et al., 2021).

4.4 AGRO WASTE-BASED EPOXY COMPOSITES

Green composites from agro wastes have received enormous interest in the scientific field due to increasing environmental concerns about the burning of crop residues (Zindani et al., 2021).

Matrix behaves as the continuous phase in composites, binds fibers together, distributes the load to fibers and protects the fibers from environmental influences (Muralidhar et al., 2020).

The compatibility between epoxy and untreated fibers was weak, as evidenced by the existence of aggregates, voids and pull-out fibers on fractured surfaces, resulting in their poor mechanical and thermal properties. Densely knitted texture was observed when fibers were treated with silane coupling agents, showing that interaction between the fibers and the epoxy enhanced. Fiber aggregates and existence of voids on fractured surfaces were also seen, indicating poor compatibility between epoxy and untreated fibers (Shih et al., 2012). Water bamboo husk fibers modified by coupling agents caused better compatibility with epoxy resin than untreated fibers. Also, mechanical properties were improved owing to loading of coupling agent-treated water bamboo husk fibers and untreated powders (Shih, 2007). Usage of epoxy as the matrix had a positive influence on wettability of organic (walnut shell) filler surface, resulting in good adhesion of component (Salasinska et al., 2018).

Polar hydroxyl groups on surface of fiber do not allow well-bonded interface with a non-polar matrix. When matrix is hydrophilic, this causes weaker mechanical properties (Bisht and Gope, 2018). The modification of fiber surface by both chemical and physical processes enhances the efficiency of natural fiber/polymeric matrix (Imoisili and Jen, 2020). Cellulose ratio in the natural fiber is increased with alkali modification, whereas lignin, oils and wax can be removed from the natural fiber structure (Boopathi et al., 2012; Arrakhiz et al., 2013). Also, alkali modification can increase fiber/matrix interface adhesion and enhance mechanical features of composite (Sareena et al., 2012; Narendar and Dasan, 2014).

The choice of production method for fabrication of polymer composites is main task. Several parameters such as the kind of matrix, type of reinforcement, i.e. long or short fibers, particulate type, non-woven mat and woven mat affect the selection of fabrication method (Verma and Goh, 2021). For the fabrication of epoxy composites, various techniques have been employed in different application areas with the necessity of final shape and size of components (Pulikkalparambil et al., 2021). Some of the fabrication techniques for the development of epoxy-based composite are explained as follows:

Hand Lay-Up: In this technique (Figure 4.1), knitted form or woven form fibers are placed in a mold and a matrix is implemented over fibers with a brush. Then, a roller is used in order to enhance interaction between matrix and fibers, ensure uniformity of resin and maintain thickness of the composite (Verma and Goh, 2021). Advantages of employing hand lay-up are freedom of design and the cheapest. Disadvantages are the need for more cycles to manufacture, the cost of skilled labor, the placement errors risk, the molding of complex parts and low surface quality (Fiore et al., 2015; Pulikkalparambil et al., 2021).

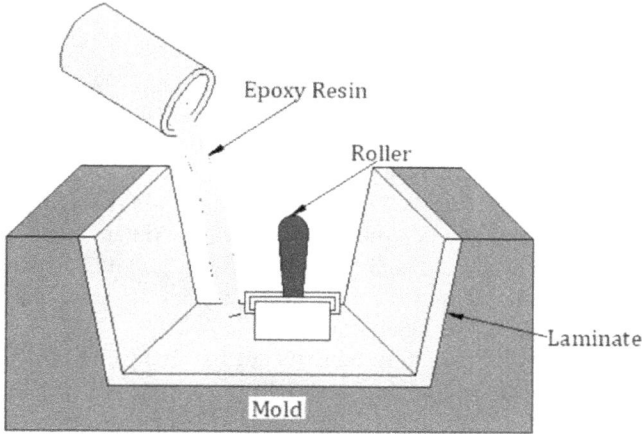

FIGURE 4.1 Hand lay-up.

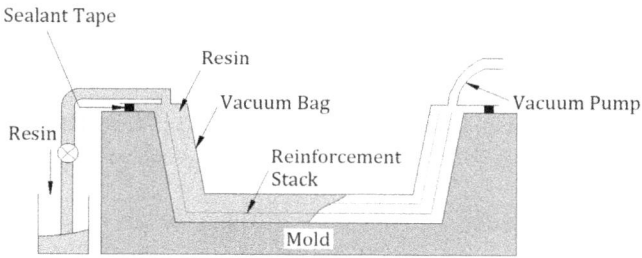

FIGURE 4.2 Vacuum infusion.

Vacuum Bagging (Vacuum Infusion): Vacuum bagging is an extended version of hand lay-up in which pressure is applied in order to enhance unification (Figure 4.2). In the method, a vacuum pump is utilized that applies a pressure of 1 atm to reinforcement (Verma and Goh, 2021). Advantages of this method are better surface finish and possibility of utilizing a heated oven to accelerated consolidation and void free. Limitations are as follows: it is more expensive and complex in comparison to hand lay-up, vacuum bag needs to be designed in accordance with component dimensions and the final component size is restricted by the mold size (Harshe, 2015; Pulikkalparambil et al., 2021).

Vacuum-Assisted Resin Transfer Molding: This infusion technique is a composite fabricating method in which vacuum pressure is employed to induce flow of liquid resin and voids in an evacuated fiber mat are infused with liquid resin. After the liquid resin is cured and converted into a solid resin matrix, the resin joins the fibers into a rigid composite. This technique is a cost-effective method to produce composite parts with better surface quality and lower void defects suitable for low volume production than conventional technique such as hand lay-up (Chun et al., 2020). Also, a high fiber-to-matrix ratio composite can be fabricated by utilizing vacuum-assisted resin infusion (Gajjar et al., 2020). Advantages of vacuum-assisted resin transfer molding are as follows: capability to produce big complex parts, could be employed to fabricate parts with different geometries, the hardener and resin could be stored separately and mixed just before infusion and there is low emission of volatile organic compounds. Limitations of this method are difficulty in reusing of bags, tubes, sealing tapes and other consumables after one cycle, limited pressure (injection and compression pressure) between atmospheric and vacuum pressure, leakage issues and less robustness (Schmachtenberg et al., 2005; Hsiao and Heider, 2012; Ouarhim et al., 2019; Pulikkalparambil et al., 2021).

Solvent Casting: Solvent casting consists of dissolving the components in a volatile solvent and after that evaporating solvent with suitable drying devices (Figure 4.3) (Cheng and Wiggins, 2017). Advantages of method are the simple manufacturing processing and no special equipment is required. Method limits the usage of any mechanical stress or high thermal processes to avoid side reactions or degradation. But the disadvantage of this technique is usage of an external solvent which can influence environmental friendliness and costs (Kong et al., 2015; Pulikkalparambil et al., 2021).

Autoclave: An autoclave is a large pressure cooker in which the prepreg is compressed employing pressurized carbon dioxide and nitrogen (Figure 4.4). It fabricates composite with closer thickness control and lower void. Disadvantage of method is that part size might be restricted by autoclave size (Eckold, 1994; Pulikkalparambil et al., 2021).

Resin Transfer Molding: Technique starts with a dry fiber preform. Preform is placed in a metal mold and mold is closed allowing preform to be compressed to specified fiber volume fraction. Then a liquid thermoset resin is injected into mold (at a pressure of 5–7 bar) (Figure 4.5). Mold and resin could be preheated prior to injection or mold could be heated after injection to cure resin. Because of high temperatures and high injection pressures, resin transfer molding tools are costly and bulky to fabricate and to

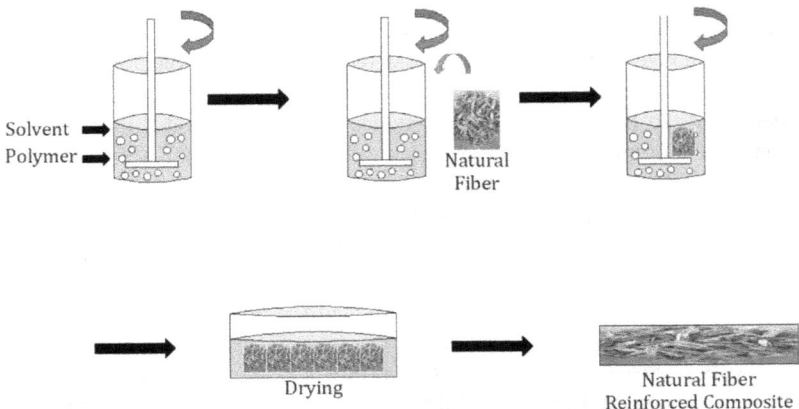

FIGURE 4.3   Solvent casting.

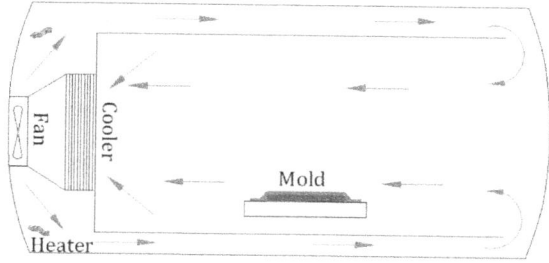

FIGURE 4.4 Autoclave.

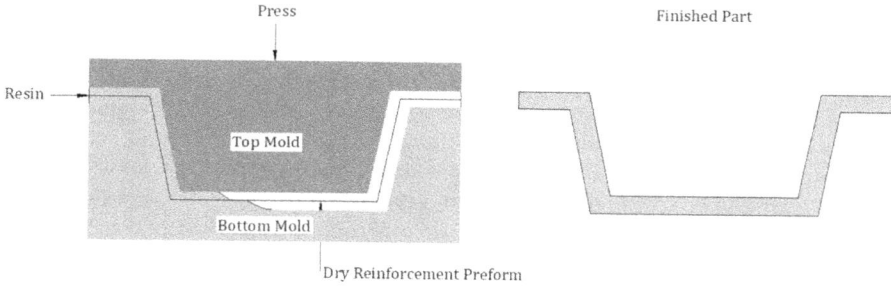

FIGURE 4.5 Resin transfer molding.

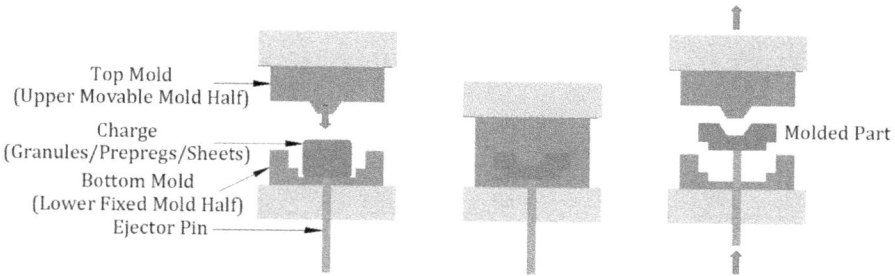

FIGURE 4.6 Compression molding.

process. The preform should be designed by choosing adequate fabric and fiber kinds and fiber volume fraction (i.e. number of plies in preform) taking into consideration cost, fiber wet-out, formability, mechanical performance and permeability to resin flow (Harshe, 2015). This technique is developed for parts made from fiber plastic composite. A reaction resin is implemented to dry, in-process fiber parts and wetted accordingly by implementing pressure in a closed vessel (Verma and Goh, 2021).

Compression Molding: This process involves a preformed filling material, a premeasured volume of powder or a viscous mixture of liquid-resin and fillers, placed into a heated cavity (Figure 4.6). Forming is made under pressure from a plunge or half of die (Abba et al., 2013). This method is one of the best techniques for manufacturing composite because of reliability and shorter cycle time (Verma and Goh, 2021). Advantages of this technique are simple processing, minimal waste and low cost. Disadvantage of employing this method is that it is restricted to small industries because of time-consuming processing. Moreover, this method needs highly trained labor to run operations (Rong et al., 2001; de Andrade Silva et al., 2008; Pulikkalparambil et al., 2021).

Pultrusion: It is a composite fabrication method in which fibers are continuously drawn through a resin bath for resin impregnation prior to entering a heated die, where an exothermic curing reaction takes place (Figure 4.7) (Santos et al., 2012). The advantages of pultrusion are continuous production, smooth surface and unlimited length of products. The disadvantages include limited size in transverse direction, being able to be reinforced in only one direction and being expensive (Chachad et al., 1995; Santos et al., 2012; Pulikkalparambil et al., 2021).

Filament Winding: It is the technique in which continuous filaments or strands of fiber are wound onto a mandrel or supporting form (Shen, 1995) (Figure 4.8). This method is especially employed for the fabrication of cylindrical composite tubes (Verma and Goh, 2021). Advantages of this method are as follows: greater reinforcement amount (70% or more to enhance the strength), adapting the orientation of the fibers and large part production is possible. Limitations are large investments and restricted shape and design (Shen, 1995; Pulikkalparambil et al., 2021).

Mixing: Mixing technique is depicted in Figure 4.9. Internal, low-shear liquid batch and rheo mixer are types of mixers. The disadvantage of these mixers is time-consuming (Cheng and Wiggins, 2017; Pulikkalparambil et al., 2021).

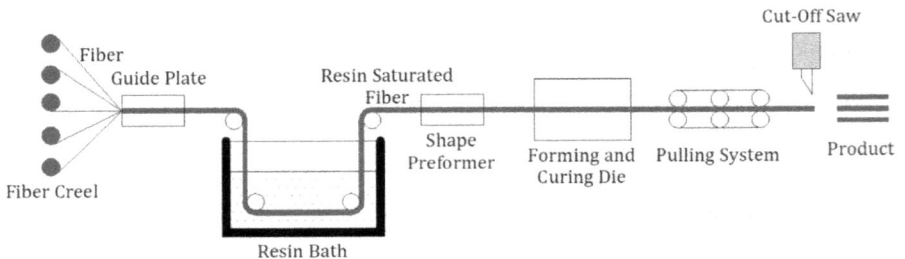

FIGURE 4.7 Pultrusion.

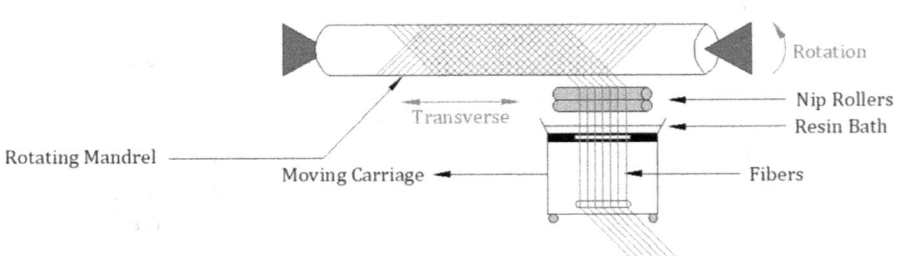

FIGURE 4.8 Filament winding.

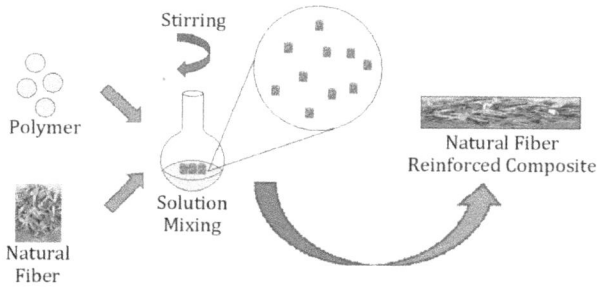

FIGURE 4.9 Mixing.

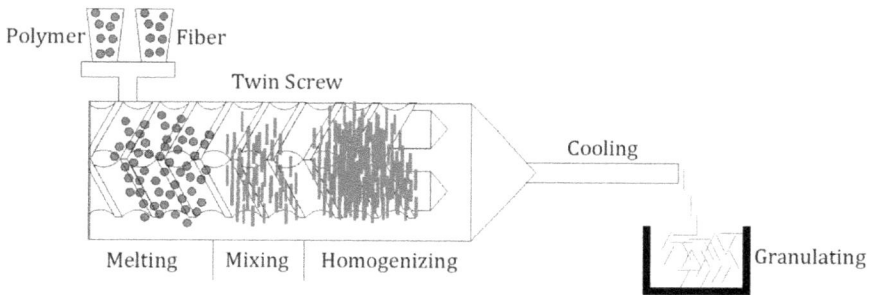

FIGURE 4.10 Extrusion.

Extrusion: Extrusion is a process that consists in forcing a material through a die with the cross section of the part to be achieved. The advantages of employing an extruder (Figure 4.10) are good heat transfer, good surface finish, high production, good mixing and low cost. Disadvantages are the size difference and part limitations (Fenouillot and Perier-Camby, 2004; Pulikkalparambil et al., 2021).

Three-Roll Mill: Three-roll mills, named as calendering mills, are an ideal continuous mixer for mixing highly viscous materials such as epoxy. It uses both extensional and shear flow produced by rollers rotating at different speeds to disperse, mix and homogenize matrices with particles or additives (Figure 4.11). Besides, this is a solvent-free method that is environmentally friendly (Cheng and Wiggins, 2017). The main advantage of this method is better dispersion. Fillers could obtain a greatly exfoliated or interca-lated structure (Gojny et al., 2004; Chang et al., 2009; Olowojoba, 2013; Pulikkalparambil et al., 2021).

Resonant Acoustic Mixer: Resonance acoustic mixer is a non-contact mix-ing approach for dispersing and distributing materials over a wide range of viscosities. It depends on application of a low-frequency and high-intensity acoustic field across entire mixing vessel to facilitate rapid mixing (Figure 4.12) (Cheng and Wiggins, 2017). Main advantage of this mixer is homogeneous mixing as opposed to continuous and batch mixers where localized mixing is observed near mixing blade tip (Rumeau et al., 2015; Pulikkalparambil et al., 2021).

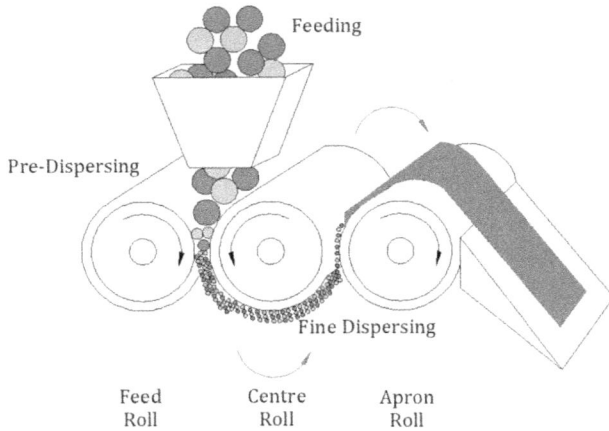

FIGURE 4.11   Three-roll mill.

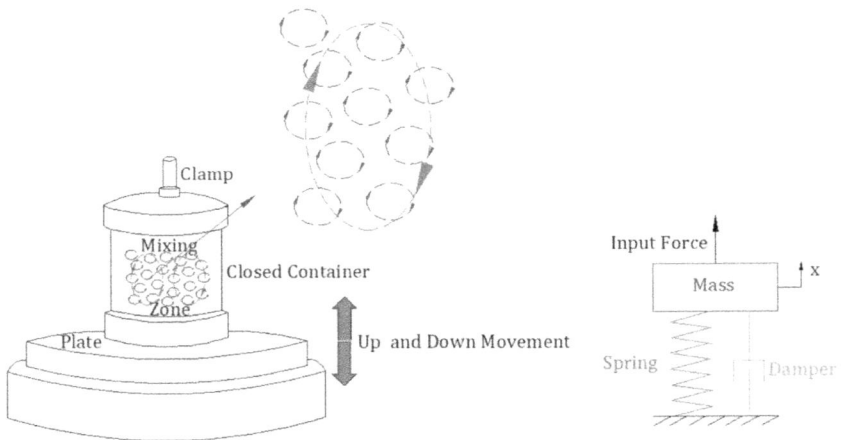

FIGURE 4.12   Resonant acoustic mixer.

By adapting autoclave molding and vacuum resin infusion, composite parts can be fabricated stronger and have less void content (Abhemanyu et al., 2019).

The fabrication techniques employed to develop epoxy composites with agro wastes in literature are presented in Table 4.1.

The potential area of application is the development of the ankle-foot orthosis employing bamboo and coir-filled epoxy composites for patients struggling with foot drop problem (Kumar and Bhowmik, 2021). The potential areas of application for the coir-reinforced epoxy composites could be decking, electronic casings, fencing and furniture (instead of timber) (Kumar and Bhowmik, 2019). Corn husk fiber-reinforced epoxy composite has potential use in non-structural applications such as seat back covers and car interior panels (Chun et al., 2020). Owing to their good electrical and mechanical properties, jute fiber-reinforced epoxy composites are useful for

TABLE 4.1

Manufacturing of Agro Waste-Based Epoxy Composites

Agro waste	Agro Waste Amount (% weight)	Manufacturing Method	References
Agave Americana Corn husk Jute	3	Hand lay-up	Karthik et al. (2020)
Arecanut husk	15	Cast molding and hand lay-up	Muralidhar et al. (2020)
Bamboo Coir	2.5, 5, 7.5, 10, 12.5	Hand lay-up	Kumar and Bhowmik 2021)
Banana fiber	10, 20, 30, 40, 50	Hand lay-up	Gairola et al. (2021)
Banana fiber/coconut sheath/Jute fiber	16/11.5/7.5 0/21.5/15.5 27/0/9 17/18/0	Hand lay-up	Abhemanyu et al. (2019)
Banana fiber Pineapple leaf fiber	20	Mixing	Shih et al. (2012)
Black rice husk ash	20, 31, 39, 43, 46	Cast molding	Hsieh et al. (2017)
Coconut shell	1, 2, 3, 4	Hand lay-up	Kaur et al. (2020)
Coir fiber/date palm fiber	100/0 50/50 75/25 25/75 0/100	Hand lay-up	Dixit et al. (2020)
Corn husk	20	Vacuum-assisted resin infusion	Chun et al. (2020)
Hazelnut shell Sunflower husk Walnut shell	15, 25, 35	Mixing	Barczewski et al. (2019)
Lagenaria siceraria fiber	15, 20, 25, 30, 35	Hand lay-up and compression molding	Nagappan et al. (2021)
Linseed flax fabric	NA	Liquid resin infusion	Vidal et al. (2021)
Oil palm ash	1, 2, 3, 4, 5	Cast molding	Abdul Khalil et al. (2013)
Oil palm boiler ash	10, 20, 30, 40, 50	Cast molding and compression molding	Rizal et al. (2020)
Oil palm empty fruit bunch/kenaf	3	Mixing and hand lay-up	Saba et al. (2016b)
Oil palm empty fruit bunch/kenaf	3	Mixing and hand lay-up	Saba et al. (2016c)
Peach palm tree fiber	50, 60, 70	Compression molding	Cordeiro et al. (2017)
Pineapple leaf fiber	2.5, 5, 7.5, 10, 12.5	Hand lay-up	Kumar et al. (2020)
Plantain	20% volume	Hand lay-up and compression molding	Imoisili and Jen (2020)
Rice husk	20	Cast molding	Bisht and Gope (2018)
Rice husk	10, 15, 20	Cast molding	Jain et al. (2019)

(*Continued*)

TABLE 4.1 (*Continued*)
Manufacturing of Agro Waste-Based Epoxy Composites

Agro waste	Agro Waste Amount (% weight)	Manufacturing Method	References
Sugarcane bagasse ash	5, 10	Compression molding	Chandradass et al. (2021)
Tamarind seed	10, 20, 30, 40% volume	Ultra-sonication	Naik et al. (2019)
Walnut shell	20, 30, 40, 50	Cast molding	Salasinska et al. (2018)

NA, not available.

electrical and electronics, low load-bearing housing and marine applications (Patel and Parsania, 2017). Pineapple leaf filler-reinforced epoxy composites can be used to develop materials for household products and casing of electronic equipment (Kumar et al., 2020). Sugarcane bagasse ash can be employed as filler to fabricate the brake pad material (Chandradass et al., 2021). Tamarind/epoxy composite can be employed in automobile interior design, coatings, dash boards, door panels, electronics components and roofing (Naik et al., 2019). The addition of high content of walnut shell into the thermoset epoxy causes the preparation of low-cost composites with suitable thermo-mechanical and mechanical features, making them viable as materials for low-demand applications and products. Considering the thermal stability and mechanical properties, optimal walnut shell filler amount for the epoxy-based composite was 30 wt%. But regarding economic issues, it is also reasonable to produce highly reinforced composites including 50 wt% of walnut shell, particularly for nonstructural applications (Salasinska et al., 2018).

Nano oil palm empty fruit bunch/kenaf/epoxy composites have good dynamic mechanical properties and thus can replace conventional building materials such as aggregates, aluminum, cement, ceramic bricks, cladding, partition, wood and steel materials to develop sustainable green buildings. The combination of low damping factor and high storage modulus shows better specific modulus and strength, so their applications can be extended to the design of bridge decks, connecting bridges, doors, door frames, plumbing components, stairs, railing and outdoor decking due to lower damping properties and lower weights in comparison to expensive wooden blogs. Their applications could also be extended to low-stress application fields such as the highway bridges support beams, housing sectors, industrial drive shafts, papermaking rollers and windmill blades. The improved properties of hybrid composites show a high possibility in advanced lightweight structural applications where renewable source and high performance are needed, such as aircraft, boats, bridges, buildings, pressure vessels and reservoirs (Saba et al., 2016c).

4.4.1 THERMAL PROPERTIES

Differential thermal analysis curve gives information regarding transformation of materials such as crystallization, melting, glass transition and sublimation during heating processing. Also it reveals information about the release and absorption of heat during the decarbonation, dehydration or burning of material (Kumar and Bhowmik, 2019).

Glass transition temperature (T_g) values of composites were lower than neat epoxy and T_g reduced with rising the fiber amount. It could be attributed to increment in free volume at fiber/matrix interface. A higher fiber content causes a higher interface and that's why a higher amount of voids, which conduces to increased polymer chain mobility (Cordeiro et al., 2017). However, it was declared that T_g of the epoxy was rose to approximately 8°C–18°C greater than pure specimen (was elevated from 58°C to 65°C–75°C) (Shih 2007). Also, T_g increased by the incorporation of nano oil palm empty fruit bunch filler into kenaf/epoxy composite (Saba et al., 2016c).

Decomposition temperatures of banana and pineapple leaf fiber-reinforced epoxy at 5% weight loss were higher than pristine epoxy. Because of nature of plant, plant fibers release the absorbed moisture at approximately 40°C–140°C, and decomposition of cellulosic materials such as cellulose and hemicelluloses continues at approximately 140°C–370°C. Generally, decomposition temperatures of the banana and pineapple leaf fiber composites were lower than neat epoxy. Thus, increased decomposition temperature of composites demonstrates that fibers were successfully bonded with epoxy in crosslinking network and degradation of the fibers was repressed compared to epoxy (Shih et al., 2012). Decomposition temperatures of treated banana fiber and pineapple leaf fiber composites (330°C and 290°C) were higher than untreated fiber-reinforced composites (264°C and 284°C), showing that treated fibers were more closely linked to epoxy than to untreated fibers. Decomposition temperature (330°C) of treated pineapple leaf fiber-reinforced epoxy was the highest in comparison to others. It meant that treated pineapple leaf fiber-reinforced composites gave better thermal properties than other specimens (Shih et al., 2012). Neat epoxy and coir fiber/epoxy composite displayed a two-step degradation phenomenon. Since there was little change in the degradation step, thermal degradation behavior was not significantly influenced by filler. Working temperature of coir-reinforced epoxy composites can be around 250°C without significant deterioration (Kumar and Bhowmik, 2019). Walnut shell filler decomposition took place in a two-stage process, while neat epoxy occurred in a single-stage manner (Salasinska et al., 2018). Thermal degradation of pure epoxy and oil palm ash-filled epoxy showed a weight loss under 100°C because of the moisture loss and evaporation of benzyl alcohol. Pure epoxy and oil palm ash-filled epoxy composites began to decompose at around 336°C. Degradation at 50% weight loss ($T_{50\%}$) of 1%–3% filler-loaded composites was greater than that of pure epoxy and tended to increase with filler loading. Beyond 3% filler content, reduction in structural destabilization was seen. It could be due to agglomeration of oil palm ash and poor compatibility between oil palm ash and epoxy. A similar result was observed for initial decomposition (T_i), final decomposition (T_f) and decomposition (T_{max}) temperatures. Residual weight percent of oil palm ash-filled epoxy composites at 800°C increased upon incorporation of filler because of inorganic residue (Abdul Khalil et al. 2013). Thermal decomposition temperatures (T_{max}, T_i and T_f) increased with the addition of oil palm boiler ash. However, thermal decomposition temperatures vaguely rose with rising temperature and diminished with size of particle. The highest thermal degradation temperature was achieved with 50 μm particle size and 50 wt% filler content (Rizal et al. 2020). Char residue increased with increasing oil palm boiler ash filler loading and particle size (Rizal et al. 2020) owing to the increment in non-volatile carbon amount with addition of oil palm boiler filler ash and with the increment in particle size (Khoshnoud 2017).

Temperature ranges from 30°C to 130°C were dehydration process where there was a 3%–6% drop in mass and could be because of moisture loss. Initial moisture amount of plantain fibers was responsible for this mass loss (Imoisili and Jen, 2020). The first step decomposition of fibers was noted at 150–320°C, which was linked to hemicellulose decomposition. The lignin decomposition was challenging and occurred at a large range of temperature with beginning from 160°C (Alemdar and Sain, 2008; Poletto et al., 2012; Imoisili and Jen, 2020). The second step of decomposition was observed at 320°C–400°C and was chiefly because of carbon combustion and cellulose decomposition (Soares et al., 1995; Kaloustian et al., 2003; Yang et al., 2006; Imoisili and Jen, 2020). The final decomposition stage began at around 400°C and continued up to 600°C (Imoisili and Jen, 2020).

Thermal gravimetric analysis showed three degradation steps for kenaf/epoxy composites and oil palm empty fruit bunch/kenaf/epoxy composites. Initial loss of weight under 100°C–150°C owing to moisture loss was found for kenaf/epoxy and oil palm empty fruit bunch/kenaf/epoxy composites. In kenaf/epoxy composites and oil palm empty fruit bunch/kenaf/epoxy hybrid composites, pyrolysis and decomposition of aromatic groups of epoxy network, coupled with decomposition of cellulosic components from the kenaf fibers, resulting in a significant weight loss in the range 300°C–400°C (Azwa and Yousif, 2013; Saba et al., 2016b). The first endothermic transitions near 100°C refer to T_g of the epoxy and second endothermic transition between 200°C and 260°C indicates removal of moisture from dehydration of the secondary alcoholic group of epoxy, coupled with decomposition of some cellulosic components from kenaf fibers successively. Third endothermic peaks are seen around 310°C solely because of complete degradation of the epoxy (Saba et al., 2016b).

Integral procedural decomposition temperature (IPDT) (Xiong et al., 2018) is a common indicator employed to evaluate thermal stability of materials. Two parameters affect IPDT: Biochar yield and initial pyrolysis temperature. High IPDT means high thermal stability. After adding 30 wt% of bagasse, IPDT of composite materials increased with respect to neat epoxy. Therefore, incorporation of bagasse improved thermal stability of epoxy. This behavior is since biochar layer occurred after pyrolysis protects epoxy and increases thermal stability of composite (Liu et al., 2021).

Thermal stability for epoxy composite determines maximum temperature at which composite is safe for applications (Pulikkalparambil et al., 2021). Black rice husk ash filler obtained from rice husks could improve thermal stability of epoxy/black rice husk composites (Hsieh et al., 2017). An increment in thermal stability was found at the 3% oil palm ash filler content condition and was owing to increment in crosslinking of epoxy in existence of oil palm ash and possessing minimum particle-to-particle interaction and well-dispersed nanoparticles (Abdul Khalil et al., 2013). Percentage of char residue and thermal stability of oil palm boiler ash/epoxy increased with rising filler amount (Rizal et al., 2020). Thermal stability improved with rising filler sizes and content (Rizal et al., 2020) because of greater silica amount and other inorganic particles in oil palm boiler ash, which improved thermal properties of epoxy composites (Khoshnoud, 2017).

Potassium permanganate ($KMnO_4$)-treated plantain fiber/epoxy composite showed better thermal stability than untreated plantain fiber/epoxy composite (Imoisili and Jen, 2020). Addition of walnut shell improved thermal stability of composite (Salasinska et al., 2018). Incorporation of nano oil palm empty fruit bunch enhanced thermal stability and char yield of kenaf/epoxy composite (Saba et al., 2016b).

Thermal resistance and flame retardancy are enhanced as water bamboo husk fibers and powders are added into epoxy matrices. The increases of char yield of the epoxy were approximately 13.5%–52.8% by incorporation of 10% water bamboo husk powder or fiber. Increased char could limit occurrence of combustible gases, reduce exothermicity of pyrolysis reaction, and inhibit thermal conductivity of burning materials (Pearce and Liepins, 1975). Therefore, with incorporation of fibers or powders, flame retardancy of the epoxy will be leveled up (Shih, 2007).

4.4.2 MECHANICAL PROPERTIES

Tensile strength of natural fibers bases on chemical composition. Tensile strength of composite obtained by increasing cellulose content of fibers increases and tensile strength reduces with increment of lignin (Abhemanyu et al., 2019). With increasing banana fiber amount, tensile strength enhanced (Abhemanyu et al., 2019). Tensile strengths of pineapple leaf and banana fiber-reinforced epoxy were higher than pure epoxy. Also, tensile strengths of treated fiber-reinforced composites were greater than untreated fiber-reinforced ones (Shih et al., 2012). Maximum tensile strength value was obtained with 3% oil palm ash filler content due to interaction between matrix and filler. Therefore, interaction caused a better stress transfer between matrix and filler particles, and as a result increased the tensile strength of composite (Abdul Khalil et al., 2013). When bond between matrix and filler is strong enough, tensile strength of a particulate composite could be greater than a pure polymer matrix (Fekete et al., 1999). Beyond 3% filler content, tensile strength reduced. It may be due to decrement in filler/matrix interaction, poor interfacial bond and agglomeration of filler particles, which resulted in early failure (Abdul Khalil et al., 2013).

With an increment in wt% of rice husk, tensile/compressive strength decreased owing to weak interfacial adhesion. Also, hydrophobic matrix and hydrophilic fiber cause an adhesion lack. At 20 wt%, rice husk coagulation occurred because of poor interfacial bonding between epoxy and rice husk, resulting in weak load transfer from matrix to reinforcement, causing failure (Jain et al., 2019). Tensile strength diminished with increasing tamarind seed fiber (Naik et al., 2019).

Arecanut husk fiber-based epoxy composites presented good elastic characteristics at higher frequencies (Muralidhar et al., 2020). At greater frequency, polymer chains are not fully deformed because of the time lack, resulting in greater modulus (Obada et al., 2020).

The highest increase in stiffness was obtained by hazelnut shell-filled composites. A gradual enhancement in the elasticity modulus was found with increment in amount of hazelnut shell filler (Barczewski et al., 2019). Maximum tensile modulus value was obtained with 3% oil palm ash filler loading. Beyond 3% filler content, tensile modulus diminished. This may be due to decrement in filler/matrix interaction, poor interfacial bond and agglomeration of filler particles (Abdul Khalil et al., 2013). Tensile modulus increased with an increment in tamarind seed filler loading up to 20% in the epoxy. It decreased after 20% of the fiber content (Naik et al., 2019).

20 wt% of corn husk fiber-reinforced epoxy gave higher tensile modulus and strength than neat epoxy (Chun et al., 2020). When pineapple leaf filler loading increased, tensile modulus and strength improved up to definite amount of filler loading (after 5 wt% of filler). Reason of this increase could be the enhanced stress transfer capacity and better mechanical interlocking with incorporation of filler particles to matrix. Source of reduction in properties with incorporation of more than 5 wt% of filler might be ineffective load transfer capacity from filler to matrix at greater filler content because of insufficient wetting and agglomeration of fillers. The loading of filler to epoxy enhanced strength of composite as compared to neat epoxy. Incorporation of pineapple leaf filler to epoxy caused an increase in strain value with respect to pure epoxy for most of filler content. This result was due to the decrease in plasticity and increased elasticity of composite with reinforcement of pineapple leaf (Kumar et al., 2020). Particulate rice husk-reinforced epoxy composites gave highest tensile strength and tensile modulus followed by hybrid (particulate mixed with rice husk in equal composition) (Jain et al., 2019).

As the coconut shell filler content increased, the tensile elongation decreased linearly from 1 wt% until 4 wt% (Kaur et al., 2020). Decrement in tensile elongation as filler amount increased might be because of formation of agglomeration, which causes low adhesion between filler and epoxy, thus causing the composite to be brittle (Shuhadah and Supri, 2009). As filler loading increases, composite becomes tougher; thus, chain mobilization in matrix decreases. Decreased chain mobility, in which the formation of the crosslinked network in chain is restricted by great attractive force, leads to an increment in brittleness of composite (Islam et al., 2017).

Elongation at break diminished with increase of oil palm boiler ash particle size in epoxy (Rizal et al., 2020) owing to decrement of deformability of interface between filler and epoxy (Husseinsyah and Mostapha, 2011). Elongation at break of oil palm boiler ash/epoxy composite diminished with increment of filler amount and with decrement of particle size (Rizal et al., 2020). Elongation at break of oil palm boiler ash/epoxy composite increased with particle size because of decrement in deformation of poor interfacial bond between matrix and filler with greater particle size (Rizal et al., 2020). This behavior increases ductility of composite, hence increased elongation (Pan et al., 2009). With an increment in filler, crack moves toward interfacial bonding and thus reduces elongation at break of composite (Rizal et al., 2020).

The material's ability to absorb energy and shock, and to withstand crack propagation, is named as impact strength. Matrix and fiber strength, resistance to crack propagation and load transfer efficiency play an important role in evaluating impact characteristics (Kumar and Bhowmik, 2019). Impact strength is energy required to fracture material (Patel and Parsania, 2017; Pulikkalparambil et al., 2021). Namely, impact strength is the capability to resist implemented stress at high speed (Pulikkalparambil et al., 2021). Filler agglomeration, poor filler/epoxy interface, sharp edges of fillers and voids can cause stress-concentrated regions that can lead to failure in cracks form with implementation of stress (DeArmitt, 2011; Jyotishkumar et al., 2013). One approach to increase impact strength is to add dispersants or coupling agents (DeArmitt and Rothon, 2011). The incorporation of dispersants decreases filler agglomeration that can otherwise behave as a stress concentration point (Pulikkalparambil et al., 2021).

Impact strength for fiber-reinforced composite bases on polymer nature, fiber and fiber/matrix interfacial adhesion. Treated fibers can absorb more energy than untreated fibers because of strong fiber/matrix interfacial adhesion (Patel and Parsania, 2017). Impact resistance usually increased with increasing the fiber content. Composites with treated fiber displayed an enhancement in the impact resistance with respect to the untreated fiber ones. This result confirmed good interfacial adhesion provided by existence of epoxy silane or isocyanate groups on fiber surface (Cordeiro et al., 2017). Loading of coir filler in epoxy increased breaking energy and impact resistance, but after 10 wt% of filler incorporation, impact strength began to decrease. This increment may be because the inclusion of particulate fillers creates a barrier to crack propagation (Kumar and Bhowmik, 2019). Impact strength of oil palm boiler ash/epoxy composite rose with increment of filler content and reduced with increasing particle size (Rizal et al., 2020). In general, filler with lower particle size possesses a low aspect ratio in comparison to filler with bigger one. But because of large particle size of the filler, stress concentration near edges and the great aspect ratio, it may behave as create initiation of epoxy composites (Ramakrishna and Rai, 2006). Therefore, it was found that impact strength of oil palm boiler ash-reinforced epoxy composite decreased with rising particle size and filler addition over 30 wt% owing to high agglomeration of filler particles with greater particle size (Rizal et al., 2020). Excessive filler loading causes inhomogeneous distribution of filler in polymeric matrix. Therefore, impact properties dropped (Chen et al., 2004).

Impact strength increased with increment in rice husk fiber weight since fiber absorbs more energy in comparison to epoxy. Existence of fiber inhibits crack growth as the probability of a higher tendency for plastic deformation in matrix increases. For rice husk-reinforced composite, impact strength was greater as extra energy was required to pull rice husk fiber out of matrix in comparison to rice husk particle (Jain et al., 2019). Impact strength improved with tamarind seed fiber amount up to 30% and then dropped with further addition (Naik et al., 2019).

Addition of all shell types (hazelnut, sunflower and walnut) resulted in a significant drop in impact strength of composite with respect to neat epoxy. While the highest impact strength was achieved with hazelnut shell-filled epoxy, the lowest impact strength was observed in sunflower shell-filled composites (Barczewski et al., 2019).

Epoxy-based composite filled with hazelnut shell was characterized by the highest flexural strength among all composites developed (composites filled with sunflower shell and walnut shell) (Barczewski et al., 2019). Maximum flexural strength value was obtained with 3% oil palm ash filler loading. Beyond 3% filler content, flexural strength dropped. This may be due to decrement in filler/matrix interaction and poor interfacial bond because of agglomeration of filler particles (Abdul Khalil et al., 2013).

Increased flexural and tensile strength by filler incorporation supports significant interaction and load transfer ability of coir fillers infused in epoxy (Kumar and Bhowmik, 2019).

Composites with higher fiber content showed higher modulus without significantly influencing the tensile strength, meaning that it has potential to produce materials with great fiber loading with high stiffness while maintaining moderate mechanical properties (Cordeiro et al., 2017). Mixing the leaf fillers in the epoxy led to an increment in flexural and rupture modulus of composite relative to the neat epoxy. Improved flexural features by filler incorporation could be attributed to good load transfer capability from filler to filler and matrix to filler through interface (Kumar et al., 2020). Composite filled with sunflower husk was characterized by the lowest flexural stiffness among all composites (composites filled with hazelnut shell and walnut shell) (Barczewski et al., 2019). Maximum flexural modulus value was obtained with 3% oil palm ash filler loading. Beyond 3% filler content, flexural modulus diminished. It may be owing to weak interface bonding or adhesion between epoxy matrix and filler or existence of agglomeration (Abdul Khalil et al., 2013). Flexural modulus of peach palm tree fiber-reinforced epoxy composite was higher than pure epoxy (Cordeiro et al., 2017).

Alkali-treated fiber-reinforced composites showed greater flexural modulus and strength than untreated fiber ones, since increment in contact area between alkali-treated fibers and matrix enhanced interfacial bonding (Boynard et al., 2003; Mylsamy et al., 2020).

Coir/epoxy composite displayed higher flexural modulus and strength than bamboo/epoxy composite. When bamboo filler content increased up to 10 wt% of content, flexural strength improved. However, for coir/epoxy composites, increment in filler content negatively influenced flexural strength. At higher filler content, considering the stiffness of material, coir-filled composite performed better under three-point bending load conditions (Kumar and Bhowmik, 2021). The highest flexural modulus and strength of produced oil palm boiler ash/epoxy were achieved with 30 wt% content and smaller particle size (50 μm). Smaller particle-size oil palm boiler ash dispersed well in the epoxy (Rizal et al., 2020). Thus, crack propagation increased and plastic deformation of matrix rose, leading to greater flexural strength (Husseinsyah and Mostapha, 2011; Firdaus and Mariatti, 2012). But flexural strength of composite decreased with increment oil palm boiler ash filler loading owing to reduced matrix and filler interactions because of agglomeration of oil palm boiler ash fibers with excessive amount of filler (Firdaus and Mariatti, 2012; Rizal et al., 2020).

Incorporation of coir in epoxy caused enhancement in flexural modulus and strength for 2.5 wt% of filler loading (Kumar and Bhowmik, 2019). This could be attributed the enhanced capability to resist implemented load because of increased interaction between filler and matrix (Nagarajan et al., 2016). Further addition of coir fillers caused lower flexural properties; however, the found values were still higher than those for pure epoxy. At higher filler content, concentration of fillers increased, causing resin engulfing and agglomeration of fillers. This was followed by a decreased efficient interface region of filler/matrix interface, therefore reducing the flexural properties (Kumar and Bhowmik, 2019).

The use of coconut shell as reinforcement filler enhanced tensile and flexural properties of composite in comparison to pure epoxy (Kaur et al., 2020). As the coconut shell filler content increased, tensile and flexural modulus, tensile and flexural strength rose linearly from 1 wt% until 3 wt% and then decreased at 4 wt%

(Kaur et al., 2020). Increment in tensile strength, elongation, impact strength, flexural and compressive properties were found by increasing the *Lagenaria siceraria* fiber length and weight percent up to 7 mm and 30 wt%, respectively. Further increment in fiber length and amount was indicative of a decrement in the properties because of poor fiber/matrix bonding (Nagappan et al., 2021). $KMnO_4$-treated plantain fiber/epoxy gave greater tensile and flexural features. However, higher concentrations of $KMnO_4$ induced a drop in flexural and tensile properties (Imoisili and Jen, 2020).

Hardness determines the resistance to penetration, abrasion, indentation and tooth decay that the composite offers (Gairola et al., 2021). It was declared that the hardness of epoxy composites was not influenced by different fibers (Abhemanyu et al., 2019). The banana fiber, coconut sheath and jute fiber-based epoxy composites gave good hardness and tensile properties values, making them suitable for utilizing as automobile body panels, replacing conventional materials currently employed (Abhemanyu et al., 2019). For all filler types, almost no influence on hardness of composites was found at 15 wt% loading. The addition of more than 15 wt% of walnut shell and hazelnut shell fillers resulted in increment in hardness of composites as compared to neat epoxy. Conversely, incorporation of higher content (25 wt% and 35 wt%) of sunflower husk led to a decrement in hardness of composite. The highest hardness was achieved by composite filled with hazelnut shell. It might be related to multilayer structure and chemical composition of organic particles dispersed in epoxy in ground shell form (Barczewski et al., 2019). In this case, composite hardness might be related to lignin amount in fillers (Arbelaiz et al., 2005). Lignin is a binder that results in a compact cell structure. Thus, the greater the lignin amount, the higher is the stiffness and compressive strength of natural fillers (Barczewski et al., 2019). As the amount of tamarind seed fiber increased, the hardness decreased (Naik et al., 2019).

Mechanical properties influenced by incorporation of natural fillers. Incorporation of particulate fillers with a low aspect ratio (walnut and hazelnut shells) made possible the production of composite with enhanced stiffness and hardness. Addition of fibrous sunflower husk as a filler caused a significant increase in viscosity of composite, leading to a drop in mechanical features of epoxy/sunflower husk composite. However, drop in mechanical strength is not a contraindication for the development of ecological and cheap parts (Barczewski et al., 2019).

The highest flexural and tensile strength values were achieved at 30 wt% banana fiber content. Rise in tensile and flexural strength was seen up to 30 wt% fiber content and then gradually diminished at 40 wt% and 50 wt% fiber amounts. Maximum hardness and impact strength values were observed at 40 wt% banana fiber content (Gairola et al., 2021). Generally, modulus, flexural, tensile and hardness of composite enhanced with increment in jute fiber content (Patel and Parsania, 2017). The highest mechanical features of produced oil palm boiler ash/epoxy composite were obtained with 30 wt% content and smaller particle size (50 μm) (Rizal et al., 2020). Smaller particle size possesses a larger surface area, which could increase interfacial adhesion between filler and matrix (Khoshnoud and Abu-Zahra 2019). As a conclusion, good load transfer is obtained; therefore, the mechanical properties of composite improve (Rizal et al., 2020). Tensile modulus and strength improved initially with rising oil palm boiler ash filler amount up to 30 wt% and then reduced with content greater than 30 wt% (Rizal et al., 2020). The increase was owing to the enhancement of adhesion bond between matrix and filler (Khoshnoud and Abu-Zahra, 2019). The decrease was owing to weak interaction between filler and matrix (Foo and Hameed, 2009), which weakened interfacial adhesion between filler and matrix (Rizal et al., 2020). Moreover, agglomeration of oil palm boiler ash particles with an excessive content is the main parameter that induced premature failure of a composite material (Evora and Shukla, 2003).

Tensile strength, modulus and hardness were obtained to be highest at 10 wt% of rice husk (Jain et al., 2019). Mechanical features (impact strength, tensile strength and modulus, flexural strength and modulus, and hardness) enhanced as NaOH solution was risen up to 8% because the treatment caused an elimination of hemicellulose, pectin, lignin and waxes from surface of rice husk. However, beyond 8%, decrement in the mechanical properties was found, which could be attributed to severe degradation of rice husk because of alkali treatment (Bisht and Gope, 2018). 10 wt% sugarcane bagasse ash/epoxy composite gave highest flexural and tensile strength. Hardness and impact strength improved by loading of 5 wt% sugarcane bagasse ash and diminished with the incorporation of 10 wt% sugarcane bagasse ash (Chandradass et al., 2021). Composites including 20 wt%, 30 wt%, 40 wt% and 50 wt% of walnut shell improved hardness and stiffness with respect to the pure epoxy. Loading of walnut shell decreased tensile and impact strength of composite (Salasinska et al., 2018).

Maximum fracture energy and toughness values were seen for 5 wt% of filler amount and further incorporation in filler amount did not occur to improve fracture properties (Kumar and Bhowmik, 2019). In other study, it was stated that coir/epoxy composite displayed maximum fracture toughness for 2.5 wt% of filler loading (Kumar and Bhowmik, 2021). Low fracture toughness of pure epoxy was due to intrinsic brittleness because of the high crosslink density. Incorporation of pineapple leaf filler in epoxy as a discontinuous or second phase caused the highest fracture toughness for 2.5 wt% of filler amount. Further increment in filler caused the decrease in fracture toughness (Kumar et al., 2020).

Fiber percentage, fiber/polymer interaction and the distribution of the fiber in matrix can affect the mechanical stability of hybrid composites (Dixit et al., 2020). Tensile and impact strength of 50% treated date palm fiber/50% untreated coir fiber/epoxy were found to be higher because strong interfacial bonding between epoxy and fibers and its higher mechanical stability make it suitable for green structural applications. Further incorporation of fiber caused a drop in mechanical stability of hybrid composites (Dixit et al., 2020).

Dynamic mechanical analysis (DMA) is utilized in order to determine the viscoelastic properties of composite. Change in the viscoelastic properties of a composite with respect to frequency or temperature is measured to determine composite behavior at dynamic condition. Loss modulus, tan delta and storage modulus are obtained from DMA (Pulikkalparambil et al., 2021).

Fiber caused an increase in storage modulus (E′) in glassy (below T_g) and rubbery region (above T_g). A sharp drop in E′ at T_g for neat epoxy was observed because of increment in chain mobility (Cordeiro et al., 2017). An enhancement of 147% in E′ was obtained with the incorporation of 46% black rice husk ash filler derived from rice husks. Black rice husk ash filler obtained from rice husk could improve the thermo-mechanical strength of epoxy/black rice husk composites (Hsieh et al., 2017). The addition of coir filler caused the enhancement in E′ (Kumar and Bhowmik, 2021). E′ increased by addition of corn husk and jute fiber (Karthik et al., 2020). Increments in E′ of the epoxy were found to be approximately 16.4% and 36.1% with loading of 10% coupling agent-treated water bamboo husk fiber and untreated powder, respectively (Shih, 2007).

Loss modulus (E″) peaks shifted to higher temperature by incorporation of water bamboo husk fibers or powders (Shih, 2007).

E′ and E″ increased with the incorporation of pineapple leaf and banana fibers, implying that fibers could improve rigidity of composites. Also, E′ and E″ of treated fiber-reinforced composite were greater than untreated fiber-reinforced ones (Shih et al., 2012). E′ and E″ increased by the incorporation of nano oil palm empty fruit bunch filler into kenaf/epoxy (Saba et al., 2016c).

The lowest damping factor among agro waste fillers was found with Agave Americana particle. This showed that Agave Americana particle had good interfacial adhesion with epoxy (Karthik et al., 2020).

Peak value in tan δ curve is utilized to find T_g and could be owing to the relaxation of polymer chains and partial segment movement (Kumar and Bhowmik, 2021). Maximum T_g for coir/epoxy composite was achieved at 2.5 wt% of filler amount, whereas maximum T_g for bamboo/epoxy composite was achieved at 5 wt% of filler amount (Kumar and Bhowmik, 2021). The damping was the highest for pure epoxy at T_g. Increment in filler content caused a continuous decrease in the damping factor due to restricted polymer mobility and decreased energy dissipation at higher filler amount. The diminished T_g in most of filler-reinforced composites indicated decreased crosslink density with respect to pure epoxy. But the increased T_g for 12.5 wt% of the filler content may be owing to improved stretching coordination in the large-scale movement of the molecular chain in amorphous region (Kumar et al., 2020). Tan δ values of composite were lower than that of pure epoxy because of decrement in the matrix content (Cordeiro et al., 2017).

Stiffness of composites at elevated temperatures increased with walnut shell content. Loading of walnut shell filler caused increment in storage modulus (Salasinska et al., 2018). This indicates that composites exhibit a greater capability to sustain loading with recoverable viscoelastic deformation at high temperature than neat epoxy (Chmielewska et al., 2014; Shakuntala et al., 2014).

Enhanced stiffness of composites with a lower loading of walnut shell is likely due to the presence of concurrent stiff filler structures and the increased crosslinking density of the epoxy. A random arrangement of filler particle and a hindered degassing of very viscous composition may be the reason why tensile modulus for composites with 40 wt% and 50 wt% walnut shell is lower in comparison to composites (30 wt% of walnut shell) revealing the greatest modulus of elasticity. The highest tan δ was obtained with neat epoxy. The maximum loss factor measured with increasing content of walnut shell filler particles decreased slightly. This phenomenon is understandable due to the addition of filler stiff domains that act as a steric barrier decreasing material attenuation (Salasinska et al., 2018).

4.5 CONCLUSIONS

Epoxy has a great range of applications owing to excellent chemical resistance, good environmental resistance, high stiffness, low shrinkage on curing and superior thermal stability. But its mechanical properties are poor. Therefore, epoxy is reinforced with fiber to increase mechanical features. The development of natural fiber-reinforced composite employing highly environmentally friendly agro waste fibers as reinforcements has received more attraction and is increasing nowadays. Therefore, this chapter gives the information about the thermal and mechanical properties when employing agro wastes as reinforcing fibers in the epoxy matrix. Below conclusions were achieved from literature works.

- Modification of fiber surface by both chemical and physical processing enhances the efficiency of natural fiber/epoxy matrix and thus improves mechanical properties of composite.
- Incorporation of agro wastes enhanced thermal stability of epoxy.
- In general, mechanical properties (tensile strength and modulus, impact strength, flexural strength and modulus, and hardness) increased as amount of agro waste fibers was increased up to a certain value. However, beyond this value decrement in mechanical properties was found.
- Incorporation of agro waste into thermoset epoxy allows the preparation of low-cost composite with suitable thermo-mechanical and mechanical features, making them viable as material for low-demand applications and parts. The potential application areas of agro waste-based epoxy composites are the ankle-foot orthosis, brake pad material, car interior panels, coatings, dash boards, decking, door panels, electronic casings, fencing, furniture, low-load bearing housing applications, marine, roofing and seat back covers.

ACKNOWLEDGMENTS

A word of "thank you" is simply not enough. But there are no words to express my feelings. Therefore, I will stick to this simple word. I would like to thank my doctor interventional neurologist Prof. Hasan Huseyin Karadeli who saved my life after the operation and treatment he performed for my brain disease in August 2019. If it had not for his treatment and operation, I would not be able to make anything I can make now. I will be grateful to him for the rest of my life. I dedicate current work to my family and Prof. Hasan Huseyin Karadeli.

REFERENCES

Abba, H.A., I.Z. Nur, and S.M. Salit. 2013. "Review of Agro Waste Plastic Composites Production." *Journal of Minerals and Materials Characterization and Engineering* 1:271–9. https://doi.org/10.4236/jmmce.2013.15041.
Abdellaoui, H., M. Raji, R. Bouhfid, and A. el kacem Qaiss. 2019. "Chapter 2: Investigation of the Deformation Behavior of Epoxy-Based Composite Materials." In *Failure Analysis in Biocomposites, Fibre-Reinforced Composites and Hybrid Composites* (Editors: M. Jawaid, M. Thariq, and N. Saba). Woodhead Publishing, Duxford, pp. 29–49.

Abdul Khalil, H.P.S., H.M. Fizree, A.H. Bhat, M. Jawaid, and C.K. Abdullah. 2013. "Development and Characterization of Epoxy Nanocomposites Based on Nano-Structured Oil Palm Ash." *Composites Part B: Engineering* 53:324–33. https://doi.org/10.1016/j.compositesb.2013.04.013.

Abhemanyu, P.C., E. Prassanth, T.N. Kumar, R. Vidhyasagar, K.P. Marimuthu, and R. Pramod. 2019. "Characterization of Natural Fiber Reinforced Polymer Composites." *AIP Conference Proceedings* 2080:020005. https://doi.org/10.1063/1.5092888.

Alemdar, A., and M. Sain. 2008. "Biocomposites from Wheat Straw Nanofibers: Morphology, Thermal and Mechanical Properties." *Composites Science and Technology* 68:557–41. https://doi.org/10.1016/j.compscitech.2007.05.044.

Alothman, O. Y., M. Jawaid, K. Senthilkumar, M. Chandrasekar, B. A. Alshammari, H. Fouad, … S. Siengchin. 2020. "Thermal Characterization of Date Palm/Epoxy Composites with Fillers from Different Parts of the Tree." *Journal of Materials Research and Technology*, 9(6), 15537–46. https://doi.org/10.1016/j.jmrt.2020.11.020

Arbelaiz, A., B. Fernández, J.A. Ramos, A. Retegi, R. Llano-Ponte and I. Mondragon. 2005. "Mechanical Properties of Short Flax Fibre Bundle/Polypropylene Composites: Influence of Matrix/Fibre Modification, Fibre Content, Water Uptake and Recycling." *Composites Science and Technology* 65:1582–92. https://doi.org/10.1016/j.compscitech.2005.01.008.

Arrakhiz, F.Z., M. Malha, R. Bouhfid, K. Benmoussa, and A. Qaiss. 2013. "Tensile, Flexural and Torsional Properties of Chemically Treated Alfa, Coir and Bagasse Reinforced Polypropylene." *Composites Part B: Engineering* 47:35–41. https://doi.org/10.1016/j.compositesb.2012.10.046.

Ashori, A. 2006. "Nonwood Fibers – A Potantial Source of Raw Material in Papermaking." *Polymer–Plastics Technology and Engineering* 45:1133–6. https://doi.org/10.1080/03602550600728976.

Ashour, T. 2017. "Chapter 9: Composites Using Agricultural Wastes." In Handbook of Composites from Renewable Materials, Design and Manufacturing (Editors: V.K. Thakur, and M.R. Kessler). John Wiley & Sons, pp. 197–240.

Asim, M., M.T. Paridah, M. Chandrasekar, R.M. Shahroze, M. Jawaid, M. Nasir, and R. Siakeng. 2020. "Thermal Stability of Natural Fibers and Their Polymer Composites." *Iranian Polymer Journal* 29:625–48. https://doi.org/10.1007/s13726-020-00824-6.

Awasthi, A., N. Hothi, P. Kaur, N. Singh, M. Chakraborty, and S. Bansal. 2017. "Elucidative Analysis and Sequencing of Two Respiratory Health Monitoring Methods to Study the Impact of Varying Atmospheric Composition on Human Health." *Atmospheric Environment* 171:32–7. https://doi.org/10.1016/j.atmosenv.2017.10.008.

Azeez, A.A., K.Y. Rhee, S.J. Park, and D. Hui. 2013. "Epoxy Clay Nanocomposites – Processing, Properties and Applications: A Review." *Composites Part B: Engineering* 45:308–20. https://doi.org/10.1016/j.compositesb.2012.04.012.

Azwa, Z.N., and B.F. Yousif. 2013. "Characteristics of Kenaf Fibre/Epoxy Composites Subjected to Thermal Degradation." *Polymer Degradation and Stability* 98:2752–9. https://doi.org/10.1016/j.polymdegradstab.2013.10.008.

Barbieri, L., F. Andreola, I. Lancellotti, and R. Taurino. 2013. "Management of Agricultural Biomass Wastes: Preliminary Study on Characterization and Valorisation in Clay Matrix Bricks." *Waste Management* 33:2307–15. https://doi.org/10.1016/j.wasman.2013.03.014.

Barczewski, M., K. Sałasińska, and J. Szulc. 2019. "Application of Sunflower Husk, Hazelnut Shell and Walnut Shell as Waste Agricultural Fillers for Epoxy-Based Composites: A Study into Mechanical Behavior Related to Structural and Rheological Properties." *Polymer Testing* 75:1–11. https://doi.org/10.1016/j.polymertesting.2019.01.017.

Bisht, N., and P.C. Gope. 2018. "Effect of Alkali Treatment on Mechanical Properties of Rice Husk Flour Reinforced Epoxy Bio-Composite." *Materials Today: Proceedings* 5:24330–8. https://doi.org/10.1016/j.matpr.2018.10.228.

Boopathi, L., P.S. Sampath, and K. Mylsamy. 2012. "Investigation of Physical, Chemical and Mechanical Properties of Raw and Alkali Treated Borassus Fruit Fiber." *Composites Part B: Engineering* 43:3044–52. https://doi.org/10.1016/j.compositesb.2012.05.002.

Boynard, C.A., S.N. Monteiro, and J.R.M. d'Almeida. 2003. "Aspects of Alkali Treatment of Sponge Gourd (*Luffa Cylindrica*) Fibers on the Flexural Properties of Polyester Matrix Composites." *Journal of Applied Polymer Science* 87:1927–32. https://doi.org/10.1002/app.11522.

Chachad,Y.R.,J.A.Roux,J.G.Vaughan,andE.Arafat.1995."Three-DimensionalCharacterization of Pultruded Fiberglass-Epoxy Composite Materials." *Journal of Reinforced Plastics and Composites* 14:495–512. https://doi.org/10.1177/073168449501400506.

Chandradass, J., M.A. Surabhi, P.B. Sethupathi, and P. Jawahar. 2021. "Development of Low Cost Brake Pad Material Using Asbestos Free Sugarcane Bagasse Ash Hybrid Composites." *Materials Today: Proceedings* 45:7050–7. https://doi.org/10.1016/j.matpr.2021.01.877.

Chandrasekar, M., I. Siva, T. S. M. Kumar, K. Senthilkumar, S. Siengchin, and N. Rajini. 2020. "Influence of Fibre Inter-ply Orientation on the Mechanical and Free Vibration Properties of Banana Fibre Reinforced Polyester Composite Laminates." Journal of Polymers and the Environment, 28(11), 2789–2800. https://doi.org/10.1007/s10924-020-01814-8

Chang, L., K. Friedrich, L. Ye, and P. Toro. 2009. "Evaluation and Visualization of the Percolating Networks in Multi-Wall Carbon Nanotube/Epoxy Composites." *Journal of Materials Science* 44:4003–12. https://doi.org/10.1007/s10853-009-3551-3.

Chaudhary, A.K., P.C. Gope, and V.K. Singh. 2015. "Water Absorption and Thickness Swelling Behavior of Almond (*Prunus Amygdalus* L.) Shell Particles and Coconut (*Cocos Nucifera*) Fiber Hybrid Epoxy-Based Biocomposite." *Science and Engineering of Composite Materials* 22:375–82. https://doi.org/10.1515/secm-2013-0317.

Chen, N., C. Wan, Y. Zhang, and Y. Zhang. 2004. "Effect of Nano-CaCO$_3$ on Mechanical Properties of PVC/Blendex Blend." *Polymer Testing* 23:169–74. https://doi.org/10.1016/S0142-9418(03)00076-X.

Chen, R., K. Hu, H. Tang, J. Wang, F. Zhu, and H. Zhou. 2019. "A Novel Flame Retardant Derived from DOPO and Piperazine and its Application in Epoxy Resin: Flame Retardance, Thermal Stability and Pyrolysis Behavior." *Polymer Degradation and Stability* 166:334–43. https://doi.org/10.1016/j.polymdegradstab.2019.06.011.

Cheng, X., and J.S. Wiggins. 2017. "Chapter 16: Novel Techniques for the Preparation of Different Epoxy/Thermoplastic Blends. In *Handbook of Epoxy Blends* (Editors: J. Parameswaranpillai, N. Hameed, J. Pionteck, and E.M. Woo)." Springer, Cham, pp. 459–486.

Chmielewska, D., T. Sterzyński, and B. Dudziec. 2014. "Epoxy Compositions Cured with Aluminosilsesquioxanes: Thermomechanical Properties." *Journal of Applied Polymer Science* 131:40672. https://doi.org/10.1002/APP.40672.

Chun, K.S., T. Maiumunah, C.M. Yeng, T.K. Yeow, and O.T. Kiat. 2020. "Properties of Corn Husk Fibre Reinforced Epoxy Composites Fabricated Using Vacuum-Assisted Resin Infusion." *Journal of Physical Science* 31:17–31. https://doi.org/10.21315/jps2020.31.3.2.

Cordeiro, E.P., V.J.R.R. Pita, and B.G. Soares. 2017. "Epoxy-Fiber of Peach Palm Trees Composites: The Effect of Composition and Fiber Modification on Mechanical and Dynamic Mechanical Properties." *Journal of Polymers and the Environment* 25:913–24. https://doi.org/10.1007/s10924-016-0841-0.

Çöpür, Y., C. Güler, M. Akgül, and C. Taşcıoğlu. 2007. "Some Chemical Properties of Hazelnut Husk and its Suitability for Particleboard Production." *Building and Environment* 42:2568–72. https://doi.org/10.1016/j.buildenv.2006.07.011.

de Andrade Silva, F., N. Chawla, and R.D. de Toledo Filho. 2008. "Tensile Behavior of High Performance Natural (Sisal) Fibers." *Composites Science and Technology* 68:3438–43. https://doi.org/10.1016/j.compscitech.2008.10.001.

DeArmitt, C. 2011. "Chapter 26: Functional Fillers for Plastics." In *Applied Plastics Engineering Handbook* (Editor: M. Kutz)." William Andrew, Oxford, pp. 455–68.

DeArmitt, C., and R. Rothon. 2011. "Chapter 25: Dispersants and Coupling Agents." In *Applied Plastics Engineering Handbook* (Editor: M. Kutz)." William Andrew, Oxford, pp. 441–54.

Dixit, S., B. Joshi, P. Kumar, and V.L. Yadav. 2020. "Novel Hybrid Structural Biocomposites from Alkali Treated-Date Palm and Coir Fibers: Morphology, Thermal and Mechanical Properties." *Journal of Polymers and the Environment* 28:2386–92. https://doi.org/10.1007/s10924-020-01780-1.

Eckold, G. 1994. *Design and Manufacture of Composite Structures.* Woodhead Publishing, Cambridge, pp. 251–304.

Edeerozey, A.M.M., H.M. Akil, A.B. Azhar, and M.I.Z. Ariffin. 2007. "Chemical Modification of Kenaf Fibers." *Materials Letters* 61:2023–5. https://doi.org/10.1016/j.matlet.2006.08.006.

Evora, V.M.F., and A. Shukla. 2003. "Fabrication, Characterization, and Dynamic Behavior of Polyester/TiO$_2$ Nanocomposites." *Materials Science and Engineering A* 361:358–66. https://doi.org/10.1016/S0921-5093(03)00536-7.

Fekete, E., S. Molnár, G.-M. Kim, G.H. Michler, and B. Pukánszky. 1999. "Aggregation, Fracture Initiation, and Strength of PP/CaCO$_3$ Composites." *Journal of Macromolecular Science–Physics* 38:885–99. https://doi.org/10.1080/00222349908248146.

Fenouillot, F., and H. Perier-Camby. 2004. "Formation of a Fibrillar Morphology of Crosslinked Epoxy in a Polystyrene Continuous Phase by Reactive Extrusion." *Polymer Engineering and Science* 44:625–37. https://doi.org/10.1002/pen.20057.

Fiore, V., T. Scalici, F. Nicoletti, G. Vitale, M. Prestipino, and A. Valenza. 2016. "A New Eco-Friendly Chemical Treatment of Natural Fibres: Effect of Sodium Bicarbonate on Properties of Sisal Fibre and its Epoxy Composites." *Composites Part B: Engineering* 85:150–60. https://doi.org/10.1016/j.compositesb.2015.09.028.

Fiore, V., T. Scalici, G. Di Bella, and A. Valenza. 2015. "A Review on Basalt Fibre and its Composites." *Composites Part B: Engineering* 74:74–94. https://doi.org/10.1016/j.compositesb.2014.12.034.

Firdaus, S.M., and M. Mariatti. 2012. "Fabrication and Characterization of Nano Filler-Filled Epoxy Composites for Underfill Application." *Journal of Materials Science: Materials in Electronics* 23:1293–9. https://doi.org/10.1007/s10854-011-0587-3.

Foo, K.Y., and B.H. Hameed. 2009. "Value-Added Utilization of Oil Palm Ash: A Superior Recycling of the Industrial Agricultural Waste." *Journal of Hazardous Materials* 172:523–31. https://doi.org/10.1016/j.jhazmat.2009.07.091.

Gajjar, T., D.B. Shah, S.J. Joshi, and K.M. Patel. 2020. "Analysis of Process Parameters for Composites Manufacturing Using Vacuum Infusion Process." *Materials Today: Proceedings* 21:1244–9. https://doi.org/10.1016/j.matpr.2020.01.112.

Gairola, S.P., Y.K. Tyagi, B. Gangil, and A. Sharma. 2021. "Fabrication and Mechanical Property Evaluation of Non-Woven Banana Fibre Epoxy-Based Polymer Composite." *Materials Today: Proceedings* 44:3990–6. https://doi.org/10.1016/j.matpr.2020.10.103.

Gojny, F.H., M.H.G. Wichmann, U. Köpke, B. Fiedler, and K. Schulte. 2004. "Carbon Nanotube-Reinforced Epoxy-Composites: Enhanced Stiffness and Fracture Toughness at Low Nanotube Content." *Composites Science and Technology* 64:2363–71. https://doi.org/10.1016/j.compscitech.2004.04.002.

Guna, V., M. Ilangovan, M.H. Rather, B.V. Giridharan, B. Prajwal, K.V. Krishna, K. Venkatesh, and N. Reddy. 2020. "Groundnut Shell/Rice Husk Agro-Waste Reinforced Polypropylene Hybrid Biocomposites." *Journal of Building Engineering* 27:100991. https://doi.org/10.1016/j.jobe.2019.100991.

Harshe, R. 2015. "A Review on Advanced Out-of-Autoclave Composites Processing." *Journal of the Indian Institute of Science* 95:207–20.

Hodd, K. 1989. "Chapter 37: Epoxy Resins." In *Comprehensive Polymer Science and Supplements* (Editors: G. Allen, and J.C. Bevington)." Pergamon Press, Oxford, pp. 667–99.

Hsiao, K.-T., and D. Heider. 2012. "Chapter 10: Vacuum Assisted Resin Transfer Molding (VARTM) in Polymer Matrix Composites." In *Manufacturing Techniques for Polymer Matrix Composites (PMCs)* (Editors: S.G. Advani, and K.-T. Hsiao)." Woodhead Publishing, Cambridge, pp. 310–47.

Hsieh, Y.-Y., Y.-C. Tsai, J.-R. He, P.-F. Yang, H.-P. Lin, C.-H. Hsu, and A. Loganathan. 2017. "Rice Husk Agricultural Waste-Derived Low Ionic Content Carbon–Silica Nanocomposite for Green Reinforced Epoxy Resin Electronic Packaging Material." *Journal of the Taiwan Institute of Chemical Engineers* 78:493–9. https://doi.org/10.1016/j.jtice.2017.06.010.

Husseinsyah, S., and M. Mostapha. 2011. "The Effect of Filler Content on Properties of Coconut Shell Filled Polyester Composites." *Malaysian Polymer Journal* 6:87–97.

Ibrahim, N.A., W.M.Z.W. Yunus, M. Othman, and K. Abdan. 2011. "Effect of Chemical Surface Treatment on the Mechanical Properties of Reinforced Plasticized Poly(lactic Acid) Biodegradable Composites." *Journal of Reinforced Plastics and Composites* 30:381–8. https://doi.org/10.1177/0731684410396595.

Imoisili, P.E., and T.-C. Jen. 2020. "Mechanical and Water Absorption Behaviour of Potassium Permanganate (KMnO$_4$) Treated Plantain (*Musa Paradisiacal*) Fibre/Epoxy Bio-Composites." *Journal of Materials Research and Technology* 9:8705–13. https://doi.org/10.1016/j.jmrt.2020.05.121.

Ishak, M.R., Z. Leman, S.M. Sapuan, M.Y. Salleh, and S. Misri. 2009. "The Effect of Sea Water Treatment on the Impact and Flexural Strength of Sugar Palm Fibre Reinforced Epoxy Composites." *International Journal of Mechanical and Engineering* 4:316–20.

Islam, M.T., S.C. Das, J. Saha, D. Paul, M.T. Islam, M. Rahman, and M.A. Khan. 2017. "Effect of Coconut Shell Powder as Filler on the Mechanical Properties of Coir-Polyester Composites." *Chemical and Materials Engineering* 5:75–82. https://doi.org/10.13189/cme.2017.050401.

Jain, N., A. Bhatia, and H. Pathak. 2014. "Emission of Air Pollutants from Crop Residue Burning in India." *Aerosol and Air Quality Research* 14:422–30. https://doi.org/10.4209/aaqr.2013.01.0031.

Jain, N., K.S. Somvanshi, P.C. Gope, and V.K. Singh. 2019. "Mechanical Characterization and Machining Performance Evaluation of Rice Husk/Epoxy an Agricultural Waste Based Composite Material." *Journal of the Mechanical Behavior of Materials* 28:29–38. https://doi.org/10.1515/jmbm-2019-0005.

Jayamani, E., S. Hamdan, M.R. Rahman, and M.K.B. Bakri. 2014. "Investigation of Fiber Surface Treatment on Mechanical, Acoustical and Thermal Properties of Betelnut Fiber Polyester Composites." *Procedia Engineering* 97:545–54. https://doi.org/10.1016/j.proeng.2014.12.282.

Jiang, D., M. Pan, X. Cai, and Y. Zhao. 2018. "Flame Retardancy of Rice Straw-Polyethylene Composites Affected by *in situ* Polymerization of Ammonium Polyphosphate/Silica." *Composites Part A: Applied Science and Manufacturing* 1–9. https://doi.org/10.1016/j.compositesa.2018.02.023.

Jones, D., G.O. Ormondroyd, S.F. Curling, C.-M Popescu, and M.-C. Popescu. 2017. "Chemical Compositions of Natural Fibres." *Advanced High Strength Natural Fibre Composites in Construction* 23–58. https://doi.org/10.1016/B978-0-08-100411-1.00002-9.

Jyotishkumar, P., J. Pionteck, P. Moldenaers, and S. Thomas. 2013. "Preparation and Properties of TiO$_2$-Filled Poly(Acrylonitrile–Butadiene–Styrene)/Epoxy Hybrid Composites." *Journal of Applied Polymer Science* 127:3159–68. https://doi. org/10.1002/app.37729.

Kaloustian, J., T.F. El-Moselhy, and H. Portugal. 2003. "Chemical and Thermal Analysis of the Biopolymers in Thyme (*Thymus Vulgaris*)." *Thermochimica Acta* 401:77–86. https://doi. org/10.1016/S0040-6031(02)00569-5.

Karthik, D., V. Baheti, J. Novotná, A. Samková, R. Pulíček, M. Venkataraman, P. Srb, K. Voleská, Y. Wang, and J. Militky. 2020. "Effect of Particulate Fillers on Creep Behaviour of Epoxy Composites." *Materials Today: Proceedings* 31:S217–S220. https://doi. org/10.1016/j.matpr.2019.11.064.

Kaur, M., N.M. Mubarak, B.L.F. Chin, M. Khalid, R.R. Karri, R. Walvekar, E.C. Abdullah, and F.A. Tanjung. 2020. "Extraction of Reinforced Epoxy Nanocomposite Using Agricultural Waste Biomass." *IOP Conference Series: Materials Science and Engineering* 943:012021. https://doi.org/10.1088/1757-899X/943/1/012021.

Kausar, A. 2017. "Chapter 1: Polyurethane/Epoxy Interpenetrating Polymer Network." In *Aspects of Polyurethanes* (Editor: F. Yılmaz)." IntechOpen, London, pp. 1–16.

Kavitha, D., S.C. Murugavel, and S. Thenmozhi. 2021. "Flame Retarding Cardanol Based Novolac-Epoxy/Rice Husk Composites." *Materials Chemistry and Physics* 263:124225. https://doi.org/10.1016/j.matchemphys.2021.124225.

Khoshnoud, P. 2017. "Polymer Foam/Fly Ash Composites: Evaluation of Mechanical, Interfacial, Thermal, Viscoelastic and Microstructural Properties." University of Wisconsin – Milwaukee Doctoral of Philosophy.

Khoshnoud, P., and N. Abu-Zahra. 2019. "The Effect of Particle Size of Fly Ash (FA) on the Interfacial Interaction and Performance of PVC/FA Composites." *Journal of Vinyl & Additive Technology* 25:134–43. https://doi.org/10.1002/vnl.21633.

Kong, I., K.Y. Tshai, and M.E. Hoque. 2015. "Chapter 16: Manufacturing of Natural Fibre-Reinforced Polymer Composites by Solvent Casting Method." In *Manufacturing of Natural Fibre Reinforced Polymer Composites* (Editors: M.S. Salit, M. Jawaid, N. Bin Yusoff, and M.E. Hoque)." Springer, Cham, pp. 331–49.

Kumar, R., and S. Bhowmik. 2019. "Elucidating the Coir Particle Filler Interaction in Epoxy Polymer Composites at Low Strain Rate." *Fibers and Polymers* 20:428–39. https://doi. org/10.1007/s12221-019-8239-x.

Kumar, R., S. Bhowmik, K. Kumar, and J.P. Davim. 2020. "Perspective on the Mechanical Response of Pineapple Leaf Filler/Toughened Epoxy Composites under Diverse Constraints." *Polymer Bulletin* 77:4105–4129. https://doi.org/10.1007/ s00289-019-02952-3.

Kumar, R., and S. Bhowmik. 2021. "Quantitative Probing of Static and Dynamic Mechanical Properties of Different Bio-Filler-Reinforced Epoxy Composite under Assorted Constraints." *Polymer Bulletin* 78:1231–52. https://doi.org/10.1007/ s00289-020-03156-w.

Kushwaha, P.K., and R. Kumar. 2010. "Effect of Silanes on Mechanical Properties of Bamboo Fiber-Epoxy Composites." *Journal of Reinforced Plastics and Composites* 29:718–24. https://doi.org/10.1177/0731684408100691.

Lazim, Y., S.M. Salit, E.S. Zainudin, M. Mustapha, and M. Jawaid. 2014. "Effect of Alkali Treatment on the Physical, Mechanical, and Morphological Properties of Waste Betel Nut (*Areca Catechu*) Husk Fibre." *BioResources* 9:7721–36. https://doi.org/10.15376/ biores.9.4.7721-7736.

Li, X., L.G. Tabil, and S. Panigrahi. 2007. "Chemical Treatments of Natural Fiber for Use in Natural Fiber-Reinforced Composites: A Review." *Journal of Polymers and the Environment* 15:25–33. https://doi.org/10.1007/s10924-006-0042-3.

Liu, S.-H., C.-Y. Ke, and C.-L. Chiang. 2021. "Thermal Stability, Smoke Density, and Flame Retardance of Ecotype Bio-Based Flame Retardant Agricultural Waste Bagasse/Epoxy Composites." *Polymers* 13:2977. https://doi.org/10.3390/polym13172977.

Lu, T., M. Jiang, Z. Jiang, D. Hui, Z. Wang, and Z. Zhou. 2013. "Thermal Stability, Smoke Density, and Flame Retardance of Ecotype Bio-Based Flame Retardant Agricultural Waste Bagasse/Epoxy Composites." *Composites Part B: Engineering* 51:28–34. https://doi.org/10.1016/j.compositesb.2013.02.031.

Mamat Razali, N.A., M.F. Ismail, and F.A. Aziz. 2021. "Characterization of Nanocellulose from *Indica* Rice Straw as Reinforcing Agent in Epoxy-Based Nanocomposites." *Polymer Engineering and Science* 61:1594–606. https://doi.org/10.1002/pen.25683.

McNeill, V.F. 2017. "Atmospheric Aerosols: Clouds, Chemistry, and Climate." *Annual Review of Chemical and Biomolecular Engineering* 8:427–44. https://doi.org/10.1146/annurev-chembioeng-060816-101538.

Mittal, G., K.Y. Rhee, V. Mišković-Stanković, and D. Hui. 2018. "Reinforcements in Multi-Scale Polymer Composites: Processing, Properties, and Applications." *Composites Part B: Engineering* 138:122–39. https://doi.org/10.1016/j.compositesb.2017.11.028.

Mittal, V., R. Saini, and S. Sinha. 2016. "Natural Fiber-Metiated Epoxy Composites – A Review." *Composites Part B: Engineering* 99:425–35. https://doi.org/10.1016/j.compositesb.2016.06.051.

Muralidhar, N., K. Vadivuchezhian, V. Arumugam, and I.S. Reddy. 2020. "Flexural Modulus of Epoxy Composite Reinforced with Arecanut Husk Fibre (AHF): A Mechanics Approach." *Materials Today: Proceedings* 27:2265–8. https://doi.org/10.1016/j.matpr.2019.09.109.

Mylsamy, B., V. Chinnasamy, S.K. Palaniaappan, S.P. Subramani, and C. Gopalsamy. 2020. "Effect of Surface Treatment on the Tribological Properties of Goccinia Indica Cellulosic Fiber Reinforced Polymer Composites." *Journal of Materials Research and Technology* 9:16423–34. https://doi.org/10.1016/j.jmrt.2020.11.100.

Nagappan, S., S.P. Subramani, S.K. Palaniappan, and B. Mylsamy. 2021. "Impact of Alkali Treatment and Fiber Length on Mechanical Properties of New Agro Waste *Lagenaria Sicararia* Fiber Reinforced Epoxy Composites." *Journal of Natural Fibers*. https://doi.org/10.1080/15440478.2021.1932681.

Nagarajan, K.J., A.N. Balaji, K.S. Basha, N.R. Ramanujam, and R.A. Kumar. 2020. "Effect of Agro Waste α-Cellulosic Micro Filler on Mechanical and Thermal Behavior of Epoxy Composites." *International Journal of Biological Macromolecules* 152:327–39. https://doi.org/10.1016/j.ijbiomac.2020.02.255.

Nagarajan, V., A.K. Mohanty, and M. Misra. 2016. "Biocomposites with Size-Fractionated Biocarbon: Influence of the Microstructure on Macroscopic Properties." *ACS Omega* 1:636–47. https://doi.org/10.1021/acsomega.6b00175.

Naik, S., B. Halemani, and G.U. Raju. 2019. "Investigation of the Mechanical Properties of Tamarind Seed Particles Reinforced Epoxy Composites." *AIP Conference Proceedings* 2057:020023. https://doi.org/10.1063/1.5085594.

Nemli, G., S. Yıldız, and E.D. Gezer. 2008. "The Potential for Using the Needle Litter of Scotch Pine (*Pinus Sylvestris* L.) as a Raw Material for Particleboard Manufacturing." *Bioresource Technology* 99:6054–8. https://doi.org/10.1016/j.biortech.2007.12.044.

Narendar, R., and K.P. Dasan. 2014. "Chemical Treatments of Coir Pitch: Morphology, Chemical Composition, Thermal and Water Retention Behavior." *Composites Part B: Engineering* 56:770–9. https://doi.org/10.1016/j.compositesb.2013.09.028.

Nasimudeen, N. A., Karounamourthy, S., Selvarathinam, J., Kumar Thiagamani, S. M., Pulikkalparambil, H., Krishnasamy, S., & Muthukumar, C. 2021. "Mechanical, Absorption and Swelling Properties of Vinyl Ester Based Natural Fibre Hybrid Composites." *Applied Science and Engineering Progress*. https://doi.org/10.14416/j.asep.2021.08.006

Nirmal, U., J. Hashim, and M.M.H.M. Ahmad. 2015. "A Review on Tribological Performance of Natural Fibre Polymeric Composites." *Tribology International* 83:77–104. https://doi.org/10.1016/j.triboint.2014.11.003.

Nirmal, U., N. Singh, J. Hashim, S.T.W. Lau, and N. Jamil. 2011. "On the Effect of Different Polymer Matrix and Fibre Treatment on Single Fibre Pullout Test Using Betelnut Fibres." *Materials and Design* 32:2717–26. https://doi.org/10.1016/j.matdes.2011.01.019.

Oanh, N.T.K, D.A. Permadi, P.K. Hopke, K.R. Smith, N.P. Dong, and A.N. Dang. 2018. "Annual Emissions of Air Toxics Emitted from Crop Residue Open Burning in Southeast Asia over the Period of 2010–2015." *Atmospheric Environment* 187:163–73. https://doi.org/10.1016/j.atmosenv.2018.05.061.

Obada, D.O, L.S. Kuburi, M. Dauda, S. Umaru, D. Dodoo-Arhin, M.B. Balogun, I. Illiyasu, and M.J. Iorpenda. 2020. "Effect of Variation in Frequencies on the Viscoelastic Properties of Coir and Coconut Husk Powder Reinforced Polymer Composites." *Journal of King Saud University – Engineering Sciences* 32:148–57. https://doi.org/10.1016/j.ksues.2018.10.001.

Olowojoba, G.B. 2013. "Assessment of Dispersion Evolution of Carbon Nanotubes in Shear-Mixed Epoxy Suspensions by Interfacial Polarization Measurement." Karlsruhe Institute of Technology Doctoral of Philosophy.

Ouarhim, W., N. Zari, R. Bouhfid, and A. el kacem Qaiss. 2019. "Chapter 3: Mechanical Performance of Natural Fibers-Based Thermosetting Composites." In *Mechanical and Physical Testing of Biocomposites, Fibre-Reinforced Composites and Hybrid Composites* (Editors: M. Jawaid, M. Thariq, and N. Saba)." Woodhead Publishing, Duxford, pp. 43–60.

Pan, P., Z. Liang, A. Cao, and Y. Inoue. 2009. "Layered Metal Phosphonate Reinforced Poly(L-Lactide) Composites with a Highly Enhanced Crystallization Rate." *ACS Applied Materials & Interfaces* 1:402–11. https://doi.org/10.1021/am800106f.

Patel, J.P., and P.H. Parsania. 2017. "Fabrication and Comparative Mechanical, Electrical and Water Absorption Characteristic Properties of Multifunctional Epoxy Resin of Bisphenol-C and Commercial Epoxy-Treated and -Untreated Jute Fiber-Reinforced Composites." *Polymer Bulletin* 74:485–504. https://doi.org/10.1007/s00289-016-1725-0.

Pearce, E.M., and R. Liepins. 1975. "Flame Retardants." *Environmental Health Perspectives* 11:59–69. https://doi.org/10.1289/ehp.751159.

Poletto, M., A.J. Zattera, M.M.C. Forte, and R.M.C. Santana. 2012. "Thermal Decomposition of Wood: Influence of Wood Components and Cellulose Crystallite Size." *Bioresource Technology* 109:148–53. https://doi.org/10.1016/j.biortech.2011.11.122.

Pulikkalparambil, H., S.M. Rangappa, S. Siengchin, and J. Parameswaranpillai. 2021. "Chapter 1: Introduction to Epoxy Composites." In *Epoxy Composites: Fabrication, Characterization and Applications* (Editors: J. Parameswaranpillai, H. Pulikkalparambil, S.M. Rangappa, and S. Siengchin). Wiley-VCH GmbH, Weinheim, pp. 1–21.

Ramakrishna, H.V., and S.K. Rai. 2006. "Effect on the Mechanical Properties and Water Absorption of Granite Powder Composites on Toughening Epoxy with Unsaturated Polyester and Unsaturated Polyester with Epoxy Resin." *Journal of Reinforced Plastics and Composites* 25:17–32. https://doi.org/10.1177/0731684406055450.

Ravindra, K., M.K. Sidhu, S. Mor, S. John, and S. Pyne. 2016. "Air Pollution in India: Bridging the Gap between Science and Policy." *Journal of Hazardous, Toxic, and Radioactive Waste* 20:A4015003. https://doi.org/10.1061/(ASCE)HZ.2153-5515.0000303.

Rizal, S., H.M. Fizree, M.S. Hossain, D.A. Gopakumar, E.C.W. Ni, and H.P.S. Abdul Khalil. 2020. "The Role of Silica-Containing Agro-Industrial Waste as Reinforcement on Physicochemical and Thermal Properties of Polymer Composites." *Heliyon* 6:e03550. https://doi.org/10.1016/j.heliyon.2020.e03550.

Rong, M.Z., M.Q. Zhang, Y. Liu, G.C. Yang, and H.M. Zeng. 2001. "The Effect of Fiber Treatment on the Mechanical Properties of Unidirectional Sisal-Reinforced Epoxy Composites." *Composites Science and Technology* 61:1437–47. https://doi.org/10.1016/S0266-3538(01)00046-X.

Rumeau, N., D. Threlfall, and A. Wilmet. 2015. "ResonantAcoustic® Mixing – Processing and Formulation Challenges for Cost Effective Manufacturing." In *2015 Europyro 41st International Pyrotechnics Seminar*, France.

Saba, N., M. Jawaid, O.Y. Alothman, M.T. Paridah, and A. Hassan. 2016a. "Recent Advances in Epoxy Resin, Natural Fiber-Reinforced Epoxy Composites and their Applications." *Journal of Reinforced Plastics and Composites* 35:447–70. https://doi.org/10.1177/0731684415618459.

Saba, N., M.T. Paridah, K. Abdan, and N.A. Ibrahim. 2016b. "Thermal Properties of Oil Palm Nano Filler/Kenaf Reinforced Epoxy Hybrid Nanocomposites." *AIP Conference Proceedings* 1787:050020. https://doi.org/10.1063/1.4968118.

Saba, N., M.T. Paridah, K. Abdan, and N.A. Ibrahim. 2016c. "Dynamic Mechanical Properties of Oil Palm Nano Filler/Kenaf/Epoxy Hybrid Nanocomposites." *Construction and Building Materials* 124:133–8. https://doi.org/10.1016/j.conbuildmat.2016.07.059.

Salasinska, K., M. Barczewski, R. Górny, and A. Kloziński. 2018. "Evaluation of Highly Filled Epoxy Composites Modified with Walnut Shell Waste Filler." *Polymer Bulletin* 75:2511–28. https://doi.org/10.1007/s00289-017-2163-3.

Santos, L.S., E.C. Biscaia Jr., R.L. Pagano, and V.M.A. Calado. 2012. "CFD – Optimization Algorithm to Optimize the Energy Transport in Pultruded Polymer Composites." *Brazilian Journal of Chemical Engineering* 29:559–66. https://doi.org/10.1590/S0104-6632201200300013.

Sareena, C., M.T. Ramesan, and E. Purushothaman. 2012. "Utilization of Peanut Shell Powder as a Novel Filler in Natural Rubber." *Journal of Applied Polymer Science* 125:2322–34. https://doi.org/10.1002/app.36468.

Scalici, T., V. Fiore, and A. Valenza. 2016. "Effect of Plasma Treatment on the Properties of *Arundo Donax* L. Leaf Fibres and its Bio-Based Epoxy Composites: A Preliminary Study." *Composites Part B: Engineering* 94:167–75. https://doi.org/10.1016/j.compositesb.2016.03.053.

Schmachtenberg, E., J.S. zur Heide, and J. Töpker. 2005. "Application of Ultrasonics for the Process Control of Resin Transfer Moulding (RTM)." *Polymer Testing* 24:330–38. https://doi.org/10.1016/j.polymertesting.2004.11.002.

Senthil Muthu Kumar, T., K. Senthilkumar, M. Chandrasekar, S. Karthikeyan, N. Ayrilmis, N. Rajini, and S. Siengchin. 2021. "Mechanical, Thermal, Tribological, and Dielectric Properties of Biobased Composites." In: Anish Khan, Sanjay M. Rangappa, Suchart Siengchin, Abdullah M. Asiri (eds.), *Biobased Composites: Processing, Characterization, Properties, and Applications*, John Wiley & Sons. Hoboken, NJ, pp. 53–73.

Senthilkumar, K., N. Saba, M. Chandrasekar, M. Jawaid, N. Rajini, S. Siengchin, … H. A. Al-Lohedan. 2021. "Compressive, Dynamic And Thermo-Mechanical Properties of Cellulosic Pineapple Leaf Fibre/Polyester Composites: Influence of Alkali Treatment on Adhesion." *International Journal of Adhesion and Adhesives*, 106, 102823. https://doi.org/10.1016/j.ijadhadh.2021.102823

Senthilkumar, K., N. Saba, N. Rajini, M. Chandrasekar, M. Jawaid, S. Siengchin, and O. Y. Alotman. 2018. "Mechanical Properties Evaluation of Sisal Fibre Reinforced Polymer Composites: A Review." *Construction and Building Materials*, 174, 713–729.

Senthilkumar, K., I. Siva, N. Rajini, and P. Jeyaraj. 2015. "Effect of Fibre Length and Weight Percentage on Mechanical Properties of Short Sisal/Polyester Composite." *International Journal of Computer Aided Engineering and Technology*, 7(1), 60. https://doi.org/10.1504/IJCAET.2015.066168

Senthilkumar, K., S. Subramaniam, T. Ungtrakul, T. S. M. Kumar, M. Chandrasekar, N. Rajini, ... J. Parameswaranpillai. 2022. "Dual Cantilever Creep and Recovery Behavior of Sisal/Hemp Fibre Reinforced Hybrid Biocomposites: Effects of Layering Sequence, Accelerated Weathering and Temperature." Journal of Industrial Textiles, 51(2_suppl), 2372S–2390S. https://doi.org/10.1177/1528083720961416

Shakuntala, O., G. Raghavendra, and A.S. Kumar. 2014. "Effect of Filler Loading on Mechanical and Tribological Properties of Wood Apple Shell Reinforced Epoxy Composite." Advances in Materials Science and Engineering 2014:538651. https://doi.org/10.1155/2014/538651.

Shahroze, R. M., M. Chandrasekar, K. Senthilkumar, T. Senthil Muthu Kumar, M. R. Ishak, N. Rajini, ... S. O. Ismail. 2021. "Mechanical, Interfacial and Thermal Properties of Silica Aerogel-Infused Flax/Epoxy Composites." International Polymer Processing, 36(1), 53–59. https://doi.org/10.1515/ipp-2020-3964

Shen, F.C. 1995. "A Filament-Wound Structure Technology Overview." Materials Chemistry and Physics 42:96–100. https://doi.org/10.1016/0254-0584(95)01554-X.

Shen, M.-Y., C.-F. Kuan, H.-C. Kuan, C.-Y. Ke, and C.-L. Chiang. 2021. "Study on Preparation and Properties of Agricultural Waste Bagasse Eco-Type Bio-Flame-Retardant/Epoxy Composites." Journal of Thermal Analysis and Calorimetry 144:525–38. https://doi.org/10.1007/s10973-020-10368-9.

Shih, Y.-F. 2007. "Mechanical and Thermal Properties of Waste Water Bamboo Husk Fiber Reinforced Epoxy Composites." Materials Science and Engineering A 445–446:289–95. https://doi.org/10.1016/j.msea.2006.9.032.

Shih, Y.-F., J.-X. Cai, C.-S. Kuan, and C.-F. Hsieh. 2012. "Plant Fibers and Wasted Fiber/Epoxy Green Composites." Composites Part B: Engineering 43:2817–21. https://doi.org/10.1016/j.compositesb.2012.04.044.

Shuhadah, S., and A.G. Supri. 2009. "LDPE-Isophthalic Acid-Modified Egg Shell Powder Composites (LDPE/ESP₁)." Journal of Physical Science 20:87–98.

Soares, S., G. Camino, and S. Levchik. 1995. "Comparative Study of the Thermal Decomposition of Pure Cellulose and Pulp Paper." Polymer Degradation and Stability 49:275–83. https://doi.org/10.1016/0141-3910(95)87009-1.

Thomas, S. K., J. Parameswaranpillai, S. Krishnasamy, P. M. S. Begum, D. Nandi, S. Siengchin, ... N. Sienkiewicz. 2021. "A Comprehensive Review on Cellulose, Chitin, and Starch as Fillers in Natural Rubber Biocomposites." Carbohydrate Polymer Technologies and Applications, 2, 100095. https://doi.org/10.1016/j.carpta.2021.100095

Van de Weyenberg, I., J. Ivens, A. De Coster, B. Kino, E. Baetens, and I. Verpoest. 2003. "Influence of Processing and Chemical Treatment of Flax Fibres on Their Composites." Composites Science and Technology 63:1241–6. https://doi.org/10.1016/S0266-3538(03)00093-9.

Verma, D., and K.L. Goh. 2021. "Effect of Mercerization/Alkali Surface Treatment of Natural Fibres and Their Utilization in Polymer Composites: Mechanical and Morphological Studies." Journal of Composites Science 5:175. https://doi.org/10.3390/jcs5070175.

Verma, S., V.K. Midha, and A.K. Choudhary. 2020. "Multi-Objective Optimization of Process Parameters for Lignin Removal of Coir Using TOPSIS." Journal of Natural Fibers. https://doi.org/10.1080/15440478.2020.1739589.

Vidal, J., D. Ponce, A. Miravete, J. Cuartero, and P. Castell. 2021. "Bio-Binders for the Improvement of the Performance of Natural Fibers as Reinforcements in Composites to Increase the Sustainability in the Transport Sector." Mechanics of Advanced Materials and Structures 28:1079–87. https://doi.org/10.1080/15376494.2019.1633447.

Vijayan, P.P., D. Puglia, P. Jyotishkumar, J.M. Kenny, and S. Thomas. 2012. "Effect of Nanoclay and Carboxyl-Terminated (Butadiene-co-Acrylonitrile) (CTBN) Rubber on the Reaction Induced Phase Separation and Cure Kinetics of an Epoxy/Cyclic Anhydride System." Journal of Materials Science 47:5241–53. https://doi.org/10.1007/s10853-012-6409-z.

Vinod, A., M.R. Sanjay, S. Siengchin, and S. Fischer. 2021a. "Fully Bio-Based Agro-Waste Soy Stem Fiber Reinforced Bio-Epoxy Composites for Lightweight Structural Applications: Influence of Surface Modification Techniques." *Construction and Building Materials* 303:124509. https://doi.org/10.1016/j.conbuildmat.2021.124509.

Vinod, A., T.G.Y. Gowda, R. Vijay, M.R. Sanjay, M.K. Gupta, M. Jamil, V. Kushvaha, and S. Siengchin. 2021b. "Novel *Muntingia Calabura Bark* Fiber Reinforced Green-Epoxy Composite: A Sustainable and Green Material for Cleaner Production." *Journal of Cleaner Production* 294:126337. https://doi.org/10.1016/j.jclepro.2021.126337.

Xiong, X., L. Zhou, R. Ren, S. Liu, and P. Chen. 2018. "The Thermal Decomposition Behavior and Kinetics of Epoxy Resins Cured with a Novel Phthalide-Containing Aromatic Diamine." *Polymer Testing* 68:46–52. https://doi.org/10.1016/j.polymertesting.2018.02.012.

Yang, H., R. Yan, H. Chen, D.H. Lee, D.T. Liang, and C. Zheng. 2006. "Pyrolysis of Palm Oil Wastes for Enhanced Production of Hydrogen Rich Gases." *Fuel Processing Technology* 87:935–42. https://doi.org/10.1016/j.fuproc.2006.07.001.

Zhang, L. 2013. "Production of Bricks from Waste Materials – A Review." *Construction and Building Materials* 47:643–55. https://doi.org/10.1016/j.conbuildmat.2013.05.043.

Zindani, D., S.R. Maity, and S. Bhowmik. 2021. "Extended TODIM Method Based on Normal Wiggly Hesitant Fuzzy Sets for Deducing Optimal Reinforcement Condition of Agro-Waste Fibers for Green Product Development." *Journal of Cleaner Production* 301:126947. https://doi.org/10.1016/j.jclepro.2021.126947.

5 Grass Fiber-Based Epoxy Composites
Thermal and Mechanical Properties

M. Ramesh, M. Tamil Selvan, A. Felix Sahayaraj
KIT-Kalaignarkarunanidhi Institute of Technology

CONTENTS

DOI: 10.1201/9781003271017-5

5.1   INTRODUCTION

Natural fiber-reinforced composite plays a vital role in engineering applications since the last decade due to their characteristics such as biodegradability, easy availability, cheap, lightweight, and good mechanical strength. Plant-based fibers are used as common reinforcing material in various applications such as automotive, aerospace, home appliances, and structural applications. Artificially made carbon, aramid and glass fibers also have very good properties but these synthetic fibers have several disadvantages, especially in terms of environment pollution, biodegradability, and emitting greenhouse gases while burning. Several studies revealed that plant-based fibers are good alternatives to synthetic fibers in terms of several composite applications. Natural fibers are derived from the various parts of the plants such as bark, stem, seed, fruit, leaf, and roots. The structure of plant fiber cellulose is determined by the origin and age of plant parts.

Several noncellulose components, including hemicellulose, lignin and wax, cover the cellulose element in the fiber. Fibers from grasses like mendong grass [10], Napier grass [11,12], elephant grass [13], snake grass [14], *Sansevieria ehrenbergii* [15], wild cane [16], *Typha domingensis* [17], kusha grass [18], sisal [19,20], *Sansevieria cylindrica* [21], *Arundo donax* L. [22], *Sansevieria trifasciata* [8, 23], and corn husk [24] are used as reinforcing elements in polymer composites. However, research into novel natural fibers must continue in order to enhance the number of alternative fiberglass alternatives used in reinforcing polymer composites. As a tropical country, Indonesia has a varied range of plants with the ability to create fibers, one of which is a grass known as belulang grass. The extraction technique, physical, chemical, and mechanical properties of several varieties of grass, including belulang grass, will be described in this chapter. The worldwide economy is rapidly shifting forward into sustainable energy and conservation. In broad sense, plant-based fibers were commonly utilized to lower the weight of the components, i.e., the fibers are reinforced with the appropriate matrix. Natural plant fibers offer several benefits over synthetic fibers in terms of affordability, regeneration, and biocompatibility. Several researchers conducted studies in the field of natural fibers.

Fibers from snake grass plant are recently identified, which is derived from grass by a conventional mechanical and/or sustainable process. This is a new fibrous plant that grows widely in the southernmost region of India [15]. Due to low tensile strength of vetiver grass fiber, composites constructed using these fibers demonstrated better tensile characteristics than materials reinforced with other plant fibers. The combination of this fiber with other fibers was shown to have superior mechanical characteristics with respect to all other combinations. It was also discovered that the fiber failure mechanism indicates better load transmission via the fibers and matrix. In future, these types of composites might be employed in semi-structural applications where cost is a primary issue. These composites might be used in a variety of applications, including roofing materials, bricks, door and window panels, interior paneling, storage tanks, and pipelines [16]. To scrutinize the interlayer bonding strength between the reinforcement and resin, research on the micro-mechanical flexural behavior of Typha biocomposite was investigated. The research revealed that the resin instantly dispersed the massive amount to the reinforced materials and that Typha fiber is strong enough to carry the massive amount stress properly. *Typha angustifolia* natural fiber biocomposite's structural and tribological properties were investigated in a study where the authors concluded that the tensile test results showed that the tensile strength increases with increasing fiber volume fraction [17].

5.2 FIBER EXTRACTION PROCESS

Fibers are separated from various parts of the plants such as bast, stem, leaves, fruits, and seed by using different processes. The most commonly used extraction processes are water retting (microbial degradation technique) and mechanical process (decorticator). Once the fibers are taken out from the plants cleaned by water to remove the unwanted elements, the cleaned fibers are then dried in sun light or oven to reduce the moisture content. Enough drying is required as the moisture content in fiber affects the fiber surface and leads to decomposition. In the mechanical process the fibers were removed either manually or by aid of machine. Then the fibers are inserted into the decorticator machine to obtain skinny fiber. Again, the fibers were cleaned with pure water to remove the dirty materials. The water retting method involves soaking the plant's stems, bark, leaves, seeds, and fruit in water for a certain length of time to release the fibers, which are then cleansed with clean water and dried.

From Table 5.1, it is observed that the grass fiber extraction methods can be grouped into three categories. The most used process is water retting because this is very simple, easy to extract, and produce good strength. In addition, it was clear that the fibers can be extracted from the plant in three ways such as water retting, mechanical process, and combined water retting and mechanical process. But mostly water retting process is preferred over the mechanical process because it is simple and easy with optimum results. The fiber properties depend upon the soaking time

TABLE 5.1
Shows Extraction Techniques of Various Grass Fibers

S. No.	Fibers	Scientific Name	Extraction Process	Ref.
1.	Snake grass fiber	*Sansevieria ehrenbergii*	Water retting	[1]
2.	Pampa grass fiber	*Cortaderia selloana*	Water retting	[2]
3.	Kusha grass fiber	*Desmostachya bipinnata*	Water retting	[3]
4.	Giant reed fiber	*Arundo donax* L.	Mechanical decorticator	[4]
5.	Cylindrical snake plant fiber	*Sansevieria cylindrica*	Mechanical decorticator	[5]
6.	Sabai grass fiber	*Eulaliopsis binata*	Water retting	[6]
7.	Napier grass fiber	*Pennisetum purpureum*	Mechanical and water retting	[7]
8.	Snake grass fiber	*Dracaena trifasciata*	Water retting	[8]
9.	Broom grass fiber	*Thysanolaena maxima*	Water retting	[9]
10.	esparto grass fiber	*Stipa tenacissima*	Water retting	[10]
11.	Sarkanda grass fiber	*Saccharum bengalense*	Water retting	[11]
12.	Ripe bulrush fiber	*Typha latifolia*	Water retting and Mechanical decorticator	[12]
13.	Wild cane grass fiber	*Saccharum spontaneum*	Water retting	[13]
14.	Mendong grass fiber	*Fimbristylis globulosa*	Mechanical decorticator	[14]

and drying time. Drying can be done naturally (sun light) or artificially (oven-dried). Plant-based fibers are primarily obtained using retting method, which would be an easy and cost-effective approach, before being exposed to chemical therapeutic approaches. Generally, lignocellulosic fibers are composed of cellulose, hemicellulose, and lignin [18]. Different techniques, such as physical separation and chemical therapeutic interventions, were implemented to reduce the lengthy time consumption. The microbial population and humidity in grasses during the retting method enable for the biological degradation components of parenchyma cells and the epoxy materials that encircle the fibers, allowing for the secession of single fibers from the grasses. While using moisture, the response time must be thoroughly reviewed because inordinate retting can prove problematic in separating each fiber or degrade the fiber strength [19]. The plant's leaves were the origin of vakka fiber. The discarded leaflets that fell into the water were dried in the shadow for 2 or 3 days. After 15 days in a pool of water, the free outer layers were separated. This was again submerged in a freshwater treating for 3 days. The fibers were then fully separated from the sheaths. Likewise, the bamboo fibers were removed in 3 days.

The date fibers were removed from the pinnate leaves, the outer layer of the stems was cut off with a blade, and the stalks were dried in the shadow for 5 days. The stems were soaked for 4 days in a freshwater retting tank before being pounded with a round mallet [20,21]. Several methods were used on the raw fibers to separate Alfa grass cellulose. The degradation of toxic pollutants in hot water solvents, subsequent using various techniques and sodium hydroxide treatment were all used to make three natural cellulose specimens. To begin, the C_7H_8, C_2H_5OH, and warm water extractives were removed using a different solvent with 10 g of dry fibers and 220 mL of C_7H_8, C_2H_5OH (1:2, v/v) combination at a constant temperature for 6 hours, treated with 5 days of observation with heated water. The purified material is then placed in the oven at 100°C [22]. *Saccharum bengalense* is also termed as "Sarkanda," whose fiber is produced from the plant's stems. A 500 g stem of the plant is gathered and submerged in water for 7 days to allow microbiological breakdown. After that, the stems were gathered and scheduled using a fine mineral brushes comb to extract the fibers. The decorticated fibers were then sun-dried for one day to eliminate the humidity [11,23].

Following the removal of the wild cane grasses at their foundation, the culms were dehydrated in the sunshine for 7 days after the leaflets were removed. The nodal parts were clipped, and the culms were divided into 20 cm pieces. These tubular sections with lignin at the center are converted into strands by peeling them in a lengthwise manner and removing the lignin. These strands were softened in water for a week before being exposed to a physical technique that involved gently pounding them with a wooden hammer to break and extract the fibers. The resultant fiber bundle is peeled and processed to get single fiber [24]. The fibers of mendong grass were obtained physically. The moist mendong grass straw was cut 600 mm long from the bottom, and the upper section was trimmed. Mendong grass straw was continuously crushed and then washed. The fibers of mendong grass were then steeped in water for 7 days. Mendong grass fibers were collected, washed, and dried in the air. Mendong grass fibers were mercerized by dipping in a 5% sodium hydroxide for 30 minutes at room temperature, and then washed, dried, wrapped in plastic wrap, and placed in a dry box with moisture of 40% [25].

5.3 PROPERTIES OF GRASS FIBERS

5.3.1 Chemical Properties

Layers of lignin, hemicellulose, and cellulose make up natural fibers in general. A layer of lignin, the next layer of hemicellulose, and the innermost cellulose make up the outermost layer of fiber. Fibers with high cellulose content usually have superior mechanical characteristics. Kusha grass fiber has cellulose content of 70.58 wt.%, lignin content of 14.35 wt.%, and moisture content of 8.01 wt.% [3]. Giant reed fiber has cellulose content of 43.2 wt.%, hemicellulose content of 20.5 (wt.%), and lignin content of 17.2 wt.% [4]. Cylindrical snake plant fiber has cellulose content of 79.7 wt.%, hemicellulose content of 10.13 wt.%, lignin content of 3.8 wt.%, and moisture content of 6.08 wt.% [5].

Napier grass fiber has cellulose content of 47.12 wt.%, hemicellulose content of 31.27 wt.%, and lignin content of 21.63 wt.% [7]. Snake plant fiber has cellulose content of 80 wt.%, hemicellulose content of 11.25 wt.%), lignin content of 7.8 wt.%, and moisture content of 10.55 wt.% [8]. Mendong grass fiber has cellulose content of 72.14 wt.%, hemicellulose content has 20.2 (wt.%) and lignin content of 3.44 (wt.%) [14]. The cellulose concentration of kusha grass fiber was determined to be 70.58 wt.%, which is greater than that of other grass fibers. In summary, higher cellulose concentration leads to excellent mechanical properties and elastic strength of kusha grass fiber. The fiber of kusha grass had 14.32 wt.% lignin. The wax and water amount of kusha grass fiber were 1.52 wt.% and usually around 10 wt.%, correspondingly. The ash percentage of kusha grass fiber is minimum (2.46%), which enhances its fire resistance [26]. A proportion of total content analysis proves that sodium hydroxide treatment alters the fiber content. The hemicellulose concentration of fiber filaments decreased from 31.27% (untreated) to 11.57% after sodium hydroxide treatment, with no major improvements in lignin amount. The cellulose amount rises as the sodium hydroxide concentration rises [27].

5.3.2 Physio-Mechanical Properties

Natural fiber density is very low compared to glass fiber, which is one of the major advantages in terms of strength-to-weight ratio. The density of fiber varies for different kinds of fibers. The density can be measured by various techniques such as pycnometer and Truong method. Snake grass fiber has a diameter of 45–250 μm and a density of 887 kg/m^3; the tensile strength of the composite made using this fiber is 278.82 MPa, tensile modulus is 9.71 GPa, and the elongation at break is 2.87% [1]. Kusha grass fiber has a diameter of 70–100 μm and a density of 1102.5 kg/m^3 [3]. Giant grass fiber has a density of 1168 kg/m^3; tensile strength of the composite made using this fiber is 248.82 MPa, the tensile modulus is 9.4 GPa, and the elongation at break is 3.24% [4]. Cylindrical snake plant fiber has a density of 915 kg/m^3; the tensile strength of the composite made using this fiber is 658 MPa, tensile modulus is 7.6 GPa, and the elongation at break is 10% [5]. Napier grass fiber has a diameter of 150–550 μm and a density of 358 kg/m^3; the tensile strength of the composite made using this fiber is 88.4 MPa, tensile modulus is 13.15 GPa, and the elongation at break is 0.99% [7]. Snake plant fiber has a diameter of 80–120 μm and a density of 1414.7 kg/m³ [3]. Broom grass fiber has a diameter of 185 to 520 μm, a density of 864 kg/m³; the tensile strength of the composite made using this fiber is 297.82 MPa, tensile modulus is 18.28 GPa, and the elongation at break is 2.87% [1]. Mendong grass fiber has a diameter of 282–394 μm and density of 892 kg/m³ the tensile strength of the composite made using this fiber is 452 MPa, tensile modulus is17.4 GPa, and the elongation at break is 2.87% [14].

The thermoplastic cassava starch and cogon grass fiber biocomposites are manufactured by employing thermoplastic cassava starch as the matrix and cogon grass fiber as reinforcement. Thermoplastic cassava starch is completely combined in a 100:30 ratio of cassava starch to glycerol (wt.%). Before extracting the fiber from the cogon grass, it was dried after being soaked for a week. The addition of 5% decreased the thickness, swell, and hydrophilicity of the biomaterials by 5.38% and 7.82%, correspondingly, when compared to the plain thermoplastic cassava starch. Due to the incorporation of cogon grass fiber, the humidity and water permeability of the mixtures did not vary significantly. Furthermore, the addition of cogon grass fiber to thermoplastic cassava starch improved the essential features of the hybrids for short-term product applications [28]. The composition of forage influences its probable behavior in the forestomach. The proclivity of grass fragments to create a fibrous rafting and stratify rumen materials is addressed by specific adaptations of grazing ruminants to regulate, utilize, and enhance this proclivity [29].

5.4   GRASS FIBER-REINFORCED EPOXY COMPOSITES

Water retting was used to separate the fibers from the grass stem. The grass composite materials were made using controlled compression molding process and LY556-rated synthetic polymers with HY951-rated bonding agent used [30]. The fibers from the grasses were extracted using a combination retting technique. The synthetic resin (Variety: LY556) and adhesive (methyl ethyl ketone peroxide) were used. Then the silane bonding compound (aminopropyl trimethoxy silane), finer solvent, and polish gelatin [31] also used.

5.4.1   EPOXY RESIN

Epoxy, polyesters, and vinyl esters are common thermoset matrix materials utilized in automotive, infrastructure, aerospace, marine, chemical, and electrical applications. When compared to polyester, epoxy has superior mechanical qualities. Vinyl ester has good adherence to a wide range of fibers and other substrates. For the most part, epoxy is employed as an adhesive and a laminating resin in engineering applications. When employed in polymer composites, it has good moisture barrier properties. It bonds to fibers exceedingly well, making fiber-reinforced polymer composites possible. A small quantity of a reactive curing agent is added immediately before inserting fibers into the mixture to initiate the polymerization reaction that turns the liquid resin into a solid. Amine-based hardener is one such curing agent that is advised by the provider for the manufacture of composites with a 9:1 mixing ratio [31,32].

5.4.2 PROCESSING TECHNIQUES

The fiber-reinforced polymer composites can be manufactured through various techniques such as lay-up, compression molding, spray-up, filament winding, sheet molding, extrusion, resin transfer molding (RTM), infusion molding, and injection molding. Cereal-based fermentation properties in colostrum can be investigated in vivo, in situ, and in vitro. Since determining rumen fermentation properties in vivo are time-consuming, costly, and incredibly hard to standardize, in situ and in vitro methods have been established [33]. Humidity, fiber orientation, fiber content, and reinforcement heating rate are the key factors affecting the preparation of grass fiber composites. Before handling, the water content of both the grass fiber and the matrix must be monitored, and any necessary adjustments must be carried out if water is present. An even more critical factor is the processing temperature. To prevent degradation of most grass fibers during the working phase, the average temperature that can be used is 200°C in less than 20 minutes [34-36].

5.4.2.1 Compression Molding

In this processing technique, fibers are arranged in a respective orientation. Before that, fiber length should be uniform respective to the mold length. Reinforced materials are to be preheated and placed in the molding cavity. Then, the inner portion of the molding presses and deforms them while applying tremendous pressure to the chamber. The high pressure is maintained until the reinforcement materials and resin to be hardened before opening the molding and removing it. The quantity of material used, warming duration, pressure used to molding, and cooling time are all crucial elements to consider when using this process [37].

5.4.2.2 Extrusion Molding

Extrusion is a one of the superior manufacturing processes that is widely utilized in the production of thermoplastic biodegradable polymers. Polymer's particles are put into the hollow tube of an extrusion, where they are mixed. The extrusion can be done by using either single or twin-screw-type extruder. The liquid-phase material flows within the extruder's cylinder and is pushed out via a die to form the exact dimension. Extrusion is a completely autonomous consistent manufacturing method. Pressure, temperature, and supply rate are critical factors. The extrusion eliminates any air that has become entrapped between the grains [38].

5.4.2.3 Sheet Molding

The method starts with a provision wheel of carrier surface on both the top and lower ends. The top and bottom carrier films are supplied underneath the top and bottom resin vessels, which deposit a pre-estimated thickness of epoxy onto the carrier film using extendable wipe brushes. The bottom carrier layer is then passed beneath the

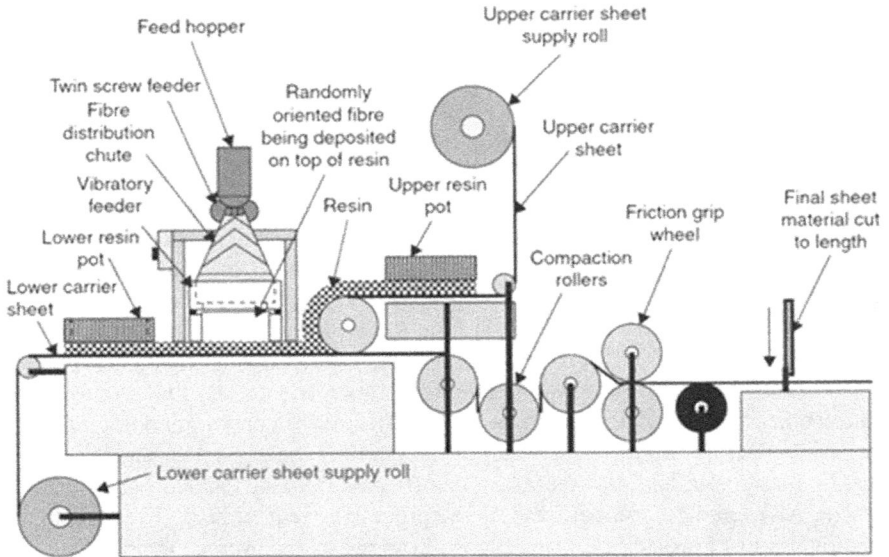

FIGURE 5.1   Workflow schematic for sheet molding method [34].

bio-fiber vibrating conveyor, which evenly distributes a measured weight percent of bio-fiber onto the resin's interface. The top and bottom carrier films are subsequently joined, resulting in a layered sheet molding composite material (Figure 5.1). The sheet material is then passed into compression wheels to assist produce consistent sheet cross-sections as well as some mixing pressures to guarantee even fiber dispersion in the epoxy. Grip rollers at the back of the journey provide the mechanical power required to draw the metal sheets at a regulated rate. After that, the sheet is trimmed to the required length [39].

When two distinct fibers were utilized to produce hybrid biomaterials, the fibers were thoroughly mixed using continuous spinning, and the resultant homogenous combination was vacuum-processed before being used to make biomaterials. Grass fibers have to be put into the sheet molding over synthetic fibers. They were fed by an extruder and vibratory conveyor arrangement. After several tries with vibratory feeding, this method of fiber insertion was perfected [40].

5.4.2.4   Injection Molding

Biomaterials were synthesized using an extruders method adopted by injection molding process. The extruder is made up of three independent temperature control sections, and double blades with a length of around 160 mm and an aspect ratio of 19. The preprocess settings were used to accomplish combination: the rotor motion was fixed at 100 rpm, the working temperature was maintained at 150°C, and the ideal duration of the substance inside the chamber was set at 2 minutes. All test specimens were produced at 30°C molding temperature, 10 bar injection pressure, and 8 seconds injection duration [41,42].

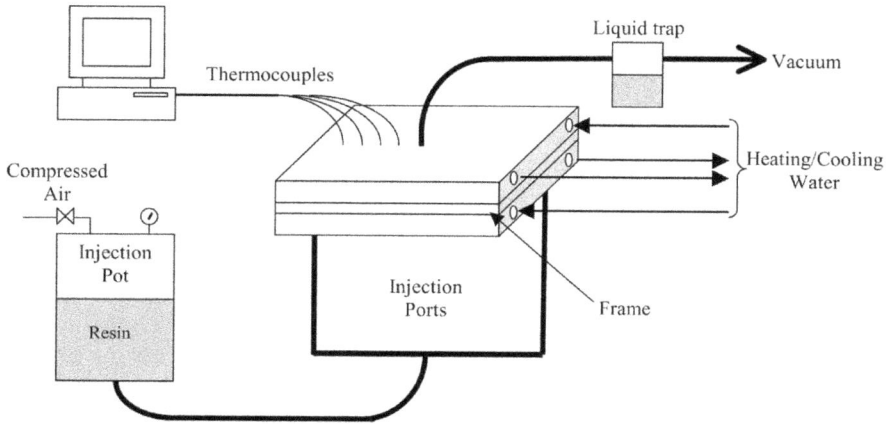

FIGURE 5.2 Workflow of the resin transfer molding process used in manufacturing of grass fiber composites [34].

5.4.2.5 Resin Transfer Molding

Designing and constructing the initial RTM for resin flow behaviors, design and manufacturing of the mold prototype, three-point flexural experiments of the resin injected then healed beams to evaluate their strengths, and finally finite element modeling to simulate the three-point flexural procedure are emphasized [43]. Rather than putting the composite material into an uncovered mold, the polymer is pre-heated and placed into the sample holder in RTM (Figure 5.2). This method is best suited to intermediate production of big parts. The fabric layers are prepared, and the resin is then injected to saturate the preforms [34].

5.5 PROPERTIES OF GRASS FIBER-REINFORCED EPOXY COMPOSITES

5.5.1 MECHANICAL PROPERTIES

The mechanical properties of grass fiber-reinforced epoxy composites such as tensile strength, bending strength, and impact strength were taken into account for the comparison of different types of natural fiber/particulate-reinforced epoxy composites. The standards like ASTM D638, ASTM D790, and ASTM D256 were followed for the conduct of above strengths. The universal testing machine was utilized for the three-point bending test, and an impact tester was used for the izod impact test to compare attributes. Plant fibers such as sisal, banana, jute, baggase, flax, ramie, and luffa were used to reinforce epoxy composites. The values in this survey were acquired using the *Lantana camera*.

5.5.1.1 Tensile Strength

In ramie-epoxy composites, a greater value of tensile strength behavior of 90 MPa was obtained [10]. Flax/epoxy composites [4] had a tensile strength behavior of

59.85 MPa, with 59 MPa produced by banana fiber-reinforced epoxy composites. *L. camera*-epoxy composites achieved the lowest tensile strength behavior of 19.08 MPa [9]. Due to the enrichment of cellulose content and stronger bonding between natural fiber and matrix, bast and leaf fiber-reinforced epoxy composites displayed higher tensile strength than other forms of fiber-reinforced epoxy composites. Broom grass alkali chemical treatment of 0.25 M concentration at fiber soaking time of 240 minutes and alkali chemical treatment of 0.5 M concentration at fiber soaking time of 240 minutes fibers had the maximum tensile strength of 82 MPa and modulus of 1 GPa. At maximal reinforcement ratio, chemical treatment of 0.5 M concentration at fiber soaking time of 240 minutes, and alkali chemical treatment of 0.25 M concentration at fiber soaking time of 480 minutes, broom grass fiber composites attained the highest tensile strength and modulus [44]. The contribution of *Cortaderia selloana* small-length fiber to organically high-density polyethylene results in a significant improvement in stiffness, but strength is unaffected even at quite large fiber loads (30 wt.%) [45]. The tensile characteristics of unmodified and modified Napier grass fiber were studied using 10% sodium hydroxide chemical treatment. After therapy, the immersion times of 6 and 24 hours give the most strength. The 24-hours immersed fiber, on the other hand, showed a greater degree of texture degradation. The sodium hydroxide treatment was found to increase the tensile strength of Napier grass fibers. The ultimate strength improved when the immersion duration was raised. Topical examination reveals that the appearance of the fibers was damaged after 18 hours of immersion time for treated fibers and above. Napier grass fibers have the capability to be employed as reinforcement materials in composite constructions [46]. Elephant grass fiber was extracted by chemical extraction method with the sodium hydroxide concentration of 4 g per liter. Those chemically extracted fiber has a greater tensile property than other fibers tested in that experiment. It has been observed that fiber volume is directly proportional to tensile strength irrespective of extraction methods and fiber modification process. However, the tensile strength of the elephant grass fiber mechanically extracted and $KMnO_4$-treated sample is less than that of other samples tested [47].

5.5.1.2 Flexural Strength

Areca-epoxy composites [11] exhibited the lowest flexural strength of 25 MPa, whereas ramie-epoxy composites [10] showed the highest value of 110 MPa. The purpose of investigating the flexural characteristics of alpha-grass strengthened starch-based biocomposites is to study the bending behavior. The alpha-fiber content of the matrix composites ranges from 5% to 35% by weight. An alkaline treatment was applied to the reinforcement materials in order to generate a proper interaction between the reinforcement and the resin. It was shown that a moderate 2.5-hour cooking procedure was sufficient to produce a decent interaction, but prolonged times produced less enhancement. Polyelectrolyte titration was used to examine the surface energies of both reinforcement and resin, and both were reported to be comparable following alkalization. The polymer nanocomposites were injection-molded and subjected to flexural testing. So many of the flexural characteristics of the polymers studied increased proportionally with reinforcing amount. The performance parameters for tensile and flexural characteristics were found to be statistically comparable. From the given data, three alternative techniques were utilized to calculate the inherent flexural strength. Finally, the Weibull theory was applied to get the best forecast of the inherent flexural modulus standard deviation [48]. Flexural characteristics improved with fiber surface modification by 5%, 6%, and 10% sodium hydroxide concentration. Combos of chemical surface modification, such as NaOH and silane, demonstrate greater flexural characteristics than untreated fiber surface modification. Under longitudinal fiber orientation, the additional mixture of alkaline and dilute epoxy therapeutic interventions increased flexural properties. Physiological techniques such as enzymatic, e-beam therapy and warm water baths degrade composite flexural characteristics. This is related to fiber degradation caused by excessive extract of fiber extractives [49]. The flexural strength of the grass fiber composites is presented in Figure 5.3 [50].

However, Sabai grass had a beneficial effect on the flexural characteristics of the biocomposites. It was heavily influenced by the density and percentage of the fiber and resin (% W/W) [50]. Then integrating wild cane grass fiber and biologically altered montmorillonite organic clay into polyester matrix. The average flexural strength of biocomposites at optimum % volume of reinforcement is enhanced to 220 MPa, and the modulus is enhanced to 4 GPa [51].

5.5.1.3 Impact Strength

The impact strength of flax fiber-reinforced epoxy composites was better, whereas the impact strength of coir fiber-reinforced epoxy composites was lower [12,14]. Snake grass fibers have tensile strength of 278.82 MPa, flexural strength of 9.71 MPa, and impact strength value of 2.87 J [1]. The tensile strength of giant grass fiber is 248 MPa, flexural strength of 9.4 MPa, and impact strength is 3.24 J [4]. Cylindrical snake plant fiber has a tensile property of 658 MPa, a flexural property of 7.6 MPa, and an impact energy of 10 J [5]. The tensile strength of Napier grass fiber is 88.4 MPa, the flexural strength is 13.15 MPa, and the impact strength is 0.99 J [7]. Broom grass fiber showed tensile strength of 297.58 MPa, flexural strength of 18.28 MPa, and impact strength of 2.87 J. Tensile strength of mendong grass fiber is observed as 452 MPa and its flexural strength is 17.4 MPa [14]. At the optimum % volume of wild crane fiber the average impact energy of organic clay-filled reinforcement biocomposites is 376.7 J/m [51].

Figure 5.4 indicates that the impact strength of unmodified *Saccharum munja* grass fiber composites is 236.1 J/m^2, whereas it is 305.4 J/m^2 for 8-hour sodium hydroxide-treated *S. munja* grass fiber composites, indicating a 29% increase. The impact strength of pure unsaturated polyester resin is 151 J/m^2, which is much lower than that of biocomposites. Reinforcement plays a major function in the impact strength of nanocomposite because it operates as a stress-transfer mechanism and prevents fracture development. The improved interface interaction between reinforcement and resin resulted in better impact strength of NaOH-treated *S. munja* grass fiber specimens [52]. The impact strength of NaOH-treated reinforcement rises with better adhesion compliance; nevertheless, extending the alkaline treatment duration to more than 120 minutes decreases these characteristics owing to fiber structural degradation [53]. Furthermore, the impact value is influenced by fiber size,

Epoxy-Based Biocomposites

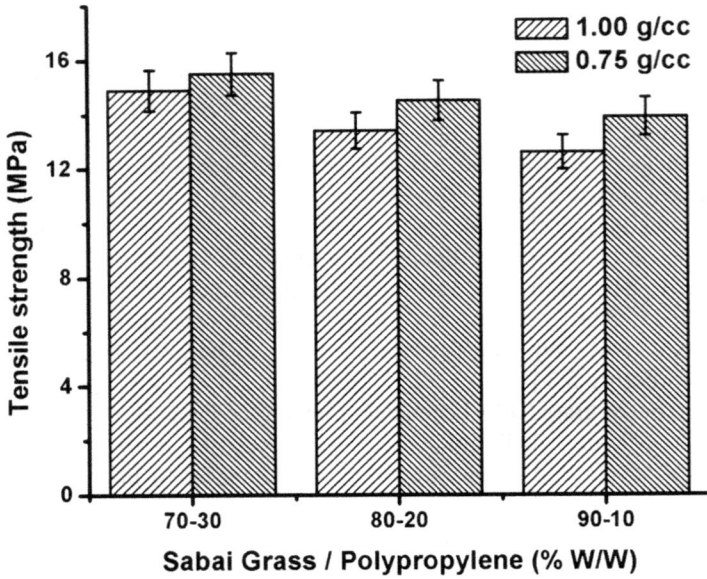

(a)

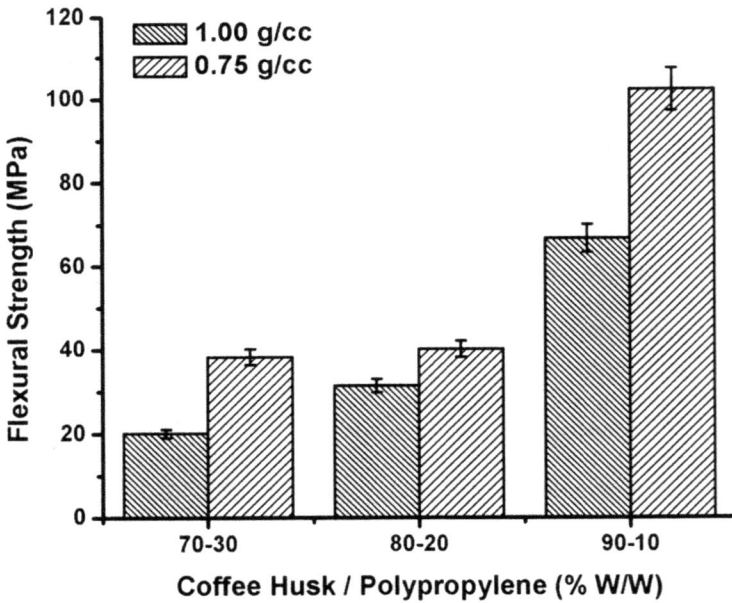

(b)

FIGURE 5.3   (a, b) Flexural strengths of the grass fiber biocomposites [50].

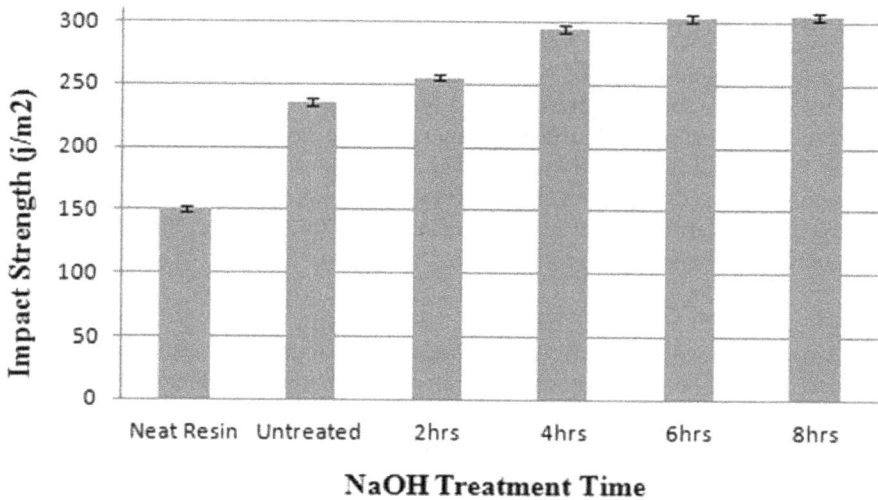

FIGURE 5.4 The influence of sodium hydroxide treatments on the izod impact strength of *S. munja* grass fibers/unsaturated polyester resin composites [52].

fiber spacing, and fiber %. Because biocomposites employ short fibers, impact energy absorption may be slightly reduced [54].

5.5.2 THERMAL PROPERTIES

Degradation of the constituents like lignin, pectin, and wax in terms of weight loss at different temperature ranges that occurs during the process is analyzed by using thermogravimetric analysis (TGA). The weight loss occurs depending on the fiber composition in the composite material. The initial weight loss that occurs in three stages is attributed to the moisture absorption, hemicellulose and non-cellulosic constituents, followed by cellulose and de-polymerization of the matrix. Snake grass fiber has a first stage of weight of 278.82 g, followed by second stage of weight 9.71 g, and then third stage of weight 2.87 g [1]. Giant grass fiber contains three weight stages: 248, 9.4, and 3.24 g [4]. Cylindrical snake plant fiber has a first stage of weight of 658 g, a second stage of weight of 7.6 g, and a third stage of weight of 1.87 g [5]. Napier grass fiber has a first stage of weight of 88.82 g, followed by second stage of weight 13.15 g and then third stage of weight 0.99 g [7]. Broom grass fiber contains three weight stages: 297.58, 18.28, and 2.87 g [9]. Mendong grass fiber has a first stage of weight of 252 g, followed by the second stage of weight of 11 g and then the third stage of weight of 4.87 g [14].

Thermal properties, crystalline structure, and tensile characteristics of Napier grass fibers of fiber surface unmodified and modified by NaOH treatment were investigated. These characteristics were discovered to be enhanced by alkali treatment. Fourier transform infrared results revealed the removal of unstructured hemicellulose from Napier grass after chemical treatment. The texture of the Napier grass following alkalization revealed surface roughening [55]. Thermal investigations on peroxide-treated straw samples and 10% NaOH-treated darbha fibers showed the degradation pattern, corresponding temperatures, and type of the degradation processes. It was also discovered that peroxide-treated straw samples were thermally more stable than darbha fibers treated with 10% alkali [56]. Physical-chemical examination, X-ray diffraction, TGA, cellulose content, and crystallinity index were used to evaluate the pampas grass. Thermal properties were measured up to 320°C, which is greater than thermoplastic polymerization temperatures [57].

5.5.2.1 Thermal Conductivity

The heat transfer of all materials rises with increasing temperature, as seen in Figure 5.5. Since the oscillation of the lattice vibrations is the thermal conductor, the humidity begins to vaporize and escape from the specimen. These findings suggest that the wild cane grass fiber composite materials studied have high thermal barrier characteristics. The inside of the fibers is porosity, allowing air to enter. This might be the cause for the composites' superior thermal insulation characteristics.

The heat transfer of the mixtures reduced, as the content level of fibers increased. Heat transfer of the polymers without organic clay ranged from 0.16 to 0.17 W/m-k in the temperature gradient of 40°C–70°C at the highest content of fiber. The heat transfer of wild cane fiber composites with 4% organic clay ranged from 0.17 to 0.18 W/m-k in the same temperature gradient range at maximum volume concentration of fiber. Heat transfer of the hybrids has enhanced by 12% with the addition of 4% organic clay at 43% volume of fiber [58].

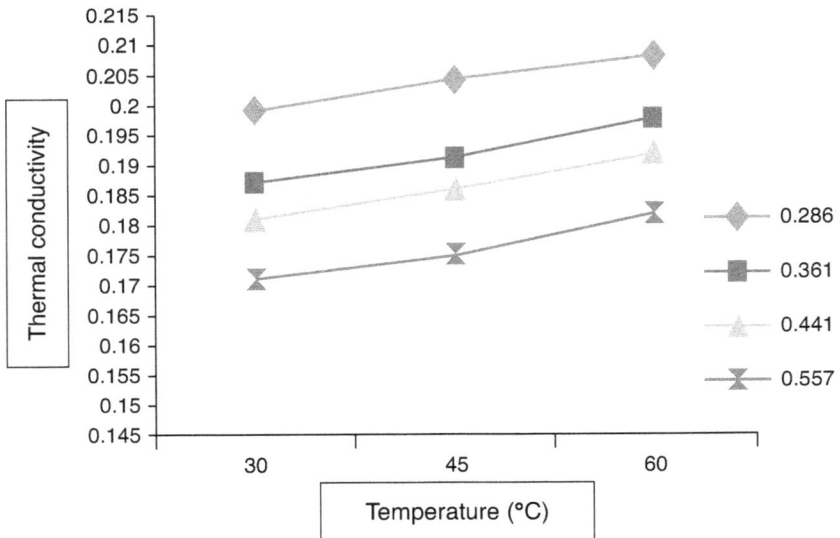

FIGURE 5.5   The heat transfer of a wild cane grass fiber composite varies with temperature at different volume percentages [58].

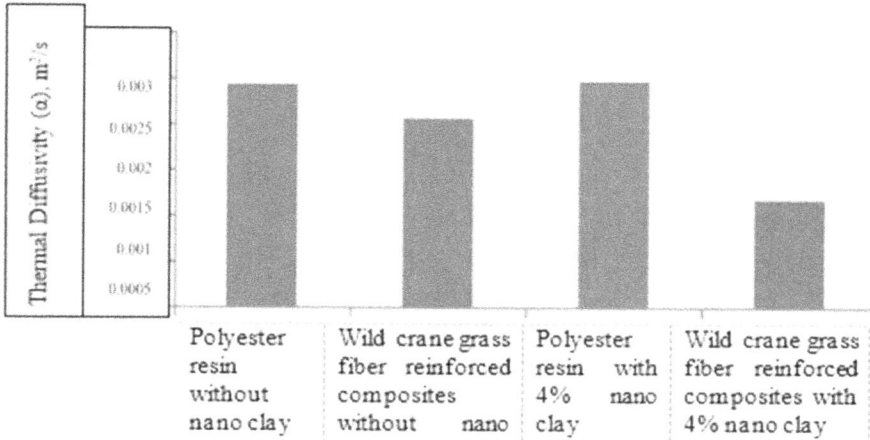

FIGURE 5.6 Comparison of thermal diffusivity of resin and reinforcement without and with 4% nano-clay [58].

5.5.2.2 Thermal Diffusivity

As sketched in Figure 5.6, the thermal diffusivity of matrix and reinforcement material is evaluated and displayed in a systematic review. A study demonstrated that the thermal diffusivity values of matrix material with and without nano-clay are almost comparable, whereas the thermal diffusivity values of reinforcement material decreased with fiber concentration and nano-clay amount.

The thermal diffusivity of reinforcement material reduces with increasing fiber volume percentage and with the addition of 4% organic clay. The proposed study's findings suggest that the reinforced material is low in weight.

5.5.2.3 Thermal Degradation

TGA was conducted to investigate the thermal degradation nature of grass fiber. Thermal degradation of grass fiber revealed that fiber withstands high temperature. The mass loss were determined by using a specimen of roughly 30 mg of trial fiber powder and a compressed gas (Argon), at a temperature rate of 10°C/min. The thermal degradation of grass fiber is presented in Figure 5.7 [59].

The thermal degradation of the grass fiber sample exhibits four phases of deterioration caused by the heating response of mendong fiber. The initial step is the start of de-volatilization, which is shown by the initial escalation in the degradation rate graph. This phase is associated with the emission of moisture and extremely light volatile molecules. The second stage is a transitional phase, which is indicated by a consistent rate of thermal degradation, indicating that the emission of volatile chemicals has begun to diminish. Mendong fiber began to deteriorate gradually. This stage occurs at temperatures ranging from 100°C to 250°C. The fiber disintegrated fast in the third stage, with all of the fiber degrading at 321° and totally degrading until temperatures of 350°C. The fourth phase is the slow burning

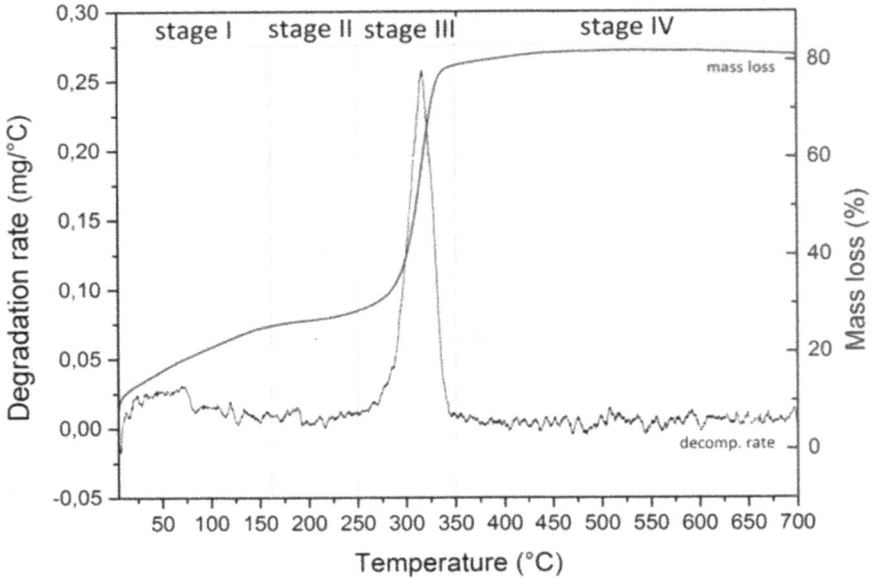

FIGURE 5.7   Temperature degradation graph for mendong fiber [59].

process residue, which is characterized by a very gradual deterioration described by a minor mass loss and a rather steady quantity of mass up to temperatures of 700°C [60].

5.6   CONCLUSION

There are lots of grass fibers naturally available and can be used as reinforcing material for the fabrication of polymer-based composites. The manufacturing process such as lay-up technique, injection molding technique, and compression molding process are commonly used fabrication methods of grass fiber composites. The constituents like cellulose and hemicellulose have major influence on the properties of these composites. The other factors such as fiber diameter, length, and fiber content also have impact on the properties. Wear resistance of these composites are influenced by increasing the fiber content. Reduction in wear has applications in automobile parts, whereas increase in friction has applied in brakes. From this chapter, it can be observed that the grass fibers have potential to replace synthetic fibers as reinforcing candidate in polymeric composites due to their excellent properties.

REFERENCES

[1] Sathishkumar, T.P., Navaneethakrishnan, P., Shankar, S., & Rajasekar, R. (2013). Mechanical properties and water absorption of snake grass longitudinal fiber reinforced isophthalic polyester composites. *Journal of Reinforced Plastics and Composites*, 32(16), 1211–1223.

[2] Jordá-Vilaplana, A., Carbonell-Verdú, A., Samper, M. D., Pop, A., & Garcia-Sanoguera, D. (2017). Development and characterization of a new natural fiber reinforced thermoplastic (NFRP) with *Cortaderia selloana* (Pampa grass) short fibers. *Composites Science and Technology*, 145, 1–9.

[3] Balaji, A.N., Karthikeyan, M.K.V., & Vignesh, V. (2016). Characterization of new natural cellulosic fiber from kusha grass. *International Journal of Polymer Analysis and Characterization*, 21(7), 599–605.

[4] Fiore, V., Scalici, T., & Valenza, A. J. C. P. (2014). Characterization of a new natural fiber from Arundodonax L. as potential reinforcement of polymer composites. *Carbohydrate Polymers*, 106, 77–83.

[5] Sreenivasan, V. S., Somasundaram, S., Ravindran, D., Manikandan, V., & Narayanasamy, R. (2011). Microstructural, physico-chemical and mechanical characterisation of *Sansevieria cylindrica* fibers–An exploratory investigation. *Materials & Design*, 32(1), 453–461.

[6] Guna, V., Ilangovan, M., Adithya, K., CV, A. K., Srinivas, C. V., Yogesh, S., & Reddy, N. (2019). Biofibers and biocomposites from sabai grass: A unique renewable resource. *Carbohydrate Polymers*, 218, 243–249.

[7] Hameem J.A.M., Majid, M. A., Afendi, M., Marzuki, H. F. A., Fahmi, I., & Gibson, A. G. (2016). Mechanical properties of Napier grass fiber/polyester composites. *Composite Structures*, 136, 1–10.

[8] Rwawiire, S., & Tomkova, B. (2015). Morphological, thermal, and mechanical characterization of *Sansevieria trifasciata* fibers. *Journal of Natural Fibers*, 12(3), 201–210.

[9] Ramanaiah, K., Prasad, A. R., & Reddy, K. H. C. (2012). Thermal and mechanical properties of waste grass broom fiber-reinforced polyester composites. *Materials & Design*, 40, 103–108.

[10] Herlina Sari, N., Wardana, I. N. G., Irawan, Y. S., & Siswanto, E. (2018). Characterization of the chemical, physical, and mechanical properties of NaOH-treated natural cellulosic fibers from corn husks. *Journal of Natural Fibers*, 15(4), 545–558.

[11] Vijay, R., Singaravelu, D. L., Vinod, A., Raj, I. F. P., Sanjay, M. R., & Siengchin, S. (2020). Characterization of novel natural fiber from saccharum bengalense grass (Sarkanda). *Journal of Natural Fibers*, 17, 1739–1747.

[12] Ramesh, M., Deepa, C., Tamil Selvan, M., & Reddy, K. H. (2020). Effect of alkalization on characterization of ripe bulrush (*Typha domingensis*) grass fiber reinforced epoxy composites. *Journal of Natural Fibers*, 19(3), 931–942.

[13] Prasad, A. R., Rao, K. M., Gupta, A. V. S. S. K. S., & Reddy, B. V. (2011). A study on flexural properties of wildcane grass fiber-reinforced polyester composites. *Journal of Materials Science*, 46(8), 2627–2634.

[14] Suryanto, H., Marsyahyo, E., Irawan, Y. S., & Soenoko, R. (2014). Morphology, structure, and mechanical properties of natural cellulose fiber from mendong grass (*Fimbristylis globulosa*). *Journal of Natural Fibers*, 11(4), 333–351.

[15] Sathishkumar, T. P., Navaneethakrishnan, P., & Shankar, O. (2012). Tensile and flexural properties of snake grass natural fiber reinforced isophthallic polyester composites. *Composites Science and Technology*, 72(10), 1183–1190.

[16] Ramesh, M., Rajeshkumar, L., Deepa, C., Tamil Selvan, M., Kushvaha, V., & Asrofi, M. (2021). Impact of silane treatment on characterization of *Ipomoea staphylina* plant fiber reinforced epoxy composites. *Journal of Natural Fibers*, 19(13), 5888–5899.

[17] Ramesh, M., Deepa, C., Rajeshkumar, L., Tamil Selvan, M., & Balaji, D. (2021). Influence of fiber surface treatment on the tribological properties of *Calotropis gigantea* plant fiber reinforced polymer composites. *Polymer Composites*, 42, 4308–4317.

[18] Madhu, P., Sanjay, M. R., Senthamaraikannan, P., Pradeep, S., Saravanakumar, S. S., & Yogesha, B. (2019). A review on synthesis and characterization of commercially available natural fibers: Part-I. *Journal of Natural Fibers*, 16, 1132–1144.

[19] Sanjay, M. R., Siengchin, S., Parameswaranpillai, J., Jawaid, M., Pruncu, C. I., & Khan, A. (2019). A comprehensive review of techniques for natural fibers as reinforcement in composites: Preparation, processing and characterization. *Carbohydrate Polymers*, 207, 108–121.

[20] Sathishkumar, T. P., Navaneethakrishnan, P., Shankar, S., Rajasekar, R., & Rajini, N. (2013). Characterization of natural fiber and composites–A review. *Journal of Reinforced Plastics and Composites*, 32(19), 1457–1476.

[21] Ramanaiah, K., Prasad, A. R., & Reddy, K. H. C. (2012). Thermal and mechanical properties of waste grass broom fiber-reinforced polyester composites. *Materials & Design*, 40, 103–108.

[22] Trache, D., Donnot, A., Khimeche, K., Benelmir, R., & Brosse, N. (2014). Physicochemical properties and thermal stability of microcrystalline cellulose isolated from Alfa fibers. *Carbohydrate Polymers*, 104, 223–230.

[23] Adeniyi, A. G., Akorede, A. S., & Ighalo, J. O. (2020). *Sansevieria Trifasciata* fibre and composites: A review of recent developments. *International Polymer Processing Journal of the Polymer Processing Society*, 35(4), 344–354.

[24] Prasad, A. R., Rao, K. M., Gupta, A. V. S. S. K. S., & Reddy, B. V. (2011). A study on flexural properties of wildcane grass fiber-reinforced polyester composites. *Journal of Materials Science*, 46(8), 2627–2634.

[25] Suryanto, H., Marsyahyo, E., Irawan, Y. S., & Soenoko, R. (2014). Morphology, structure, and mechanical properties of natural cellulose fiber from mendong grass (*Fimbristylis globulosa*). *Journal of Natural Fibers*, 11(4), 333–351.

[26] Balaji, A. N., Karthikeyan, M. K. V., & Vignesh, V. (2016). Characterization of new natural cellulosic fiber from kusha grass. *International Journal of Polymer Analysis and Characterization*, 21(7), 599–605.

[27] Kommula, V. P., Reddy, K. O., Shukla, M., Marwala, T., & Rajulu, A. V. (2013). Physico-chemical, tensile, and thermal characterization of Napier grass (native African) fiber strands. *International Journal of Polymer Analysis and Characterization*, 18(4), 303–314.

[28] Jumaidin, R., Saidi, Z. A. S., Ilyas, R. A., Ahmad, M. N., Wahid, M. K., Yaakob, M. Y., & Osman, M. H. (2019). Characteristics of cogon grass fiber reinforced thermoplastic cassava starch biocomposite: Water absorption and physical properties. *Journal of Advanced Research in Fluid Mechanics and Thermal Sciences*, 62(1), 43–52.

[29] Clauss, M., Lechner-Doll, M., & Streich, W. J. (2003). Ruminant diversification as an adaptation to the physicomechanical characteristics of forage. A reevaluation of an old debate and a new hypothesis. *Oikos*, 102(2), 253–262.

[30] Ramesh, M., Deepa, C., Selvan, M. T., Rajeshkumar, L., Balaji, D., & Bhuvaneswari, V. (2021). Mechanical and water absorption properties of *Calotropis gigantea* plant fibers reinforced polymer composites. *Materials Today: Proceedings*, 46, 3367–3372.

[31] Sahayaraj, A. F., Muthukrishnan, M., Ramesh, M., & Rajeshkumar, L. (2021). Effect of hybridization on properties of tamarind (*Tamarindus indica* L.) seed nano-powder incorporated jute-hemp fibers reinforced epoxy composites. *Polymer Composites*, https://doi.org/10.1002/pc.26326.

[32] Vimal, R., Subramanian, K. H. H., Aswin, C., Logeswaran, V., & Ramesh, M. (2015). Comparisonal study of succinylation and phthalicylation of jute fibres: Study of mechanical properties of modified fibre reinforced epoxy composites. *Materials Today: Proceedings* 2, 2918–2927.

[33] Cone, J. W., Van Gelder, A. H., Soliman, I. A., De Visser, H., & Van Vuuren, A. M. (1999). Different techniques to study rumen fermentation characteristics of maturing grass and grass silage. *Journal of Dairy Science*, 82(5), 957–966.

[34] Gholampour, A., & Ozbakkaloglu, T. (2020). A review of natural fiber composites: Properties, modification and processing techniques, characterization, applications. *Journal of Materials Science*, 55(3), 829–892.

[35] Adesogan, A. T., Arriola, K. G., Jiang, Y., Oyebade, A., Paula, E. M., Pech-Cervantes, A. A., & Vyas, D. (2019). Symposium review: Technologies for improving fiber utilization. *Journal of Dairy Science*, 102(6), 5726–5755.

[36] Chauhan, V., Kärki, T., &Varis, J. (2019). Review of natural fiber-reinforced engineering plastic composites, their applications in the transportation sector and processing techniques. *Journal of Thermoplastic Composite Materials*, 35(8), 1169–1209.

[37] Vignesh, V., Balaji, A. N., Nagaprasad, N., Sanjay, M. R., Khan, A., Asiri, A. M., & Siengchin, S. (2021). Indian mallow fiber reinforced polyester composites: Mechanical and thermal properties. *Journal of Materials Research and Technology*, 11, 274–284.

[38] Liu, W., Mohanty, A. K., Drzal, L. T., & Misra, M. (2005). Novel biocomposites from native grass and soy based bioplastic: Processing and properties evaluation. *Industrial & Engineering Chemistry Research*, 44(18), 7105–7112.

[39] Mehta, G., Mohanty, A. K., Thayer, K., Misra, M., & Drzal, L. T. (2005). Novel biocomposites sheet molding compounds for low cost housing panel applications. *Journal of Polymers and the Environment*, 13(2), 169–175.

[40] Huang, Z., Ge, H., Yin, J., & Liu, F. (2017). Effects of fiber loading and chemical treatments on properties of sisal fiber-reinforced sheet molding compounds. *Journal of Composite Materials*, 51(22), 3175–3185.

[41] Muthuraj, R., Misra, M., & Mohanty, A. K. (2015). Injection molded sustainable biocomposites from poly (butylene succinate) bioplastic and perennial grass. *ACS Sustainable Chemistry & Engineering*, 3(11), 2767–2776.

[42] Liu, W., Mohanty, A. K., Askeland, P., Drzal, L. T., & Misra, M. (2004). Influence of fiber surface treatment on properties of Indian grass fiber reinforced soy protein based biocomposites. *Polymer*, 45(22), 7589–7596.

[43] Jiang, L., Walczyk, D., McIntyre, G., Bucinell, R., & Li, B. (2019). Bioresin infused then cured mycelium-based sandwich-structure biocomposites: Resin transfer molding (RTM) process, flexural properties, and simulation. *Journal of Cleaner Production*, 207, 123–135.

[44] Srinivasababu, N., Kumar, J. S., & Reddy, K. V. K. (2014). Mechanical and dielectric properties of *Thysanolaena maxima* (broom grass) long fiber reinforced polyester composites. *Procedia Materials Science*, 6, 1006–1016.

[45] Jordá-Vilaplana, A., Carbonell-Verdú, A., Samper, M. D., Pop, A., & Garcia-Sanoguera, D. (2017). Development and characterization of a new natural fiber reinforced thermoplastic (NFRP) with Cortaderiaselloana (Pampa grass) short fibers. *Composites Science and Technology*, 145, 1–9.

[46] Ridzuan, M. J. M., Majid, M. A., Afendi, M., Azduwin, K., Kanafiah, S. A., & Dan-Mallam, Y. (2015). The effects of the alkaline treatment's soaking exposure on the tensile strength of napier fiber. *Procedia Manufacturing*, 2, 353–358.

[47] Rao, K. M. M., Prasad, A. R., Babu, M. R., Rao, K. M., & Gupta, A. V. S. S. K. S. (2007). Tensile properties of elephant grass fiber reinforced polyester composites. *Journal of Materials Science*, 42(9), 3266–3272.

[48] Espinach, F. X., Delgado-Aguilar, M., Puig, J., Julian, F., Boufi, S., & Mutjé, P. (2015). Flexural properties of fully biodegradable alpha-grass fibers reinforced starch-based thermoplastics. *Composites Part B: Engineering*, 81, 98–106.

[49] Sood, M., & Dwivedi, G. (2018). Effect of fiber treatment on flexural properties of natural fiber reinforced composites: A review. *Egyptian Journal of Petroleum*, 27(4), 775–783.

[50] Guna, V., Ilangovan, M., Adithya, K., CV, A. K., Srinivas, C. V., Yogesh, S., & Reddy, N. (2019). Biofibers and biocomposites from sabai grass: A unique renewable resource. *Carbohydrate Polymers*, 218, 243–249.

[51] Prasad, A. R., Rao, K. B., Rao, K. M., Ramanaiah, K., & Gudapati, S. K. (2015). Influence of nanoclay on the mechanical performance of wild cane grass fiber-reinforced polyester nanocomposites. *International Journal of Polymer Analysis and Characterization*, 20(6), 541–556.

[52] Singh, G. P., Madiwale, P. V., Jagtap, R. N., & Adivarekar, R. V. (2014). Extraction of fibers from *Saccharum munja* grass and its application in composites. *Journal of Applied Polymer Science*, 131(19), https://doi.org/10.1002/app.40829.

[53] Kamali Moghaddam, M. (2021). Typha leaves fiber and its composites: A review. *Journal of Natural Fibers*, https://doi.org/10.1080/15440478.2020.1870643.

[54] Ramasamy, S., Natesan, V. T., Balasubramanian, K., Justin, J. M., Samrot, A. V., & Jayaraj, J. J. (2021). Study on effect of fiber loading natural *Coccinia grandis* fiber epoxy composite. *Journal of Natural Fibers*, https://doi.org/10.1080/15440478.2021.19 52136.

[55] Reddy, K. O., Maheswari, C. U., Reddy, D. J. P., & Rajulu, A. V. (2009). Thermal properties of Napier grass fibers. *Materials Letters*, 63(27), 2390–2392.

[56] Rakesh, K. M., Ramachandracharya, S., Gokulkumar, S., & Nithin, K. S. (2021). Investigation on acoustical properties, thermal stabilities and water sorption abilities of finger millet straw fibers, darbha fibers and ripe bulrush fibers. *Materials Today: Proceedings*, 47, 5268–5275.

[57] Khan, A., Vijay, R., Singaravelu, D. L., Sanjay, M. R., Siengchin, S., Verpoort, F., & Asiri, A. M. (2021). Characterization of natural fibers from *Cortaderia selloana* grass (pampas) as reinforcement material for the production of the composites. *Journal of Natural Fibers*, 18(11), 1893–1901.

[58] Gudapati, S. K., Chidambaranathan, S., & Prasad, A. R. (2020). Influence of nanoclay on thermal properties of wild cane grass fiber reinforced polyester composites. *Materials Today: Proceedings*, 23, 632–636.

[59] Suryanto, H. (2015). Thermal degradation of mendong fiber. In *6th International Conference on Green Technology*. Universitas Islam Negeri Malang, Malang (pp. 306–309).

[60] Sahayaraj, A. F., Muthukrishnan, M., Kumar, R. P., Ramesh, M., & Kannan, M. (2021). PLA based bio composite reinforced with natural fibers–Review. In *IOP Conference Series: Materials Science and Engineering*. IOP Publishing (Vol. 1145, No. 1, p. 012069).

6 Wood Fibre-Based Epoxy Composites

Harsha Negi
IIT Roorkee

Ajitanshu Vedrtnam
Invertis University
Escuela de Arquitectura - Universidad de Alcalá

Dheeraj Gunwant
Invertis University
Apex Institute of Technology

Kishor Kalauni and Shashikant Chaturvedi
Invertis University

CONTENTS

DOI: 10.1201/9781003271017-6

6.1 INTRODUCTION

The scientific research community is looking toward the design of better-engineered eco-friendly materials. The growing need for a sustainable, environment-friendly, cheap, biodegradable material has led to using natural fibres and replacing synthetic fibres [1]. Wood fibres are the most abundantly used in a vast category of cellulose-based natural fibres. The application of wood as reinforcement in composites dates back to 1200 years ago in south China, where wood was used for making wood-reinforced clay walls and buildings like 'Tulou' [2]. Wood fibres possess good specific stiffness and good strength, have lower cost, possess lower density, and have abundant availability. Wood fibres show less variability than natural fibres from annual/seasonal crops [3]. The fibrous shape of wood fibres also helps them in dissipating wave energy and converting sound energy into heat energy; thus, wood fibre-based composites show a better coefficient of sound absorption. The load-bearing capability of wood fibres is also encouraging manufacturers to look for advancement in wood polymer composites (WPCs). However, some of the drawbacks of wood fibre over synthetic fibres include their affinity towards water absorption, lower thermal resistance, swelling on water uptake, and low thermal degradation of about 200°C [2].

6.1.1 WOOD POLYMER COMPOSITES

Figure 6.1 shows a schematic presentation of wood from macro to molecular levels, i.e., macroscopic, mesoscopic, microscopic, ultrastructural, nanoscopic, and molecular levels [2]. Wood species are categorized broadly into softwood and hardwood. Indian rosewood, Teak Oak, Sal, Beech, Mango, Maple, and Poplar are some of the hardwoods-yielding deciduous trees. Some conifer trees yield softwood, including Spruce, Fir, Larch, and Pine [4]. The surface property of wood fibres plays a vital role in determining the interfacial bond between the fibres and the matrix surface and ultimately influencing the composite's properties. Chemical composition, fibre morphology, extractive chemicals, and processing conditions influence the surface property of fibre, and the important compositions of some of the wood types are listed in Table 6.1.

WPCs consist of wood fibre or wood flour/dust as reinforcement and polymer as the matrix. The WPCs provide recyclability, reduced density and cost, increased stiffness, and good mechanical properties. However, the hydrophilic wood flour/fibres with high polar character are less compatible with hygroscopic, non-polar polymer matrix; thus, there is a poor interfacial adhesion present between the flour/fibres and polymer in WPCs [2]. In addition to this, the wood flour/fibres consist of hydroxyl groups that form hydrogen bonds, ensuing the accumulation of dust and irregular dispersion of fibres in the polymer [5]. To improve interfacial adhesion and provide better dispersion processes, physical treatments, chemical treatments, and coupling agents are used (Figure 6.2) [2]. The type of wood species, wood dust content, type of coupling agent, and type of matrix material greatly influence WPC's properties. Some typical applications of WPCs are observed in automotive, construction, and infrastructure applications [5].

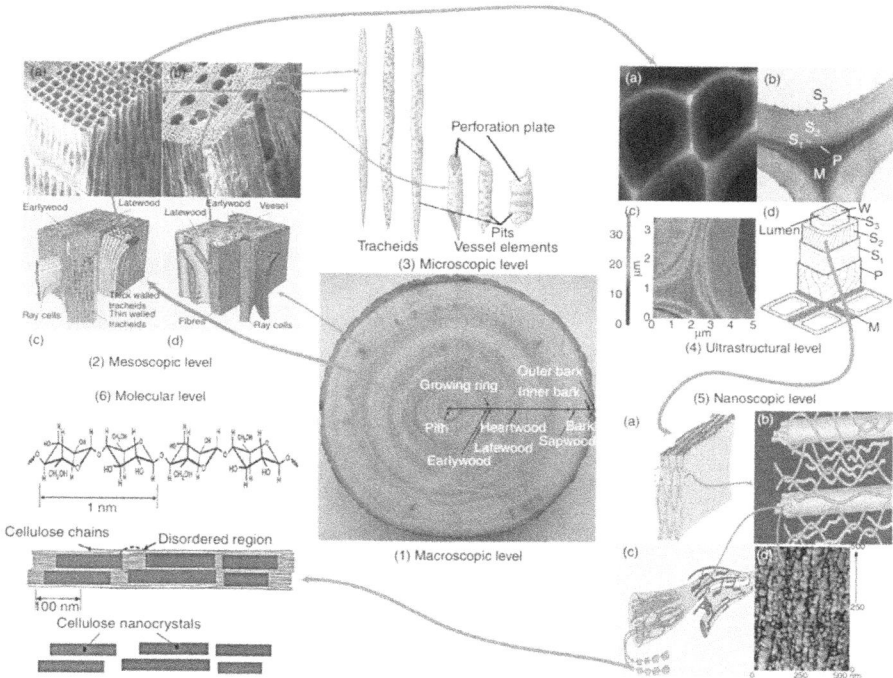

FIGURE 6.1 Macro- to molecular-level representation of wood fibre [2]. (Licence number: 5427511169663.)

TABLE 6.1
Different Chemical Compositions in Wood Fibres [2]

Type of Wood Fibre	Cellulose (%)	Hemicellulose (%)	Lignin (%)	Extractives (%)
Softwood	40–45	25–30	26–34	0–5
Hardwood	45–50	21–35	22–30	0–10
Thermo-mechanical pulp wood	37.07 ± 0.6	29.2 ± 0.1	13.8 ± 0.7	0.8 ± 0.6
Unbleached softwood	69.0 ± 2.5	22.0 ± 0.7	8.8 ± 1.8	0.2 ± 0.1
Unbleached hardwood	78 ± 0.5	19.3 ± 0.1	2.4 ± 0.4	0.3 ± 0.2
Bleached softwood	79.2 ± 0.2	20.0 ± 0.1	0.8 ± 0.1	0 ± 0
Bleached hardwood	78 ± 0.2	20.3 ± 0.1	1.3 ± 0.1	0.5 ± 0.1

Licence number: 5427511169663.

Among several thermoset matrices used, epoxy, vinyl esters, and polyesters are primarily used for aircraft, marine, automotive, infrastructure, chemical, and electrical applications. Good mechanical properties, high impact resistance, moisture resistance, chemical resistance, good adhesion with substrates, and good damage

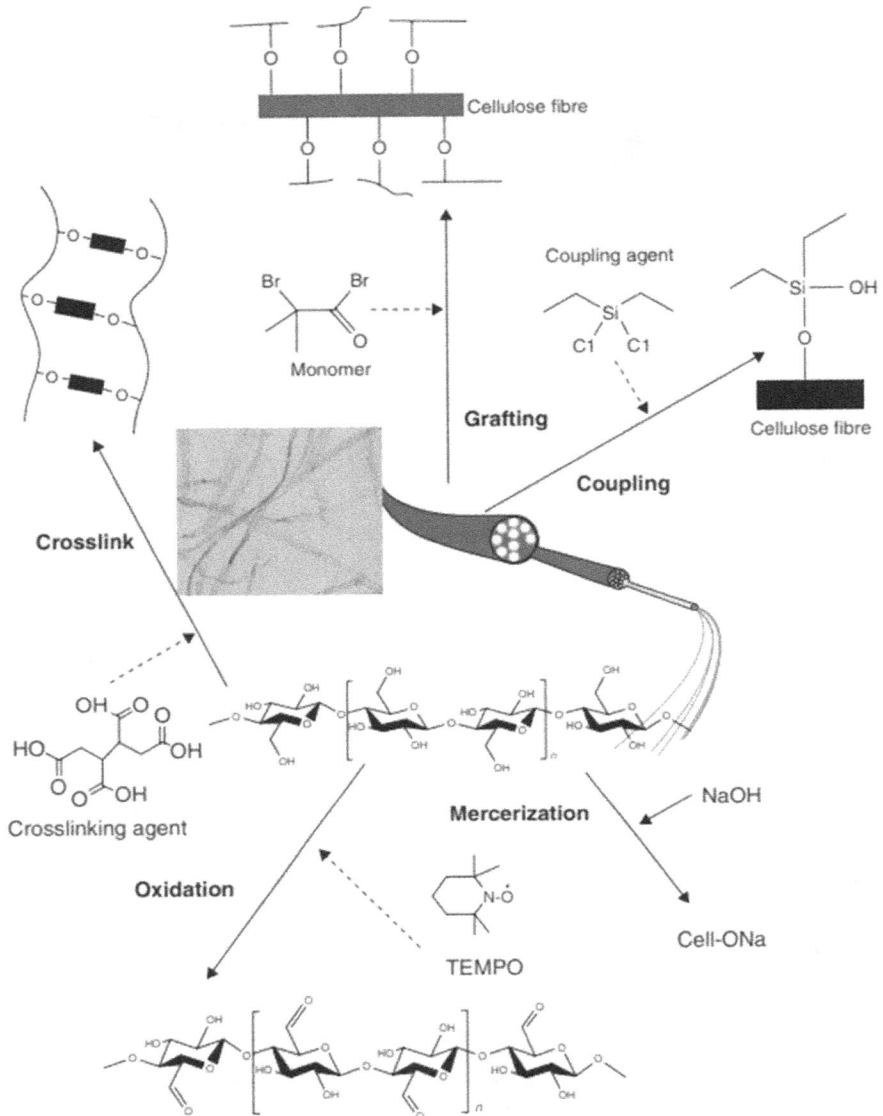

FIGURE 6.2 Chemical treatments and modification mechanism [2]. (Licence number: 5427511169663.)

tolerance make epoxy an excellent choice for use as a matrix. Epoxy acts as an effective laminating resin and adhesive for various applications. It bonds excellently with fibres and possesses great moisture barrier properties when used as a matrix in a composite. Moreover, epoxy composites are easily cured in the absence of any pressure and at ambient temperature with a curing agent or heat-cured [6,7].

6.2 MECHANICAL PROPERTIES OF WPCs

The WPCs are used in various structural as well as non-structural fields, encompassing a wide range from outdoor decking to product modelling. The mechanical characteristics of WPCs significantly influence their utilization in applications [8]. The next section discusses the mechanical properties of wood/epoxy composites:

6.2.1 WOOD FIBRE-REINFORCED EPOXY COMPOSITES

Kumar [9] developed composites with eucalyptus fibre and epoxy resin. The results have reported that the increase in fibre content has increased the mechanical behaviour (tensile and flexural strength). 25/75 eucalyptus/epoxy composites showed maximum tensile and flexural strength, whereas 20/80 eucalyptus/epoxy composites showed maximum compression strength. Huang et al. [10] reported Moso bamboo fibre-reinforced epoxy laminate (BEL) composites categorized as first-class and second-class BEL composites based on their Young's modulus. First-class BEL composites had $E \geq 12,000\,MPa$, and second-class BEL composites had $12,000\,MPa$ <E> $1000\,MPa$. On comparison, BEL shows equal or higher compression strength, shear strength, and tensile properties than other conventional composites (wood-epoxy laminates). However, BEL composites showed a slightly higher density than conventional wood/epoxy laminate composites. Naik et al. [11] developed screw pine roots fibre-reinforced epoxy composites. The composites have shown improved mechanical properties with the increment of fibre contents. Sarikaya et al. [12] evaluated the properties of birch fibre and bleached sulphate eucalyptus fibre-reinforced epoxy composites. Since the bleached sulphate fibre showed higher bonding than untreated fibres, eucalyptus fibre/epoxy composites showed better results than birch fibre/epoxy composites. Tables 6.2 and 6.3 summarize the mechanical properties of wood-epoxy composites.

Oliviera et al. [13] reported the mechanical behaviour of eucalyptus bark fibre-reinforced epoxy composites. Results indicated a reduction in tensile strength (TS) and an increment in tensile modulus with the addition of eucalyptus fibres. The decline in strength was attributed to the poor fibre–matrix adhesion. Veettil et al. [22] fabricated chemical-treated eucalyptus fibre-epoxy composite. Fibres were arranged in uniaxial, biaxial, and criss-cross patterns. Uniaxial composites showed maximum flexural strength, and biaxial composites showed maximum flexural modulus. Dinesh et al. [14] investigated the impact behaviour of jute fibre-epoxy composite. The composite showed good mechanical properties. Saxena and Gupta [13] studied the properties of Sal wood flour-epoxy and mango wood flour-epoxy composites. Sal wood flour has higher strength, stiffness, and cellulose percentage than mango wood flour. Hence, mango wood flour/epoxy composites reported lower tensile properties than Sal wood flour/epoxy composites.

6.2.2 WOOD PARTICULATE-REINFORCED EPOXY COMPOSITES

Valasek and Chocholous [15] investigated the mechanical characterization of wood flour-epoxy composites, and results showed a decrease in impact resistance and

TABLE 6.2
Mechanical Properties of Wood-Epoxy Composites Based on Tensile Load

Composite	Tensile Strength (MPa)	Young Modulus (GPa)	Ref.
Wood-fibre/Epoxy Composite			
10/90, 15/85, 20/80, and 25/75 eucalyptus/epoxy	58.4, 60.4, 66, and 70.08	10.6, 137.2, 117.8, and 104.4	[9]
30/70, 40/60, and 50/50 eucalyptus/ epoxy	25, 23, and 15	0.36×10^{-3}, 0.38×10^{-3}, and 0.45×10^{-3}	[13]
Jute fibre/epoxy	22		[14]
10/90, 20/80, and 30/70 screw pine roots/epoxy	23, 29, and 26		[11]
First-class BEL, second-class BEL	185, 172	13.8, 11.6	[10]
Birch/epoxy, eucalyptus/epoxy	29.53, 45.28		[12]
Wood Particulate/Epoxy Composite			
Mango wood flour/epoxy, Sal wood flour/epoxy	12.07, 13.52	1.28×10^{-3}, 1.29×10^{-3}	[13]
5/95, 10/90, 15/85, and 20/80 wood/ epoxy	21.65, 20.77, 20.53, and 18.27		[15]
5/95, 10/90, 15/85, and 20/80 Kayu Malam wood/epoxy	27.60–29.50, 30.00–31.50, 32.90–35.50 and 31.90–32.90		[16]
5/95, 10/90, 15/85 and 20/80 Keranji wood/epoxy	31.50–34.30, 33.90–36.10, 34.90–36.80 and 31.70–36.10		[16]
Coarse, rough, and soft sawdust/ epoxy	24.77, 28.92, and 10.92	0.86, 0.89, and 0.39	[17]
Coarse, rough, and soft chip wood/ epoxy	17.92, 21.95, and 19.89	0.78, 0.90, and 0.85	[17]
Palm tree trunk fibre/epoxy	28.44	1.66×10^{-3}	[18]
5/95, 20/80 wood/epoxy	12, 14		[19].
Hybrid Wood/Epoxy Composite			
10/50/40 rosewood/jute/epoxy, 10/50/40 padauk/jute/epoxy, and 5/5/50/40 rosewood/padauk/jute/ epoxy	24.00, 43.00, and 32.50		[14]
24.75/8.25/67, 16.5/16.5/67, and 8.25/24.75/67 mango/Sal/epoxy	13.83, 15.26, and 15.01	1.32, 1.36, and 1.33	[20]
2.5/25, 5/25, 7.5/25, and 10/25 (wood%/glass%) wood/glass/epoxy	46.33, 35.33, 56.00, and 44.42		[21]

TABLE 6.3

Mechanical Properties of Wood-Epoxy Composites Based on Flexural Load

Composite	Flexural Strength (MPa)	Flexural Modulus (GPa)	Ref.
Wood Fibre/Epoxy Composite			
10/90, 15/85, 20/80, and 25/75 eucalyptus/ epoxy	57.34, 57.74, 58.67, and 60.00	10.75, 10.83, 8.80, and 8.18	[9]
Uni, bi, and criss-cross eucalyptus/epoxy	245.59, 154.14, and 154.44	34.94, 33.62, and 22.44	[22]
Jute fibre/epoxy	3.25		[14]
Birch/epoxy, eucalyptus/epoxy	58.83, 79.92		[12]
Wood Particulate/Epoxy Composite			
Mango wood flour/epoxy, Sal wood flour/ epoxy	54.15, 43.47	2.21, 2.44	[13]
5/95, 10/90, 15/85, and 20/80 wood/epoxy			[15]
5/95, 10/90, 15/85, and 20/80 Kayu Malam wood/epoxy	2.53–2.80, 2.85–3.01, 3.15–3.30, and 2.72–2.95		[16]
5/95, 10/90, 15/85, and 20/80 Keranji wood/epoxy	3.01–3.45, 3.25–3.61, 3.49–3.90, and 3.10–3.59		[16]
Coarse, rough, and soft sawdust/epoxy	0.47, 0.28, and 0.47	1.32, 1.02, and 2.25	[23]
Coarse, rough and soft Chip wood/epoxy	0.60, 0.32 and 0.48	1.37, 2.68 and1.92	[23]
Palm tree trunk fibre/epoxy	89.43	5.12	[18]
5/95, 20/80 wood/epoxy	27.00, 35.00		[19]
Hybrid Wood/Epoxy Composite			
Uni, bi, and criss-cross eucalyptus/glass fibre/epoxy	290.10, 167.99, and 171.65	32.73, 36.91, and 26.12	[22]
10/50/40 rosewood/jute/epoxy, 10/50/40 padauk/jute/epoxy, and 5/5/50/40 rosewood/padauk/jute/epoxy	3.6, 4.2, and 4.1	2.72, 3.32, and 3.08	[14]
24.75/8.25/67, 16.5/16.5/67, and 8.25/24.75/67 mango/Sal/epoxy	50.36, 62.67 and 61.89		[20]
(14.73%/9.42% glass/fir wood) fir wood/ glass fibre/epoxy	141.4	2.90	[24]

TS, and no significant change in abrasion wear and hardness. Alzomor et al. [25] reported wood powder and flake-reinforced synthetic epoxy and bio-epoxy foam matrix-based composites. Composites were placed in ultraviolet (UV) radiation in the region 280–320 nm, using an array of fluorescent lamps at 50°C for 2000 and

4000 hours in a weatherometer apparatus. Results indicated that cell structure size was reduced with UV exposure time. Also, the filler diameter directly influenced the pore size in the composite. Hisham et al. [23] developed waste wood sawdust and chip wood-reinforced epoxy resin composites. The reinforcements were divided into three size categories, i.e., coarse, rough, and soft fibre. The flexural strength of the composite increases with the addition of filler content. Kumar et al. [26] reported the mechanical characterization of sundi wood dust-reinforced epoxy composite. The results confirmed that the mechanical behaviour had been improved with an increase in fibre percentage up to a limit, after which the properties gradually decreased. This is due to the agglomeration of fibres at higher fibre content, leading to poor curing of the composite.

Jayamani et al. [16] reported the properties of hydrogen peroxide and boric acid-treated Kayu Malam (diospyros) and Keranji (dialium) wood sawdust-reinforced epoxy composites. The mechanical behaviour (tensile and flexural strength) increased with filler concentration of up to 15% fibre; however, further increase in filler content caused a reduction in mechanical properties. Also, the impact strength was found to be less than virgin epoxy. Further, the chemical treatment improved the tensile and flexural properties of the composites. Hisham et al. [17] developed sawdust and chip wood/epoxy resin composites. The reinforcements were categorized into coarse, rough, and soft fibre. Composites showed increased tensile modulus with increased wood dust content from coarse to rough but decreased filler size from rough to soft. Alshammari et al. [18] reported palm tree trunk fibre-epoxy composite. The increase in filler content improved the properties of the composite. Hoque et al. [19] developed meranti wood powder-reinforced epoxy composites by the solution casting process. Adding fibre content improved the composites' mechanical properties (tensile, flexural, and impact strength).

Samoilenko et al. studied the mechanical properties of a modified epoxy matrix containing isocyanate, and sodium silicate reinforced with hemp wood core. An insignificant change in flexural properties was observed in the composites after weathering. This could be due to, firstly, epoxy ring opening caused by temperature, secondly, surface activation by UV irradiation, promoting crosslinking of composites, and lastly, reduction of gaps promoted by fibre swelling [27].

6.2.3 Hybrid Wood-Reinforced Epoxy Composites

Veettil et al. [22] fabricated an alkali-treated eucalyptus and glass fibre-reinforced epoxy composites. The uniaxial, biaxial, and criss-cross fibre patterns were used. Further, the flexural strength of the composite increased with the inclusion of E-glass fibres. The fibre orientation influenced the flexural strength in the following order: Uniaxial > Criss-cross > Biaxial. Dinesh et al. [14] reported the influence of padauk wood and rosewood dust on the properties of jute fibre-epoxy composites. The results concluded that padauk/jute/epoxy composites had better tensile, flexural, compressive strength, hardness, and energy absorption in impact analysis. Thus, the dust of padauk wood helped increase the composite's mechanical properties due to the dust's fine distribution, which further improved adhesion with epoxy. Rosewood dust improved the thermal stability of composite due to its coarse structure. Padauk wood/epoxy composites showed better shore D hardness than rosewood/epoxy composites. Ranakoti et al. [21] discussed the mechanical properties of *Dalbergia sissoo* wood flour/filler glass fibre-epoxy composites. A rise in wood flour content (2.5–10 wt%) improved the wear rate of the composites. Further, the impact strength increased with wood flour up to 7.5 wt%, and thereafter a reduction was observed. Saxena and Gupta [20] studied the effect of the hybridization of Sal wood flour on the mechanical properties of mango wood flour/epoxy composites. The composite containing an equal percentage of Sal and wood filler showed the best mechanical properties.

6.3 MORPHOLOGICAL CHARACTERIZATION OF WPCs

The study of morphology helps provide important information about the composites that help establish structure–property relationships. Scanning electron microscope (SEM) image shows the interfacial adhesion between fibre and polymer, pore cell size, fibre distribution, fibre agglomeration, composite degradation, and presence of deformities in the composite. This helps to provide a purpose for the increase or decrease in the properties of the composites. The morphology of wood/epoxy composites is discussed in this section.

6.3.1 WOOD PARTICULATE-REINFORCED EPOXY COMPOSITES

A review of wood powder and flake-reinforced synthetic epoxy and bio-epoxy foam matrix-based composites reveals a decrease in pore cell size after increased UV irradiation exposure at higher absorption frequencies. Composites with bio-epoxy matrix and flakes as reinforcement show a decrease in cell size. Cell structure size is reduced with increased exposure to UV irradiation [25]. SEM images examined the fractured samples after flexural test of waste wood sawdust and chip wood-reinforced epoxy resin composites. Poor filler and matrix adhesion in the composites is visible due to voids and filler pull-out in untreated wood/epoxy composites [23]. Morphology of sundi wood dust-reinforced epoxy matrix composites showed an absence of agglomeration, revealing good distribution and dispersion of filler in the composite. An increase in filler content above 10% shows epoxy entrapped between bundled filler and unwetted epoxy, leading to poor properties in the composites [26]. Morphology of fractured surfaces of sawdust and chip wood-reinforced epoxy resin composites is reported. Chip wood/epoxy composite show cracks and voids, indicating poor adhesion at the interface. Images show a small amount of fibre pull-out. Untreated composites showed voids and bubble formation, leading to a decline in the strength of the composites [17]. SEM of fractured surfaces of palm tree trunk fibre/epoxy composites reported fibre pull-out, voids, delamination, and fibre breakage. This resulted in decreased properties of the composite [18]. The morphology of meranti wood powder-reinforced epoxy composites showed little deformation after machining and grinding. However, wood still maintained its cellular structure. The interface between wood and matrix showed no voids, but polymer was absent on the wood lumen [19]. Hemp wood core/modified epoxy composites showed smooth and undamaged surfaces before weathering process. After weathering process, the samples showed a more yellowish colour [27].

6.3.2 HYBRID WOOD-REINFORCED EPOXY COMPOSITES

Morphology of alkali-treated eucalyptus and glass fibre-reinforced epoxy composites showed the highest fibre pull-out in biaxial composites, resulting in weak interfacial bonding. On the other hand, the adhesion between eucalyptus fibre and epoxy matrix was visible in the SEM images [22]. Morphology of padauk wood and rosewood dust-reinforced jute fibre/epoxy composites was reported. Padauk wood/jute fibre/epoxy composites showed less pull-out of wood dust leading to good interfacial adhesion. Rosewood dust/jute fibre/epoxy composites showed agglomeration caused by the interlocking structure of rosewood [14]. SEM images of jute-epoxy composites showed fibre tear, fibre pull-out, fibre bending, fibre cracks, and resin fracture leading to poor mechanical properties of the composite. SEM of the unfractured and fractured specimen of Sal wood flour-reinforced mango wood flour/epoxy composites shows wood fractures and interfacial bonding in the composite [20]. The morphological investigation of eucalyptus/epoxy composites showed more fibre pull-out for 10/90 eucalyptus/epoxy composites and less for 25/75 eucalyptus/epoxy composites. This indicated poor fibre–matrix adhesion in the 10/90 eucalyptus/epoxy composites [9]. In addition, SEM of tensile-fractured specimens of eucalyptus bark fibre/epoxy composites showed the presence of voids. This explains the poor fibre–matrix adhesion leading to the detachment of fibre from the matrix and to a reduction in the strength of the eucalyptus bark fibre/epoxy composite [13]. Some of the SEM images of the WPCs discussed in the previous section are given in Table 6.4.

6.4 EFFECT OF WEATHERING

Effects of weathering include combined effects of heat, moisture, and UV radiation. Among these three, UV radiation and moisture have shown a substantial impact on the composite properties. Degradation after weathering is due to thermal degradation, hydrolysis, photo-oxidation, and photo radiation, which affect the physical, chemical, and mechanical properties of composites. For example, WPCs exposed to sunlight show weight loss, colour fading, yellowing, surface roughening, embrittlement, and reduced mechanical properties. In addition, after weathering, the composites have reduced TS because of degradation between reinforcement and matrix [28,29]. The global mapping of the average annual radiation dosage for outdoor trial locations is given in Figure 6.3.

WPCs have the most significant share of applications in outdoor decking and automotive components [8]. Hence, the effect of moisture, heat, and UV radiation significantly impacts their properties. Some of the WPCs and their corresponding weathering test available in the literary works are given in Table 6.5.

The water absorption mechanism in WPC is displayed in Figure 6.4. Table 6.6 lists the composites' immersion time and water absorption percentage. A rise in wood content due to the affinity of wood towards water is observed as a result of water absorption that occur in the composites [21]. Hybrid composites with 50%/50% concentration of reinforcement showed the highest sorption coefficient and minimum water uptake and diffusion coefficient [20].

TABLE 6.4

The SEM Images of WPC

S. No.	Composite	SEM Image	Ref.
1.	Date palm fibre-epoxy composite		[18] Open access
2.	Rosewood dust-based jute-epoxy composite		[14] License Number- 5427691199851
3.	Padauk wood-based jute-epoxy composite		

(Continued)

TABLE 6.4 (*Continued*)
The SEM Images of WPC

S. No.	Composite	SEM Image	Ref.
4.	Chip wood/epoxy composite		[23] Order number-1 289697
5.	Sawdust/epoxy composite		
6.	Miranti/epoxy composite		[19] Licence number: 5427620044736

(Continued)

TABLE 6.4 (*Continued*)
The SEM Images of WPC

S. No.	Composite	SEM Image	Ref.
7.	Sundi dust/epoxy composite		[26] Licence number: 5427570180177
8.	Hybrid wood-epoxy composite		[20] Order number: 501773438

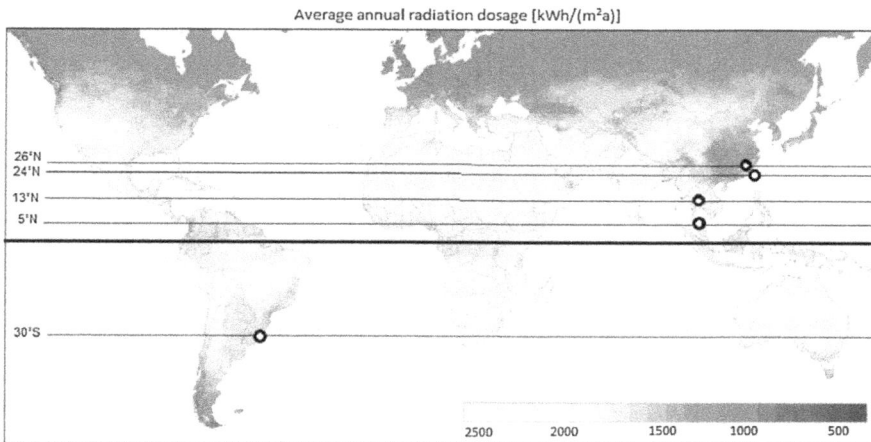

FIGURE 6.3 Outdoor trial locations and global radiation of dosage from analyzed papers [29]. (Licence number: 5427550546468.)

TABLE 6.5
WPCs and Their Corresponding Weathering Test

S. No.	Wood Flour/Fibre	Polymer	Weathering Test	Effect of Weathering	Ref.
1.	Beech (Fagus)	Polypropylene	Accelerated weathering	Filler protrusion on surface, higher colour change, and anti-fungal efficiency in heat treated filler/composites.	[30]
2.	Lodgepole pine, Western red cedar and Maple Trembling aspen	Polypropylene	Accelerated UV weathering	Lowest surface chemistry, reduced cracks, protrusion of filler, and good colour stability in western red cedar extractive composites.	[31]
3.	Douglas fir	High-density polyethylene	Accelerated weathering	Mechanical properties reduced, dimensional stability, and reduced colour change in biochar and carbon--containing composites.	[32]
4.	Spruce/fir	Polypropylene	Cyclic weathering (moisture, frost, and heat)	Mechanical properties reduced and addition of fly ash lowered the properties further after weathering.	[33]
5.	*Pinus radiata*	Polypropylene	Accelerated weathering	Decreased mechanical properties, carboxylic acid increased, fibre swelling, and fibre pull-out.	[34]
6.	*P. radiata*	Polyhydroxyalkanonate	Natural weathering	Decreased mechanical properties, fungal attacks, and colour change.	[35]
7.	Rubberwood	Polypropylene, low-density polyethylene, polyvinyl chloride, high-density polyethylene, and polystyrene	Natural weathering	Reduction in mechanical performance and lowest performance for wood-low density polyethyene (LDPE) composites after weathering.	[36]
8.	Pine mixture	High-density polyethylene	Accelerated weathering	Wood-polymer delamination and cracks on the surface, and water uptake.	[37]
9.	Pine	High-density polyethylene	UV weathering and Moisture absorption	Water uptake, reduced mechanical properties, and coupled composites delayed water absorption.	[38]

(Continued)

TABLE 6.5 (*Continued*)
WPCs and Their Corresponding Weathering Test

S. No.	Wood Flour/Fibre	Polymer	Weathering Test	Effect of Weathering	Ref.
10.	Spruce	Polypropylene	Accelerated weathering	Wood fibre exposed, cracks, reduced mechanical properties and carbon black-based composites delayed cracking.	[39]
11.	Spruce	Polypropylene	Accelerated weathering, natural weathering, cyclic test, and water immersion test	Reduced cracks in carbon black and wollastonite-based composites.	[40]
12.	Eucalyptus and pine	Polypropylene and ethylene vinyl acetate	Natural weathering	Surface cracks, decreased mechanical properties, colour change, and lowered intensity of carbonyl absorbance.	[41,42]
13.	P. radiata	High-density polyethylene	Accelerated cyclic of water immersion, followed by freeze thaw	Decreased mechanical properties, reduced interfacial adhesion, and colour change. Better performance of recycled polymer-based composites and maleic anhydride-grafted polypropylene (MAPP)-containing composite.	[43]
14.	Pine	High-density polyethylene	Accelerated and outside weathering	Colour change and crystallinity increased.	[44]
15.	Pine	Polyvinyl chloride	Accelerated and outside weathering	Water absorption, degradation, and change in carbonyl concentration.	[45]
16.	Pine, and bamboo	Polypropylene, polyvinyl chloride, and high-density polyethylene	Natural weathering	Decrease in flexural strength.	[46]
17.	Rubberwood	Polypropylene	Natural weathering	Photodegradation and decreased mechanical properties. UV stabilizer-based composites resisted surface crack and lightness.	[47]
18.	Birch and Aspen	Polypropylene, and maleic anhydride-grafted linear low-density polyethylene	Accelerated weathering	Slight increase in flexural modulus of elasticity (MOE) and impact. Reduced water absorption in coupled composites.	[48]

(*Continued*)

TABLE 6.5 (*Continued*)
WPCs and Their Corresponding Weathering Test

S. No.	Wood Flour/Fibre	Polymer	Weathering Test	Effect of Weathering	Ref.
19.	*Hevea brasiliensis* (Rubber tree)	Polyvinyl chloride	UV weathering	Decrease in flexural properties, onset decomposition temperature decreased, surface cracks.	[49]
20.	Mango, sheesham, mahogany, and babool	Polypropylene	Natural weathering	Tensile and flexural properties decreased and wear increased.	[5]
21.	Pine, poplar, and bamboo	Ethylene vinyl acetate	UV weathering	Water uptake and mass loss.	[50]
22.	Maple, and kenaf fibre	High-density polyethylene	Accelerated weathering	Loss of mechanical properties, photodegradation, and mass loss.	[51]
23.	Beech	Polypropylene	UV radiation, water-spray, and frost	Cracks formation and decreased mechanical properties.	[52]
24.	Poplar	Polypropylene	UV weathering	Surface cracks, protrusion, colour change, and photodegradation.	[53,54]
25.	Padauk	Poly(butylene succinate)	Water absorption and sunlight exposure	Increased lightness, inflated wood fillers, and decreased mechanical properties.	[55]
26.	Ponderosa pine	High-density polyethylene	Accelerated weathering	Mechanical properties reduced and carbonyl index increased. Pigment and photostabilizers reduced weathering effect.	[56]
27.	Pine	High-density polyethylene	Accelerated weathering	Surface erosion, wood fibre swelling, surface cracking, photodegradation.	[57]
28.	Poplar	High-density polyethylene	Accelerated weathering	Colour and lightness change. Dimensional and colour stability in esterified composites.	[58]
29.	Spruce, pine, and fir	Polypropylene	Natural weathering	Lightness change, cracking, mass loss, and reduced mechanical properties.	[59]

(*Continued*)

TABLE 6.5 (*Continued*)
WPCs and Their Corresponding Weathering Test

S. No.	Wood Flour/Fibre	Polymer	Weathering Test	Effect of Weathering	Ref.
30.	Mango, sheesham, mahogany, and babool	Polypropylene	Natural weathering	Wear increased, tensile properties decreased, and matrix degradation.	[60]
31.	Oak	Poly(lactic acid), Polyhydroxybutyrate, bioflex, and solanyl	UV weathering	Mechanical properties reduced, crystallinity increased, and surface degradation.	[61]

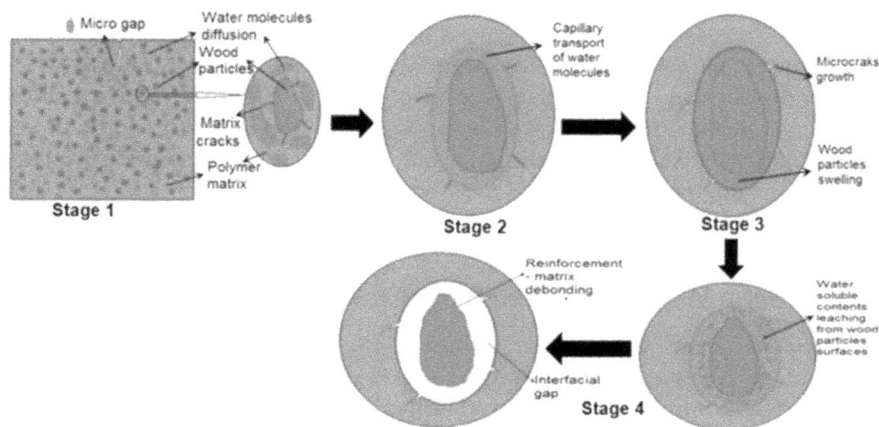

FIGURE 6.4 The water absorption mechanism in WPC [4]. (Licence number: 5427550832730.)

TABLE 6.6
Water Absorption of Wood Fibre-Reinforced Epoxy Composites

Composite	Immersion Time (hours)	Water Absorption Percentage (%)	Ref.
Wood Fibre/Epoxy Composite			
10/90, 20/80, and 30/70 screw pine roots/epoxy	24	0.79, 1.03, and 1.62	[11]
Hybrid Wood/Epoxy Composite			
24.75/8.25/67, 16.5/16.5/67, and 8.25/24.75/67 mango/Sal/epoxy	-	4.77, 3.82, and 4.50	[20]
2.5/25, 5/25, 7.5/25, and 10/25 (wood%/glass%) wood/glass/epoxy	216	1.25, 2.60, 2.80, and 2.90	[21]
Fir wood/glass fibre/epoxy	1177, 3048, and 6572	1.25, 1.81, and 1.90	[24]

6.4.1 Wood Fibre-Reinforced Epoxy Composites

The screw pine roots/epoxy composites are immersed in water for 24 hours. Results show that with an increase in fibre content, there is more affinity towards water absorption [11]. Among all the jute/wood/epoxy composites, the unfilled jute fibre/epoxy composites showed increased water absorption due to the hygroscopic property of jute fibres [14].

6.4.2 Wood Particulate-Reinforced Epoxy Composites

Sal wood flour/epoxy composites displayed maximum water absorptivity because of the high cellulose content of the fillers. In comparison to mango wood flour-epoxy, the Sal wood flour-epoxy composites had a higher diffuse coefficient and lower sorption coefficient [20]. Water absorption of palm tree trunk fibre/epoxy composites was reported. With increased absorption time, the hydrophilic fibres showed increased water uptake and thickness swelling [18]. Water absorption of meranti wood powder-reinforced epoxy composites was carried out in both water and soil. Reports suggested that increased wood filler content increased water absorption of the composites [19]. Hemp wood core/modified epoxy composites showed weight gain and swelling after weathering process. However, the degree of water absorption was low because of the high lignin content in the filler [27].

6.4.3 Hybrid Wood-Reinforced Epoxy Composites

Dalbergia sissoo wood flour filler glass fibre-reinforced epoxy composites were immersed in Alaknanda River water every 24 hours for a total of 216 hours. Reports show minimum water absorption for 25/75 glass fibre/epoxy composites and maximum absorption for 10/25/65 wood flour/glass/epoxy composites. Hence, with the addition of the filler contents, the water uptake of composites increased [21]. Water absorption for wood dust (padauk and rose)-reinforced jute-epoxy composites was studied. The padauk wood/jute fibre/epoxy showed the least water uptake compared to rosewood-jute fibre-epoxy composites and rosewood/padauk-jute fibre-epoxy composites. The coarse rosewood structure caused the development of improper wetting, air pockets, and voids in the composite, leading to a rise in water absorptivity [14].

The water absorption of Sal wood flour/mango wood flour/epoxy composites revealed that the composite containing an equal percentage of Sal and mango wood filler showed minimum water uptake, lowest diffuse coefficient, and highest sorption coefficient after saturation. This was due to the resistance of voids caused by good interfacial bonding between the filler and the matrix [20]. Cerbu and Cosereanu studied the consequences of moisture on the properties of fir wood flour-glass fibre-epoxy composites. Composites developed by hand lay-up method were immersed in water for 1177, 3048, and 6572 hours. Reduced flexural properties after 3048 hours were reported, which recovered after 6572 hours [24].

6.5 BIODEGRADABILITY OF WPC

Biodegradation causes the individual composite constituents to degrade and a reduction in interfacial strength between the constituents of a composite. Biodegradation oxidizes the composite surface, alters the crystallinity of the matrix phase and degradation of interfacial strength. The chemical composition of natural fibres majorly influences the degradation characteristics of the fibres. Figure 6.5 displays the significance of water and soil biodegradability on the mass of WPC [19]. Degradation is affected by the temperature, moisture, UV radiation, and microorganism activities to which a natural fibre/polymer composite is subjected. Since wood-epoxy composites have been used in various outdoor applications such as furniture, fencing, window parts, door panels, roofline products, and decking, the biodegradability of composites plays a significant role in determining its applications [5,8,28]. The biodegradability depends on the concentration of wood in the WPCs. This trend is evident as a higher percentage of wood will lead to its larger metabolization owing to increased fungal activity. The biodegradation is further accelerated in the presence of moisture compared to dry conditions [5]. It has been established that biodegradability is directly proportional to the loss of mechanical properties. Unfortunately, very little research on wood/epoxy composite biodegradability has been reported. The biodegradability of some of the wood/epoxy composites is given in this section.

Figure 6.6 shows the SEM micrographs of the samples after zero and 12 months of burial at 100× and 1000× magnifications. The growth of fungal hyphae cells and fungal spores is evident from the micrographs, especially at 1000× magnification [35]. The micrographs also depict depression in the middle of particles manifesting a site with simultaneous degradation of wood and poly(3-hydroxybutyrate-co-3-hydroxyvalerate) (PHBV). The mechanism of fungal action is explained in Figure 6.7. After burial for 1 year, the wood/PHBV interface was loosened due to the action of moisture. This loosened interface provides pathways for the fungi and bacteria, permitting them to enter the bulk composite matrix and initiate degradation. This network of cracks and interconnecting pores would result in crack propagation under

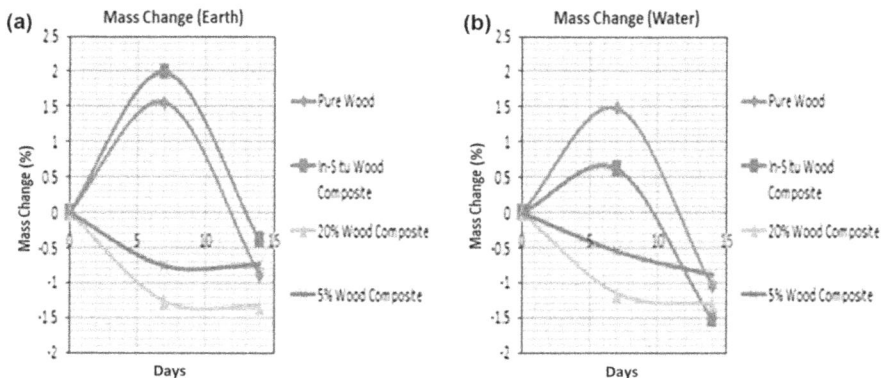

FIGURE 6.5 Change in the mass of WPC during 14 days of (a) soil degradation and (b) water degradation [19]. (Licence number: 5427620044736.)

	Initial (100x)	After 12 months (100x)	After 12 months (1000x)
PHA0W			
PHA20W			
PHA50W			
PLA50W			
PE50W			

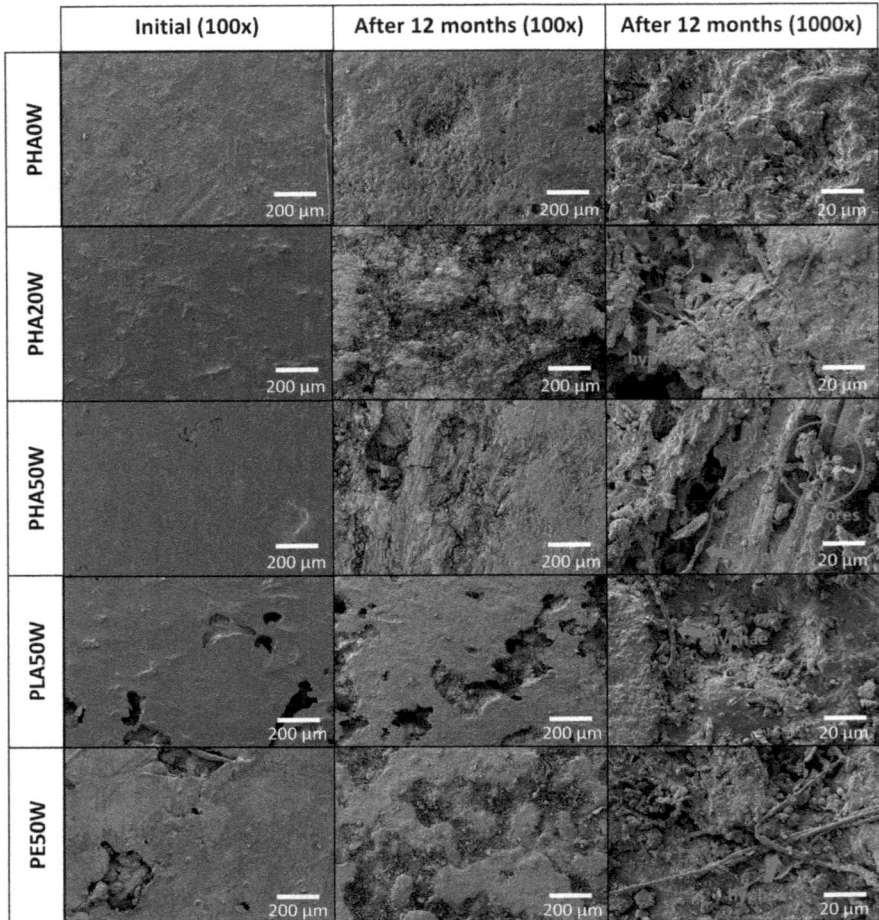

FIGURE 6.6 SEM micrographs showing fungal hyphae cell growth and fungal spores on the surface of samples [35]. (Licence number: 5427560821088.)

the applied load. The decay mechanism initiates from the surface of the composite sample in the form of surface erosion. This erosion results in the degradation of the superficial wood particles. In the presence of pathways, the hyphae would advance further. Without adequate fibre/matrix interfacial strength, the hyphae penetrate through the cracks, directly attacking the matrix polymer chains.

6.5.1 WOOD PARTICULATE-REINFORCED EPOXY COMPOSITES

The experimentations were done by immersing samples in water and soil for 14 days. Biodegradability reports of meranti wood powder/epoxy composites suggested that the change in mass was about $\pm 1.5\%$ because of the short period of the experimentation [19]. The meranti wood powder/epoxy composites had good water resistance in both water and soil. Since no impurities were absorbed along with water, this

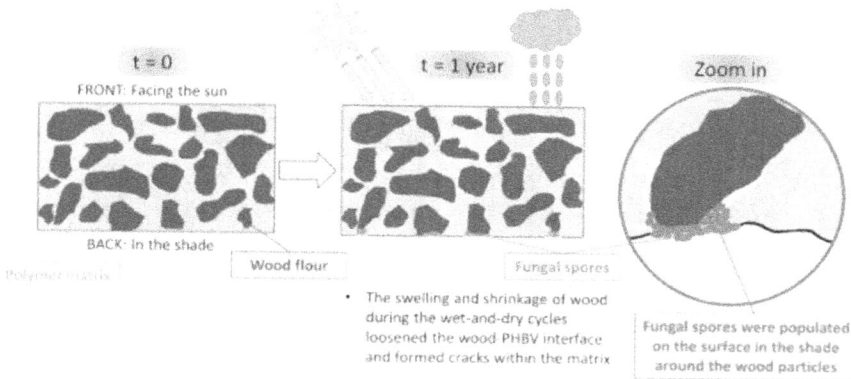

FIGURE 6.7 Description of the fungal action on wood-epoxy composites [35]. (Licence number: 5427560821088.)

decreased the composites' biodegradability. Significant degradation of hemp wood core/modified epoxy composites was reported after 1 year of soil burial. Composites with high epoxidized soybean oil (ESO) content had higher degradability by microorganisms due to the natural origin of the ESO component [27].

6.5.2 Hybrid Wood-Reinforced Epoxy Composites

The biodegradability of the padauk wood and rosewood dust-reinforced jute fibre/epoxy composites was studied [14]. Since padauk wood/jute fibre/epoxy composites had the highest resistance towards water absorption, they are the least susceptible to bacterial intake and, thus, are the least biodegradable. The biodegradability of composites in descending order is as follows: jute fibre-epoxy > rosewood-jute fibre-epoxy > rosewood-padauk-jute fibre-epoxy > padauk-wood-jute fibre-epoxy composites.

6.6 CASE STUDY: BIODEGRADATION OF WPCs AND EFFECTS ON THE THERMO-MECHANICAL PROPERTIES

Deodar (*Cedrus deodara*) wood powder of size 600 μm and polypropylene composites were fabricated via injection moulding technique by the method explained in Ref. [5]. Biodegradation tests of the composite samples were investigated according to the ASTM D5338-11 by burying them in the earth. Additionally, the buried samples were subjected to thermo-mechanical characterization to determine the effects of biodegradation on thermal and mechanical performances. Biodegradation exposure degrades the mechanical properties of WPCs. Figure 6.8 shows the variation of the TS fracture toughness of mode I (K_{IC}) of exposed and unexposed WPCs.

The single edge-notched bending test has been used to determine K_{IC} values of fabricated WPCs as per ASTM D-5045. In addition, the samples were subjected to the 3PB loading standard on a universal testing machine (UTM). The K_{IC} was determined by equation (6.1):

(a)   ■Unexposed ▨Exposed

(b)   ■Unexposed ▨Exposed

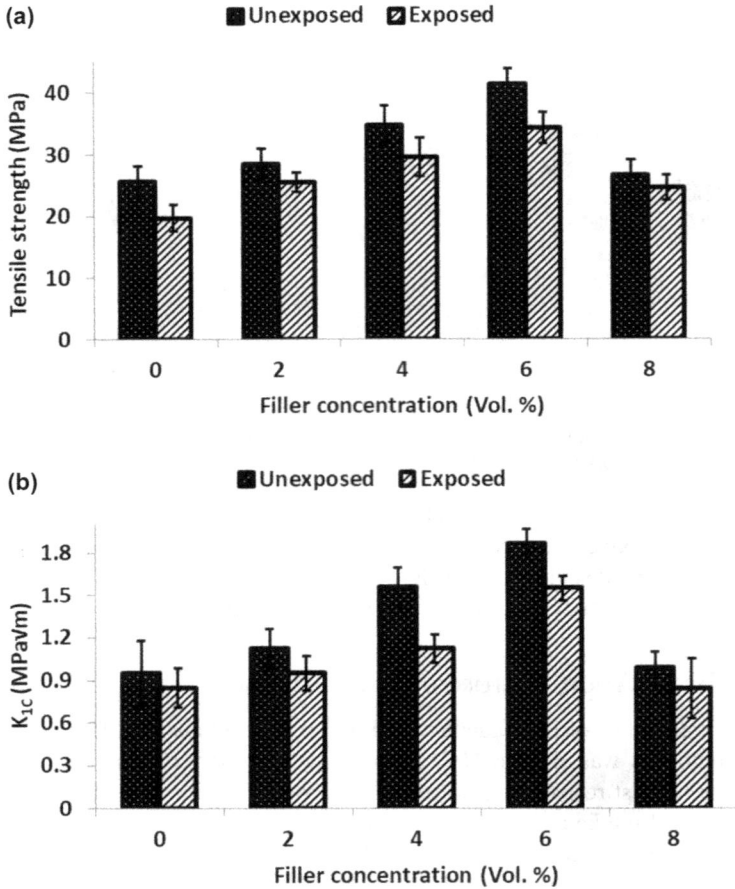

FIGURE 6.8 Variation of (a) TS and (b) K_{IC} values of unexposed and exposed WPC speci-mens vs filler concentration.

$$K_{IC} = \left(\frac{F}{BW^{\frac{1}{2}}} \right) f(x), \tag{6.1}$$

where 'F' is the maximum applied load, 'B' is the thickness of the specimen, 'W' is the width of the specimen, and 'a' is the initial crack length.

Factor 'f (x)' is the geometry function determined using the equation (6.2):

$$f(x) = 6\sqrt{x} \left[\frac{1.99 - x(1-x)(2.15 - 3.93x + 2.7x^2)}{(1+2x)(1-x)^{\frac{3}{2}}} \right]. \tag{6.2}$$

Here, x = a/W.

As shown in Figure 6.8a, the unexposed neat epoxy samples displayed TS of 25.56 MPa, continuously improving up to 41.23 MPa for 6 vol.% filler concentration.

In contrast, the exposed pure epoxy exposed samples showed a TS of 19.63 MPa, which improved to 34.21 MPa for 6 vol.% filler concentration. At a filler concentration of 6 vol.%, the TS of exposed samples deteriorated by 17.02% compared to the unexposed samples. Similar trends were obtained for K_{IC} values. The exposure resulted in a 16.66% reduction in the K_{IC} value for exposed samples compared to the unexposed samples.

The inferior tensile properties of exposed specimens are because of the degradation of the WPC samples due to biodegradation. The microorganisms (fungi and bacteria) enter the WPCs through the surface micro-cracks damaging the internal structure of the matrix. The network of cracks thus formed results in rapid crack advancement under applied load. As shown in Figure 6.8b, similar trends were observed for the fracture toughness of the exposed WPC samples. In addition, post-exposure, the K_{IC} values decreased significantly because of the degradation of the

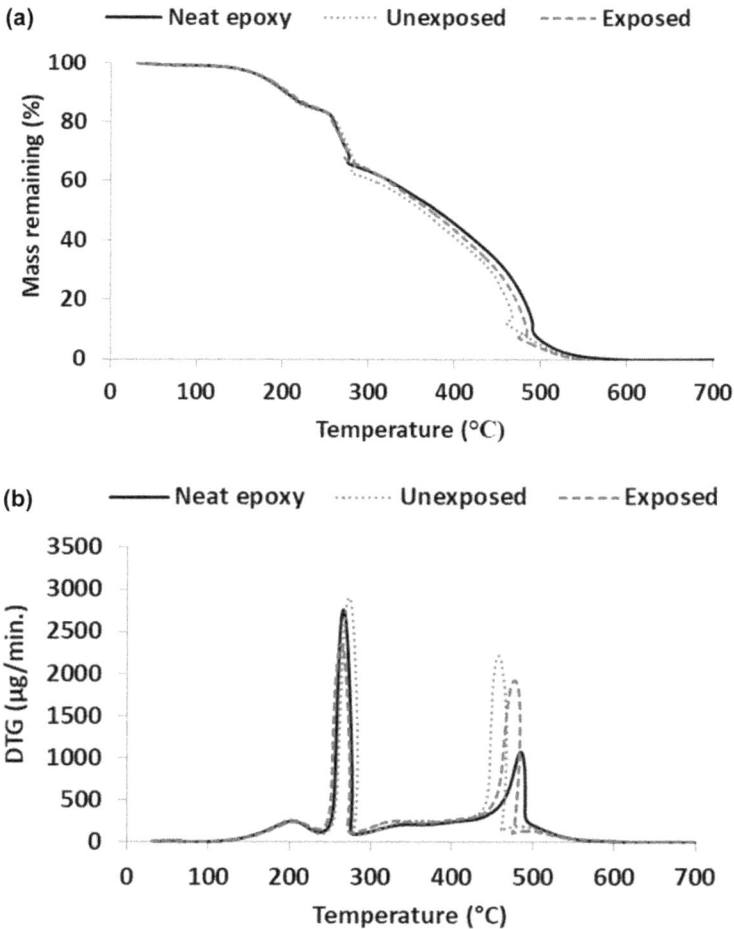

FIGURE 6.9 (a) Remaining mass percentage, and (b) DTG curve for neat epoxy and WPCs at 6% by volume filler concentration under diverse conditions.

TABLE 6.7
Results of Thermogravimetric Analysis

S. No.	Specimen	T_5 (°C)	T_{50} (°C)	T_{80} (°C)	R_{max} (µg/min)	T_I (°C)	T_{II} (°C)
1	Neat epoxy	177.89	379.39	480.33	2759.37	265.94	484.86
2	Unexposed	176.35	361.69	460.30	2904.56	273.22	476.77
3	Exposed	180.41	371.27	470.82	2352.80	263.62	457.64

epoxy polymer. Thermal properties of pure epoxy, unexposed and exposed samples were evaluated via thermogravimetric analysis. Figure 6.9a and b shows the percentage mass remaining and differential thermogravimetry(DTG) curves for the WPCs, respectively. The thermal behaviour of these samples is summarized in Table 6.7. Parameters T_5, T_{50}, and T_{80} refer to the 5%, 50%, and 80% degradation temperatures, respectively. All the samples were decomposed in two stages. On the other hand, parameter R_{max} represents the maximum degradation rate. Temperatures T_I and T_{II} represent the two DTG peaks. The neat epoxy and unexposed/exposed samples displayed two-stage decomposition. The neat epoxy displayed 5% degradation at 177.89°C against the 176.35°C and 180.41°C displayed by unexposed and exposed specimens, respectively. The 50% degradation of neat epoxy occurred at 379.39°C against 361.69°C for the unexposed samples and 371.27°C for exposed samples. Similarly, the 80% degradation temperatures were recorded as 480.33°C, 460.30°C, and 470.82°C for epoxy, unexposed, and exposed samples, respectively. The R_{max} values are 2759.37, 2904.56, and 2352.80 µg/min for epoxy, unexposed, and exposed samples, respectively. It can be concluded from the results that the exposed samples exhibited a significantly higher decomposition rate in the second decomposition step.

6.7 CONCLUSIONS

Mechanical properties, morphology, biodegradability, and weathering effect of wood/epoxy composites were reviewed.

- In general, the mechanical performance of WPCs improved with the increase in reinforcement up to a specific limit based on constituents. However, a further increase in reinforcement resulted in the agglomeration of wood and, ultimately, a decrease in properties. Treatment of filler/fibre helped to improve the properties of WPCs.
- The morphological characterization helped deduce the interfacial adhesion, fibre distribution, fibre wetting, and defects on the composite surface. The presence of voids, fibre pull-out, fibre tear, fibre bending, fibre cracks, and resin fracture were primarily observed in composites having poor interfacial adhesion and hence poor mechanical properties. SEM of UV-weathered samples reported a decrease in cell structure size with increased exposure to UV irradiation.
- The wood fibre/filler's hygroscopic property caused increased composites' affinity towards water absorption after weathering and water absorption

tests. The literature indicated increased weight gain and swelling because of water absorption.

- Water uptake exhibited an increasing trend with the rise in filler or fibre content in the composite. The coarse reinforcer caused a rise in water absorption due to improper wetting and air pockets and voids making way for water to enter the composite. Low water absorption was observed for composites with high lignin reinforcers and those with good interfacial adhesion.
- Biodegradability of composites tested in soil or both soil and water resulted in mass change, degradability by microorganisms, and bacterial attack.
- The case study concludes that WPC biodegradation damages the internal structure of the matrix by the entry of microorganisms through the surface micro-cracks. The cracks propagate under applied load and lead to a reduction in the mechanical performance of the composite.

REFERENCES

[1] V. Mittal, R. Saini, S. Sinha, Natural fiber-mediated epoxy composites – A review, *Compos. Part B Eng.* 99 (2016) 425–435. https://doi.org/10.1016/j.compositesb.2016.06.051.

[2] D. Dai, M. Fan, 1- Wood fibres as reinforcements in natural fibre composites: Structure, properties, processing and applications, in: A. Hodzic, R. Shanks (Eds.), *Natural Fibre Composites*, Woodhead Publishing, 2014: pp. 3–65. https://doi.org/10.1533/9780857099228.1.3.

[3] RC Neagu, E.K. Gamstedt, F. Berthold, Stiffness contribution of various wood fibers to composite materials, *J. Compos. Mater.* 40 (2006) 663–699. https://doi.org/10.1177/0021998305055276.

[4] M.Z.R. Khan, S.K. Srivastava, M.K. Gupta, A state-of-the-art review on particulate wood polymer composites: Processing, properties and applications, *Polym. Test.* 89 (2020) 106721. https://doi.org/10.1016/j.polymertesting.2020.106721.

[5] S. Kumar, A. Vedrtnam, S.J. Pawar, Effect of wood dust type on mechanical properties, wear behavior, biodegradability, and resistance to natural weathering of wood-plastic composites, *Front. Struct. Civ. Eng.* 13 (2019) 1446–1462. https://doi.org/10.1007/s11709-019-0568-9.

[6] R. Jeyapragash, V. Srinivasan, S. Sathiyamurthy, Mechanical properties of natural fiber/particulate reinforced epoxy composites – A review of the literature, *Mater. Today Proc.* 22 (2020) 1223–1227. https://doi.org/10.1016/j.matpr.2019.12.146.

[7] N. Saba, M. Jawaid, O.Y. Alothman, M. Paridah, A. Hassan, Recent advances in epoxy resin, natural fiber-reinforced epoxy composites and their applications, *J. Reinf. Plast. Compos.* 35 (2016) 447–470. https://doi.org/10.1177/0731684415618459.

[8] M. Schwarzkopf, M. Burnard, Wood-plastic composites—performance and environmental Impacts, in A. Kutnar, S. Muthu (Eds.), Environmental Impacts of Traditional and Innovative Forest-based Bioproducts. Environmental Footprints and Eco-design of Products and Processes, Springer, Singapore, 2016: pp. 19–43. https://doi.org/10.1007/978-981-10-0655-5_2.

[9] A. Kumar, Mechanical properties evaluation of eucalyptus fiber reinforced epoxy composites, *J. Mater. Environ. Sci* 6 (2015): 1400–1410.

[10] X.-D. Huang, C.-Y. Hse, T.F. Shupe, Evaluation of the performance of the composite bamboo/epoxy laminated material for wind turbine blades technology, *BioResources.* 10 (2014) 660–671. https://doi.org/10.15376/biores.10.1.660-671.

[11] V. Naik, M. Kumar, V. Kaup, Study on the mechanical properties of alkali treated screw pine root fiber reinforced in epoxy matrix composite material, *AIP Conf. Proc.* 2317 (2021) 020023. https://doi.org/10.1063/5.0036136.

[12] E. Sarikaya, H. Çallioğlu, H. Demirel, Production of epoxy composites reinforced by different natural fibers and their mechanical properties, *Compos. Part B Eng.* 167 (2019) 461–466. https://doi.org/10.1016/j.compositesb.2019.03.020.

[13] C.G. de Oliveira, F.M. Margem, S.N. Monteiro, F.P.D. Lopes, Comparison between tensile behavior of epoxy and polyester matrix composites reinforced with eucalyptus fibers, *J. Mater. Res. Technol.* 6 (2017) 406–410. https://doi.org/10.1016/j.jmrt.2017.08.002.

[14] S. Dinesh, P. Kumaran, S. Mohanamurugan, R. Vijay, D.L. Singaravelu, A. Vinod, M.R. Sanjay, S. Siengchin, K.S. Bhat, Influence of wood dust fillers on the mechanical, thermal, water absorption and biodegradation characteristics of jute fiber epoxy composites, *J. Polym. Res.* 27 (2019) 9. https://doi.org/10.1007/s10965-019-1975-2.

[15] P. Valasek, P. Chocholous, Mechanical properties of epoxy resins with organic filler - Wood flour, *Eng. Rural. Dev.* 16 (2013) 232–237.

[16] E. Jayamani, P. V. S. H. Prashanth, K. H. Soon, Y. C. Wong, processing and characterization of chemically modified wood sawdust reinforced (diospyros, dialium) epoxy composites, *Mater. Werkst.* 52 (2021) 1048–1056. https://doi.org/10.1002/mawe.202000306.

[17] S. Hisham, A.A. Faieza, N. Ismail, S.M. Sapuan, M.S. Ibrahim, Tensile Properties and micromorphologies of sawdust and chipwood filled epoxy composites, *Key Eng. Mater.* 471–472 (2011) 1070–1074. https://doi.org/10.4028/www.scientific.net/KEM.471-472.1070.

[18] BA. Alshammari, N. Saba, M.D. Alotaibi, M.F. Alotibi, M. Jawaid, O.Y. Alothman, Evaluation of mechanical, physical, and morphological properties of epoxy composites reinforced with different date palm fillers, *Materials.* 12 (2019) 2145. https://doi.org/10.3390/ma12132145.

[19] M. Enamul Hoque, M.A.M. Aminudin, M. Jawaid, M.S. Islam, N. Saba, M.T. Paridah, Physical, mechanical, and biodegradable properties of meranti wood polymer composites, *Mater. Des.* 64 (2014) 743–749. https://doi.org/10.1016/j.matdes.2014.08.024.

[20] M. Saxena, M. Gupta, Mechanical, thermal, and water absorption properties of hybrid wood composites, *Proc. Inst. Mech. Eng. Part J. Mater. Des. Appl.* 233 (2019) 1914–1922. https://doi.org/10.1177/1464420718798661.

[21] L. Ranakoti, M.K. Gupta, P.K. Rakesh, Analysis of mechanical and tribological behavior of wood flour filled glass fiber reinforced epoxy composite, *Mater. Res. Express.* 6 (2019). https://doi.org/10.1088/2053-1591/ab2375.

[22] A. Antham Veettil, A. Dharmarajan Thayyil, N.B. Thamba, R.V. Mangalaraja, Study on flexural properties of lotus, eucalyptus composites reinforced with epoxy and e-glass fibers in different orientations, *J. Nat. Fibers.* 19 (2020) 3001–3014. https://doi.org/10.1080/15440478.2020.1838991.

[23] S. Hisham, A.A. Faieza, N. Ismail, S.M. Sapuan, M.S. Ibrahim, Flexural mechanical characteristic of sawdust and chipwood filled epoxy composites, *Key Eng. Mater.* 471–472 (2011) 1064–1069. https://doi.org/10.4028/www.scientific.net/KEM.471-472.1064.

[24] C. Cerbu, C. Cosereanu, Moisture effects on the mechanical behavior of fir wood flour/glass reinforced epoxy composite, *BioResources.* 11 (2016) 8364–8385. https://doi.org/10.15376/BIORES.11.4.8364-8385.

[25] A. Alzomor, A.Z.M. Rus, H.A. Wahab, N.S.M. Salim, N. Marsi, M.A. Zulhakimie, M.M. Farid, Dynamic mechanical analysis and morphology of petroleum-based and bio-epoxy foams with wood filler, *AIP Conf. Proc.* 2339 (2021) 020042. https://doi.org/10.1063/5.0045146.

[26] R. Kumar, K. Kumar, P. Sahoo, S. Bhowmik, Study of mechanical properties of wood dust reinforced epoxy composite, *Procedia Mater. Sci.* 6 (2014) 551–556. https://doi.org/10.1016/j.mspro.2014.07.070.

[27] T. Samoilenko, L. Yashchenko, N. Yarova, O. Brovko, Behaviour of modified epoxy-urethanes reinforced with hemp wood core under accelerated weathering and soil burial test, *Mater. Today Commun.* 31 (2022) 103508. https://doi.org/10.1016/j.mtcomm.2022.103508.

[28] Z.N. Azwa, B.F. Yousif, A.C. Manalo, W. Karunasena, A review on the degradability of polymeric composites based on natural fibres, *Mater. Des.* 47 (2013) 424–442. https://doi.org/10.1016/j.matdes.2012.11.025.

[29] D. Friedrich, Comparative study on artificial and natural weathering of wood-polymer compounds: A comprehensive literature review, *Case Stud. Constr. Mater.* 9 (2018) e00196. https://doi.org/10.1016/j.cscm.2018.e00196.

[30] D. Aydemir, M. Alsan, A. Can, E. Altuntas, H. Sivrikaya, Accelerated weathering and decay resistance of heat-treated wood reinforced polypropylene composites, *Drv. Ind.* 70 (2019) 279–285. https://doi.org/10.5552/drvind.2019.1851.

[31] Y. Peng, N. Yan, J. Cao, Utilization of three bark extractives as natural photostabilizers for the photostabilization of wood flour/polypropylene composites, *Fibers Polym.* 21 (2020) 1488–1497. https://doi.org/10.1007/s12221-020-9694-1.

[32] X. Wang, Z. Yu, A.G. McDonald, Effect of different reinforcing fillers on properties, interfacial compatibility and weatherability of wood-plastic composites, *J. Bionic Eng.* 16 (2019) 337–353. https://doi.org/10.1007/s42235-019-0029-0.

[33] A. Benešová, J. Vanerek, R. Drochytka, K. Havlíčková, Cyclic weathering of wood - polymer composite modified by a fly ash admixture, *Adv. Mater. Res.* 1000 (2014) 145–149. https://doi.org/10.4028/www.scientific.net/AMR.1000.145.

[34] K.B. Adhikary, S. Pang, M.P. Staiger, Accelerated ultraviolet weathering of recycled polypropylene—sawdust composites, *J. Thermoplast. Compos. Mater.* 22 (2009) 661–679. https://doi.org/10.1177/0892705709096550.

[35] C.M. Chan, S. Pratt, P. Halley, D. Richardson, A. Werker, B. Laycock, L.-J. Vandi, Mechanical and physical stability of polyhydroxyalkanoate (PHA)-based wood plastic composites (WPCs) under natural weathering, *Polym. Test.* 73 (2019) 214–221. https://doi.org/10.1016/j.polymertesting.2018.11.028.

[36] T. Ratanawilai, K. Taneerat, Alternative polymeric matrices for wood-plastic composites: Effects on mechanical properties and resistance to natural weathering, *Constr. Build. Mater.* 172 (2018) 349–357. https://doi.org/10.1016/j.conbuildmat.2018.03.266.

[37] BK. Segerholm, R.E. Ibach, M.E.P. Wålinder, Moisture sorption in artificially aged wood-plastic composites, *BioResources.* 7 (2012) 1283–1293.

[38] S.C. Pech-Cohuo, I. Flores-Cerón, A. Valadez-González, C.V. Cupul-Manzano, F. Navarro-Arzate, R.H. Cruz-Estrada, Interfacial shear strength evaluation of pinewood residue/high-density polyethylene composites exposed to UV radiation and moisture absorption-desorption cycles, *BioResources.* 11 (2016) 3719–3735.

[39] S. Butylina, M. Hyvärinen, T. Kärki, Weathering of wood-polypropylene and wood-wollastonite-polypropylene composites containing pigments in Finnish climatic conditions, *Pigment Resin Technol.* 44 (2015) 313–321. https://doi.org/10.1108/PRT-08-2014-0066.

[40] S. Butylina, T. Kärki, Resistance to weathering of wood-polypropylene and wood-wollastonite-polypropylene composites made with and without carbon black, *Pigment Resin Technol.* 43 (2014) 185–193. https://doi.org/10.1108/PRT-03-2013-0019.

[41] A.L. Catto, L.S. Montagna, S.H. Almeida, R.M.B. Silveira, R.M.C. Santana, Wood plastic composites weathering: Effects of compatibilization on biodegradation in soil and fungal decay, *Int Biodeterior Biodegrad.* 109 (2016) 11–22. https://doi.org/10.1016/j.ibiod.2015.12.026.

[42] A.L. Catto, L.S. Montagna, R.M.C. Santana, Abiotic and biotic degradation of post-consumer polypropylene/ethylene vinyl acetate: Wood flour composites exposed to natural weathering, *Polym. Compos.* 38 (2017) 571–582. https://doi.org/10.1002/pc.23615.

[43] K.B. Adhikary, S. Pang, M.P. Staiger, Effects of the accelerated freeze-thaw cycling on physical and mechanical properties of wood flour-recycled thermoplastic composites, *Polym. Compos.* 31 (2010) 185–194. https://doi.org/10.1002/pc.20782.

[44] J.S. Fabiyi, A.G. McDonald, D. McIlroy, Wood modification effects on weathering of HDPE-based wood plastic composites, *J. Polym. Environ.* 17 (2009) 34–48. https://doi.org/10.1007/s10924-009-0118-y.

[45] J.S. Fabiyi, A.G. McDonald, Weathering performance of delignified pine-based poly-vinyl chloride composites, *J. Reinf. Plast. Compos.* 32 (2013) 547–563. https://doi.org/10.1177/0731684412472382.

[46] D. Friedrich, Effects from natural weathering on long-term structural performance of wood-polymer composite cladding in the building envelope, *J. Build. Eng.* 23 (2019) 68–76. https://doi.org/10.1016/j.jobe.2019.01.025.

[47] C. Homkhiew, T. Ratanawilai, W. Thongruang, Effects of natural weathering on the properties of recycled polypropylene composites reinforced with rubberwood flour, *Ind. Crops Prod.* 56 (2014) 52–59. https://doi.org/10.1016/j.indcrop.2014.02.034.

[48] H. Kallakas, T. Poltimäe, T.-M. Süld, J. Kers, A. Krumme, The influence of acceler-ated weathering on the mechanical and physical properties of wood-plastic composites, *Proc. Est. Acad. Sci.* 64 (2015) 94. https://doi.org/10.3176/proc.2015.1S.05.

[49] A. Kositchaiyong, V. Rosarpitak, H. Hamada, N. Sombatsompop, Anti-fungal perfor-mance and mechanical–morphological properties of PVC and wood/PVC composites under UV-weathering aging and soil-burial exposure, *Int. Biodeterior. Biodegrad.* 91 (2014) 128–137. https://doi.org/10.1016/j.ibiod.2014.01.022.

[50] DF Li, F.R. Xing, L. Li, JZ Li, Effects of EVA on the weathering resistance performance of wood plastic composite, *Adv. Mater. Res.* 741 (2013) 7–10. https://doi.org/10.4028/www.scientific.net/AMR.741.7.

[51] T. Lundin, S.M. Cramer, R.H. Falk, C. Felton, Accelerated weathering of natural fiber-filled polyethylene composites, *J. Mater. Civ. Eng.* 16 (2004) 547–555. https://doi.org/10.1061/(asce)0899-1561(2004)16:6(547).

[52] A. Naumann, I. Stephan, M. Noll, Material resistance of weathered wood-plastic com-posites against fungal decay, *Int. Biodeterior. Biodegrad.* 75 (2012) 28–35. https://doi.org/10.1016/j.ibiod.2012.08.004.

[53] Y. Peng, R. Liu, J. Cao, X. Guo, Effects of vitamin E combined with antioxidants on wood flour/polypropylene composites during accelerated weathering, *Holzforschung.* 69 (2015) 113–120. https://doi.org/10.1515/hf-2014-0044.

[54] Y. Peng, X. Guo, J. Cao, W. Wang, Effects of two staining methods on color stabil-ity of wood flour/polypropylene composites during accelerated UV weathering, *Polym. Compos.* 38 (2017) 1194–1205. https://doi.org/10.1002/pc.23683.

[55] N. Petchwattana, J. Sanetuntikul, P. Sriromreun, B. Narupai, Wood plastic compos-ites prepared from biodegradable poly(butylene succinate) and Burma padauk sawdust (*Pterocarpus macrocarpus*): Water absorption kinetics and sunlight exposure investiga-tions, *J. Bionic Eng.* 14 (2017) 781–790. https://doi.org/10.1016/S1672-6529(16)60443-2.

[56] NM. Stark, L.M. Matuana, Surface chemistry and mechanical property changes of wood-flour/high-density-polyethylene composites after accelerated weathering, *J. Appl. Polym. Sci.* 94 (2004) 2263–2273. https://doi.org/10.1002/app.20996.

[57] NM. Stark, L.M. Matuana, Characterization of weathered wood–plastic composite surfaces using FTIR spectroscopy, contact angle, and XPS, *Polym. Degrad. Stab.* 92 (2007) 1883–1890. https://doi.org/10.1016/j.polymdegradstab.2007.06.017.

[58] L. Wei, A.G. McDonald, C. Freitag, J.J. Morrell, Effects of wood fiber esterification on properties, weatherability and biodurability of wood plastic composites, *Polym. Degrad. Stab.* 98 (2013) 1348–1361. https://doi.org/10.1016/j.polymdegradstab.2013.03.027.

[59] T.-H. Yang, T.-H. Yang, W.-C. Chao, S.-Y. Leu, characterization of the property changes of extruded wood–plastic composites during year round subtropical weathering, *Constr. Build. Mater.* 88 (2015) 159–168. https://doi.org/10.1016/j.conbuildmat.2015.04.019.

[60] A. Vedrtnam, S. Kumar, S. Chaturvedi, Experimental study on mechanical behavior, biodegradability, and resistance to natural weathering and ultraviolet radiation of wood-plastic composites, *Compos. Part B Eng.* 176 (2019) 107282. https://doi.org/10.1016/j.compositesb.2019.107282.

[61] NS. Yatigala, D.S. Bajwa, S.G. Bajwa, Compatibilization improves performance of biodegradable biopolymer composites without affecting UV weathering characteristics, *J. Polym. Environ.* (2018). http://dx.doi.org/10.1007/s10924-018-1291-7.

7 Palm Fiber-Based Epoxy Composites

Vellaichamy Parthasarathy
Rajalakshmi Institute of Technology

Arumugam Senthil
SRM Institute of Science and Technology

Deivanayagampillai Nagarajan
Rajalakshmi Institute of Technology

Balakrishnan Sundaresan
Ayya Nadar Janaki Ammal College

CONTENTS

7.1 INTRODUCTION

The utilization of plant-based biomass resources has been increased worldwide for the sake of environmental benefit and economic growth[1,2]. Natural fibers (NFs) derived from plants have been employed as reinforcing agents due to low cost, biodegradability, low density, easy availability, high strength, toughness, reduced tool wear, good thermal stability, recyclability, and environmental benefits for developing NF-based polymer composites[3–6]. NFs such as jute, kenaf, sisal, hemp, pineapple leaf, oil palm, beach palm, sugar palm (SP) and date palm are abundantly available as waste products in major parts of the world[7,8]. The production cost of NFs is economically viable as compared to synthetic fibers such as aramid, carbon and glass fibers[9,10]. Therefore, the NFs are considered as promising alternatives for synthetic fibers in semi-structural and structural applications. NF-based polymer composites are promising alternatives to synthetic fiber-based composites in various applications due to their enhanced properties[11,12]. NF-reinforced polymer composites find

DOI: 10.1201/9781003271017-7

151

applications in automotive industries to make structural parts for the car, car bumper beam, ship, airplane and some packaging materials[13]. However, poor fiber–matrix adhesion hinders the usage of NF-based composites in various applications. In NFs-based polymer composites, the polymer matrix either thermoset or thermoplastic transfers the applied load to the stiff NFs at the interface through shear stress[14] while using them for tribological applications. Therefore, the properties of the composite depend on the fiber–matrix interfacial adhesion[15]. Epoxy resin is a thermosetting polymer that is used to develop NF-based epoxy composites for outdoor applications owing to its chemical resistance and good thermal properties[16,17]. The interaction of NF and matrix creates an interfacial bonding at the interface that depends on the compatibility of fiber and matrix. The possible interfacial bonding between the NF and matrix is illustrated in Figure 7.1.

The surface of the NFs can be modified either by physical treatments such as corona treatment, steam explosion and cold plasma treatment or chemical treatments like plasma treatment[18], alkali treatment[19,20], maleic anhydride and organosilanes[21,22] to improve adhesion between epoxy matrix and NFs for developing composites with improved properties. Among the chemical methods, the alkali treatment is economically viable to modify the surface of NFs.

The oligomers of diglycidyl ether of bisphenol A become a thermosetting epoxy resin while interacting with hardener[23]. The cured epoxy resin suffers from low resistance to crack propagation, low impact resistance and fracture toughness[24–27]. The high-performance applications of epoxy are restricted due to its inadequate properties[28,29]. The properties of epoxy can be improved by incorporating various fillers and additives such as nanoparticles, plasticizers and carbon nanotubes into the epoxy matrix[30]. The reinforcement of epoxy resin with NFs is also an ideal way to extend its applications in the field of engineering and technology, as there is an improvement in the mechanical and physical properties of epoxy after reinforcement[31–33]. The epoxy-based composite can be prepared by spray-up and lay-up techniques. This chapter intends to summarize the reinforcing effect of various palm fibers such as oil palm fiber (OPF), sugar palm fiber (SPF), peach palm fiber (PPF) and date palm fiber (DPF) on the mechanical, thermal and viscoelastic properties of epoxy composites.

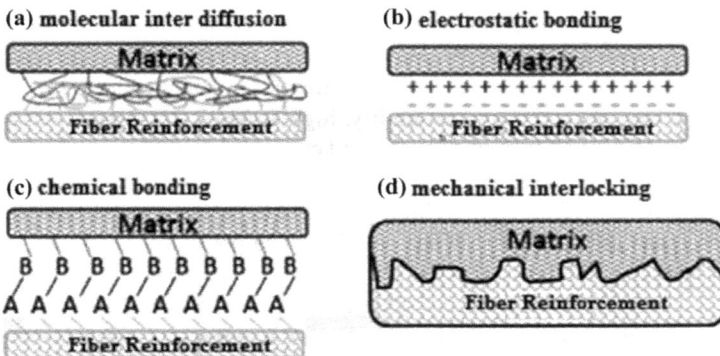

FIGURE 7.1 Types of interfacial bonding between NF and matrix[34] (a) molecular inter diffusion (b) electrostatic bonding (c) chemical bonding and (d) mechanical interlocking.

7.2   OIL PALM FIBER-REINFORCED EPOXY COMPOSITES

Oil palm tree (Figure 7.2a) is cultivated in 42 countries that include south East Asia, West Africa and Latin America for producing edible oil, and the extraction of each ton of oil yields 1.1 tons of OPF waste[35]. The chemical composition of the extracted OPF from the different parts of the tree is listed in Table 7.1. OPF is used for reinforcing polymer matrices to meet the requirement of various applications. OPF is suitable for composite fabrication because it enables mechanical interlocking with resin matrix due to its porous morphology.

The OPF was treated with sodium hydroxide (NaOH) to improve the surface roughness[36]. The different weight percentages (5, 10, 15 and 20 wt%) of OP fibers were mixed with epoxy resin along with hardener to develop an OPF/epoxy composite. The reinforcing effect of treated OPF on the mechanical and acoustical properties of epoxy composite was analyzed and compared with untreated epoxy composite. The average yield and tensile strength of the treated OPF/epoxy composite increased while increasing fiber content. The sound absorption coefficient of the treated OPF/epoxy composite was reported to be higher as compared to the untreated epoxy composite. The morphological analysis revealed that the surface of OPF became cripple, rough and torn out after the chemical treatment. The high tensile strength of the epoxy composite was due to the addition of treated OPF, since the rough surface of OPF created friction that restricted the agglomeration of fiber within the resin matrix. It was also concluded that alkaline treatment improved the interaction of OPF with the epoxy matrix.

The different volume percentages of short random OP fibers (5 to 20 vol%@5 vol%) were loaded into epoxy resin to study the reinforcing effect of short OPF on the mechanical properties of the epoxy composites by Yusoff and his research group[37]. The fabrication of epoxy composites was carried out by hand lay-up process using mold size of 200 mm × 150 mm × 3 mm. The mold was loaded with epoxy containing evenly dispersed OPF, and that was compressed to obtain the composite plates with a thickness of 3 mm, and these composite plates were used to perform tensile and flexural analyses as per ASTM standards. The addition of short random OPF to epoxy resin did not enhance the tensile properties of the composites, as tensile strength (60 MPa) was higher for pristine epoxy. The epoxy composite reinforced with 5 vol% of OPF showed the highest tensile strength (29.9 MPa) among the prepared composites. There was not much difference in the tensile strength values of fabricated composites even while increasing the volume fractions of OPF. This was due to the randomly distributed fibers in the matrix. Hence, the epoxy composite failed to hold the applied load, resulting in a decrement in the tensile strength. The fiber length also played a major role in determining the mechanical properties of the fabricated composites. The flexural strength of the pristine epoxy resin (app. 98 MPa) was higher as compared to the fabricated epoxy composites loaded with different volume fractions of OPF because the short random-OPF reinforcement failed to improve the flexural strength of the composites. The epoxy composite reinforced with 10 vol% of OPF exhibited the highest flexural strength (51 MPa) among the prepared composites. The flexural strength of the epoxy composite was found to be decreased with the increasing OPF content. The poor fiber alignment, weak interfacial fiber–matrix adhesion

TABLE 7.1

The Chemical Composition of the Extracted OPF from Various Parts of Tree[38]

Chemical Composition (%)	OPF from Empty Fruit Bunches	OPF from Kernel Shell	OPF from Mesocarp	OPF from Frond	OPF from Trunk
Cellulose	43–65	27–35	43–44	45–50	29–37
Holocellulose	68–86	40–47	70–71	80–83	42–45
Hemicellulose	17–33	15–19	33–35	34–38	12–17
Lignin	13–37	48–55	22–24	20–21	18–23

and the formation of bubbles during the composite fabrication were suggested for the decrement in the flexural modulus. The fiber and matrix interaction was studied by analyzing the scanning electron micrographs (SEM) of the fractured surface of the epoxy composites. According to this study, dispersion of fiber, fiber length and interfacial matrix–fiber adhesion need to be considered to enhance the mechanical properties of the composites.

7.3   DATE PALM FIBER-REINFORCED EPOXY COMPOSITES

The date palm tree is illustrated in Figure 7.2b. Alothnam and his research team[39] extracted DPF from different parts of matured palm trees such as tree trunks, fruit bunch stalk, leaf stalk and leaf sheath, and used them as reinforcing agents to develop DPF/epoxy composites. The DPF extracted from various parts of the trees differs in terms of morphological features which lead to variation in mechanical and thermal properties. The chemical composition of DPF is given in Table 7.2.

The epoxy resin was reinforced with DPF extracted from different parts of the palm tree to study its viscoelastic properties and thermal expansion. Among the prepared DPF/epoxy composites, the tree trunk fiber-reinforced epoxy composite showed excellent thermal stability with a higher onset and inflection temperature, and lower weight loss (22.45%). The DPF/epoxy composites were subjected to dynamic mechanical analysis (DMA) to assess tan delta and storage modulus. The storage modulus is associated with the stiffness and load-bearing capacity of a material. The epoxy composite reinforced with fruit bunch stalk fiber exhibited the least constant value (C) of 0.0021 as compared with composites reinforced with other DP fibers. The lower value of the constant affirmed the excellent stress transfer between epoxy and fiber, better interfacial adhesion and uniform dispersion of fibers in the epoxy matrix. The tan delta describes the fiber–matrix adhesion and impact resistance of the material. The tan delta was observed to be increased till glass transition temperature (T_g) and then it declined in the rubbery region while increasing the temperature for the DPF/epoxy composites. The fruit bunch stalk fiber-reinforced epoxy composite exhibited a lower tan delta peak, which proved better interfacial adhesion between epoxy matrix and fiber. The nature of fiber

TABLE 7.2
Chemical Constituents of DPF[40]

Chemical Composition	Value (%)
Hemicellulose	43.21
Cellulose	26.92
Extractives	1.75
Lignin	27.42
Others	0.70

dispersion either uniform or nonuniform in the polymer matrix is understood from the Cole–Cole plot. The obtained plot was semi-circular in shape for the leaf stalk fiber/epoxy composite, which indicated poor interfacial bonding between epoxy and fiber. The Cole–Cole plot was elliptical in shape for the composites reinforced with fruit bunch stalk and leaf sheath fibers, which represented good interfacial bonding and heterogeneous fiber dispersion in the epoxy matrix. The SEM micrographs of the fractured fruit bunch stalk/epoxy composite exhibited only fewer fiber breakage and fiber pull-outs, which affirmed the enhanced bonding strength between epoxy and fiber. The fruit bunch stalk/epoxy composite showed high tensile modulus (2.88 GPa) and tensile strength (40.12 MPa) owing to good interfacial bonding between the fiber and epoxy matrix. The effectiveness of the fibers can be estimated using equation (7.1). The efficiency of fiber is higher only when the constant value (reinforcing coefficient) is low.

$$C = \frac{\left(E_g'/E_r'\right)composite}{\left(E_g'/E_r'\right)resin}, \tag{7.1}$$

where E_g' is the storage modulus in the glass region and E_g' is the storage modulus in the rubbery region.

Gheith et al.[40] also employed DPF as a reinforcing agent and epoxy as a matrix to fabricate DPF/epoxy composites with different DPF content (40, 50 and 60 wt%) by hand lay-up process and studied their dynamic mechanical, flexural and thermal properties to conclude the effectiveness of the DPF reinforcement. The flexural modulus and strength of the epoxy composite containing 50 wt% of DPF were estimated as 3.28 GPa and 32.64 MPa, respectively, and these values were observed to be higher as compared to that of pure epoxy (flexural modulus – 2.26 GPa and flexural strength – 26.15 MPa). However, at higher DPF loading (60 wt%), there was a decline in their values due to the availability of insufficient epoxy matrix to cover all reinforced DPF[41], resulting in poor fiber–matrix interfacial adhesion. The epoxy composite reinforced with 50% of DPF exhibited a two-step degradation process with maximum decomposition temperature (T_d) at 316.9°C and higher residual content of 12.51%, and also its thermal stability was reported to be higher as compared to pristine epoxy (T_d – 30.02°C and residue content – 9.58%).

FIGURE 7.2 Images of (a) oil palm tree[42], (b) date palm tree[43], (c) sugar palm tree[44], and (d) peach palm tree[45].

7.4 SPF-REINFORCED EPOXY COMPOSITES

SP tree (Figure 7.2c) is found in tropical regions of southeast Asian nations such as Malaysia, Indonesia and Philippines[46]. SP cultivation yields SPFs as waste products. SPF is found naturally in Indonesian and Malaysian rainforests[47]. The chemical composition of SPF is listed in Table 7.3. The SP fibers can be used as reinforcement for composites owing to their significant stiffness and fiber strength. The polymer composites reinforced with SPF find applications in structural and automobile sectors due to their excellent mechanical properties.

The SP fibers were immersed in different concentrations of sodium hydroxide solution (0.25 M and 0.5 M) to carry out alkaline treatment by Bachitar et al.[48] The alkaline-treated SPF (10 wt%) was incorporated into the epoxy matrix to study the flexural properties of alkaline-treated SPF/epoxy composites. The alkaline treatment enhanced the interfacial bonding between the SPF and epoxy matrix.

The alkaline-treated SPF/epoxy composite with the dimension of 127 mm × 12.7 mm and a thickness of 3.2 mm was subjected to flexural analysis by using the universal tensile machine at a crosshead speed of 5 mm/min. The bending stress was calculated by using equation (7.2):

$$\text{Flexural strength } (\sigma_{max}) = \frac{3PL}{2bd^3},$$
(7.2)

where L is the support span (mm), P is the load at yield, d is the thickness (mm) and b is width (mm).

The epoxy composite reinforced with SPF treated with 0.25 M NaOH and 1 hour soaking time showed a maximum flexural strength of 96.7 MPa, which was observed to be increased by 24.4% as compared to the untreated epoxy composite. The maximum flexural strength was estimated as 6948 MPa for the epoxy composite reinforced with SPF treated with 0.5 M NaOH at 4 hours soaking time, and the increase in flexural modulus was 148% as compared to the untreated epoxy composite. The alkaline treatment improved the crystallinity of the SP fibers, resulting in improved flexural modulus of the treated epoxy composite. There was a removal of lignin and hemicellulose contents, resulting in an improvement in the crystallinity of cellulose content in SP fiber[49]. The improvement in the modulus of the treated composite was due to the above-said reason. The presence of holes on the SEM micrograph of the fracture surface of untreated composite was due to the pull-out of the fiber from matrix locking, which revealed a poor bonding between matrix and fiber. Good interfacial bonding was understood by the presence of fibers on the SEM micrographs of the fracture surface of the treated composite even after breakage. The surface of the epoxy composite with higher flexural strength was seen with fewer holes as compared to the composite with low flexural modulus.

Ishak and his co-workers studied the impact and flexural strength of the seawater-treated SPF/epoxy composites[50]. An improvement in the surface characteristics of treated SPF was due to the removal of pectin and the outer layer of hemicellulose during treatment. The seawater treatment improved the interfacial adhesion of SPF with the matrix, resulting in improved impact and flexural strength. The epoxy composites were subjected to flexural and impact tests according to the ASTM D790 and ASTM D256 standards, respectively, to conclude the effectiveness of the seawater-treated SPF. The flexural modulus was estimated as 18.46 MPa with 5.06% of improvement for the epoxy composite reinforced with 30 wt% treated SPF, and 14.16 MPa with 4.27% of improvement for 20 wt% treated SPF-reinforced epoxy composite as compared to the untreated epoxy composite. The epoxy composite loaded with 30 wt% treated SPF showed a higher impact value of 53.87 MPa with an improvement of 7.35% as compared to the untreated epoxy composite.

In another study, Bachtiar et al.[51] investigated the reinforcing effect of alkaline-treated SPF on the impact properties of epoxy composites. The treated-SPF/epoxy composite was prepared by hand lay-up process. The energy required to break the specimen is termed as impact strength. The treated epoxy composite with the dimension of 63.5 × 12.7 × 3.2 mm was used to carry out the Izod impact test according to

TABLE 7.3
The Chemical Composition of the Extracted SPF from Various Parts of Tree[53]

Biomass Waste	Cellulose (%)	Holocellulose (%)	Lignin (%)	Moisture (%)	Extractive (%)	Ash (%)
SPF from Frond	66.5	61.2	18.9	2.7	2.5	3.1
SPF from bunch	61.8	71.8	23.5	2.7	2.2	3.4
SPF from ijuk	52.3	65.6	31.5	7.4	4.4	4.0
SPF from trunk	40.6	61.1	46.4	1.5	6.3	2.4

ASTM D256 standard. The velocity of the striking nose was maintained at 3.46 m/s at the moment of impact. The employed pendulum energy was 5 J for testing. The impact strength was estimated using equation (7.3):

$$\text{Impact strength (E)} = \frac{J}{I}, \tag{7.3}$$

where J is the value from machine and t is the thickness.

The impact response of fiber-reinforced composite depends on interfacial bond strength, fiber and matrix properties[52]. The impact energy is dissipated by the matrix or fiber fracture, fiber pull-out and debonding[51]. The fiber fracture is associated with the dissipation of less impact energy in comparison with fiber pull-out. The fiber fracture represents a strong interfacial bonding in the composite, while the fiber pull-out represents poor interfacial bonding. The epoxy composite reinforced with treated SPF at a higher NaOH concentration (0.5 M) and 8 hours of soaking time exhibited higher impact strength of 60 J/m, and the impact strength was improved by 12.85% as compared to the untreated composite. The strong alkali treatment enabled the better bonding between matrix and SPF through the exposed OH groups on the fiber surface due to the removal of cementing materials (hemicellulose and pectin) resulting in the improved impact strength for the prepared composite.

7.5   PEACH PALM FIBER-REINFORCED EPOXY COMPOSITES

The peach palm (PP) tree (Figure 7.2d) is cultivated in Brazil and Central America, and it also yields two food crops such as fruits and the heart of palm (the inner core of the tree). The extraction of the heart of palm (edible part) generates a huge amount of agricultural waste since only one-fifth of the stem tip is used for food purposes[54]. The produced agricultural waste creates environmental issue which is addressed by extracting NFs from the agricultural waste. Cordeiro et al.[55] studied the reinforcing effect of the extracted PP fibers from the trunk of PP trees on the mechanical properties of the epoxy-based composites. PPFs were functionalized with glycidyloxy-propyl-trimethoxy silane (GPTMS) to improve fiber–matrix interactions. The epoxy composite reinforced with 70 wt% of GPTMS-treated PP fibers showed significantly higher mechanical performance as compared to pristine epoxy resin. An increase in modulus in the rubbery region was understood from the DMA for the developed epoxy composite. The efficient interaction between the GPTMS-treated PP fibers and epoxy resin was affirmed by the SEM micrographs. The fracture surface of the surface-modified epoxy composite with untreated PPF was seen to be irregular, and the presence of voids and fiber debonding due to the pull-out of fiber from the matrix affirmed a poor fiber–matrix interfacial adhesion. However, there was no void on the fracture surface of the epoxy composite reinforced with the GPTMS-treated PPF, which affirmed the better adhesion between fiber and matrix.

7.6 CONCLUSION

The various reported works on epoxy composites reinforced with different palm fibers such as PP, DP, OP and SP fibers were discussed to conclude the important findings of these works. The palm fibers extracted from four different parts of palm trees such as tree trunk, fruit bunch stalk, leaf stalk and leaf sheath were used for reinforcing epoxy composites. The mechanical properties of epoxy resin were reported to be increased after reinforcing it with palm fibers. However, the epoxy composite reinforced with treated palm fibers showed excellent mechanical properties as compared to the untreated epoxy composites. The fiber dispersion, fiber length and matrix–fiber adhesion were concluded as important factors in determining the mechanical properties of palm fiber-reinforced epoxy composites. The Cole–Cole plot was used to conclude the nature of fiber dispersion (uniform or nonuniform) in the epoxy matrix.

REFERENCES

[1] Petrone, G.; Meruane, V. Mechanical properties updating of anon-uniform natural fibre composite panel by means of a parallel genetic algorithm. *Compos A: Appl. Sci. Manuf.* **2017**, 94, 226–233.
[2] Bambach, M. Compression strength of natural fibre composite plates and sections of flax, jute and hemp. *Thin. Walled. Struct.* **2017**, 119, 103–113.
[3] Chollakup, R.; Smitthipong, W.; Suwanruji, P. Environmentally friendly coupling agents for natural fibre composites. *Nat. Polym.* **2012**, 1, 161–182.
[4] Lee, S. H.; Wang, S. Biodegradable polymers/bamboo fiber biocomposite with bio-based coupling agent. *Compos A: Appl. Sci. Manuf.* **2006**, 37, 80–91.
[5] Chand, N.; Dwivedi, U. K. Effect of coupling agent on abrasive wear behaviour of chopped jute fibre-reinforced polypropylene composites. *Wear.* **2006**, 261, 1057–1063.
[6] Sgriccia, N.; Hawley, M. C.; Misra, M. Characterization of natural fiber surfaces and natural fiber composites. *Compos A: Appl. Sci. Manuf.* **2008**, 39, 1632–1637.
[7] Saba, N.; Jawaid, Paridah, M.; Al-othman, O. A review on flammability of epoxy polymer, cellulosic and non-cellulosic fiber reinforced epoxy composites. *Polym. Adv. Technol.* **2016**, 27 (5), 577–590.
[8] Saba, N.; Jawaid, M.; Hakeem, K.; Paridah, M.; Khalina, A.; Alothman, O. Potential of bioenergy production from industrial kenaf (*Hibiscus cannabinus* L.) based on Malaysian perspective. *Renew. Sustain. Energy Rev.* **2015**, 42, 446–459.

[9] Al-Oqla, F. M.; Alothman, O. Y.; Jawaid, M.; Sapuan, S. M.; Es-Saheb, M. H. Processing and properties of date palm fibers and its composites. *Biomass Bioenerg.* **2014**, 1–25.

[10] Witayakran, S.; Kongtud, W.; Boonyarit, J.; Smitthipong, W.; Chollakup, R. Development of oil palm empty fruit bunch fiber reinforced epoxy composites for bumper beam in automobile. *Key Eng. Mater.* **2017**, 751, 779–784.

[11] Kalia, S.; Kaith, B.; Kaur, I. Pretreatment of natural fibers and their application as reinforcing material in polymer composites-a review. *Polym. Eng. Sci.* **2009**, 49, 1253–1272.

[12] Li, X.; Tabil, L. G.; Panigrahi, S. Chemical treatments of natural fiber for use in natural fiber-reinforced composites: a review. *J. Polym. Environ.* **2007**, 15, 25–33

[13] Reddy, N.; Yang, Y. Biofibers from agricultural byproducts for industrial applications. *Trends Biotechnol.* **2005**, 23, 22–27.

[14] Bachtiar, D.; Sapuan, S. M.; Hamdan, M. M.; Flexural properties of alkaline treated sugar palm fibre reinforced epoxy composites, Int. *J. Automot. Mech. Eng.* **2010**, 1, 79–90.

[15] Suwanruji, P.; Smitthipong, W.; Chollakup, R. Chapter 5. Purpose of natural fiber surface treatment and coupling agent in bio-based composites, in: W. Smitthipong, R. Chollakup, M. Nardin (Eds.), *Bio-Based Composites for High-Performance Materials From Strategy to Industrial Application*, CRC Press, Taylor & Francis Group, Boca Raton, FL, **2014**, pp. 59–86.

[16] Mittal, V.; Saini, R.; Sinha, S. Natural fiber-mediated epoxy composites-a review. *Compos. B. Eng.* **2016**, 99, 425–435.

[17] Saba, N.; Jawaid, M.; Alothman, O. Y.; Paridah, M. T.; Hassan, A. (2016) Recent advances in epoxy resin, natural fiber-reinforced epoxy composites and their applications. *J Reinf. Plast. Compos.* **2016**, 35, 447–470. doi:10.1177/0731684415618459

[18] Scalici, T.; Fiore, V.; Valenza, A. Effect of plasma treatment on the properties of *Arundo donax* L. leaf fibres and its bio-based epoxy composites: a preliminary study. *Compos. B. Eng.* **2016**, 94, 167–175.

[19] Li, X.; Tabil, L. G.; Panigrahi, S. Chemical treatments of natural fiber for use in natural fiber-reinforced composites: a review. *J. Polym. Environ.* **2007**, 15, 25–33.

[20] Edderozey, A. M. M.; Akil, H. M.; Azhar, A. B.; Ariffin, M. I. Z. Chemical modification of kenaf fibers. *Mater. Lett.* **2007**, 61, 2023–2025.

[21] Lu, T.; Jiang, M.; Jiang, Z.; Hui, D.; Wang.; Z, Zhou.; Z. (2013) Effect of surface modification of bamboo cellulose fibers on mechanical properties of cellulose/epoxy composites. *Compos. B. Eng.* **2013**, 51, 28–34.

[22] Kushwaha, P. K.; Kumar, R. (2010) Effect of silanes on mechanical properties of bamboo fiber–epoxy composites. *J. Reinf. Plast. Compos.* **2010**, 29, 718–724.

[23] Zhang, J.; Dong, H.; Tong, L. Investigation of curing kinetics of sodium carboxymethyl cellulose/epoxy resin system by differential scanning calorimetry. *Thermochim. Acta.* **2012**, 549, 63–68.

[24] Alamri, H.; Low, I. M. Effect of water absorption on the mechanical properties of nano-filler reinforced epoxy nanocomposites. *Mater. Des.* **2012**, 42, 214–222.

[25] Njuguna, J.; Pielichowski, K.; Alcock, J. R. Epoxy-based fibre reinforced nanocomposites. *Adv. Eng. Mater.* **2007**, 9, 835–847.

[26] Mirmohseni, A.; Zavareh, S. Preparation and characterization of an epoxy nanocomposite toughened by a combination of thermoplastic, layered and particulate nano-fillers. *Mater. Des.* **2010**, 31, 2699–2706.

[27] Comas-Cardona, S.; Groenenboom, P.; Binetruy, C. A generic mixed FE-SPH method to address hydro-mechanical coupling in liquid composite moulding processes. *Compos. Part A: Appl. Sci. Manuf.* **2005**, 36, 1004–1010.

[28] Raquez, J.; M, Dele'glise, M.; Lacrampe, M. F. Thermosetting (bio)materials derived from renewable resources: a critical review. *Prog. Polym. Sci.* **2010**, 35, 487–509.

[29] Njuguna, J.; And K. P.; Alcock, J. R. Epoxy-based fibre reinforced nanocomposites. *Adv. Eng. Mater.* **2007**, 9, 835–847.

[30] Saba, N.; Jawaid, M.; Alothman, O. Y.; Paridah, M. T.; Hassan, A. Recent advances in epoxy resin, natural fiber-reinforced epoxy composites and their applications. *J Reinf. Plast. Compos.* **2016**, 35 (6), 447–470.

[31] Abdellaoui, H.; Bensalah, H.; Echaabi, J. Fabrication, characterization and modelling of laminated composites based on woven jute fibres reinforced epoxy resin. *Mater. Des.* **2015**, 68, 104–113.

[32] Masoodi, R.; Pillai, K. M.; A study on moisture absorption and swelling in bio-based jute-epoxy composites. *J. Reinf. Plast. Compos.* **2012**, 31, 285–294.

[33] Jawaid, M.; Abdul Khalil, H. P. S.; Abu Bakar, A.; Mechanical performances of oil palm empty fruit bunches/jute fibres reinforced epoxy hybrid composites. *Mater. Sci. Eng A.* **2010**, 527, 7944–7949.

[34] Rao, J.; Zhou, Y.; Fan, M. Revealing the interface structure and bonding mechanism of coupling agent treated WPC. *Polymers.* **2018**, 10 (3), 266.

[35] Kakou, C. A.; Arrakhiz, F. Z.; Trokourey, A.; Bouhfid, R.; Qaiss, A.; Rodrigue, D. Influence of coupling agent content on the properties of high density polyethylene composites reinforced with oil palm fibers. *Mater. Des.* **2014**, 63, 641–649.

[36] Bakri, M. K.; Jayamani E.; Heng S. K.; Hamdan, S. Reinforced oil palm fiber epoxy composites: an investigation on chemical treatment of fibers on acoustical, morphological, mechanical and spectral properties. *Mater. Today: Proc.* **2015**, 1, 2747–2756.

[37] Yusoff, M. Z.; Salit. M. S.; Ismail, N.; Wirawan, R. Mechanical properties of short random oil palm fibre reinforced epoxy composites. *Sains Malaysiana.* **2010**, 39 (1), 87–92.

[38] Maluin, F. N.; Hussein, M. Z.; Idris, A. S. An overview of the oil palm industry: challenges and some emerging opportunities for nanotechnology development. *Agronomy.* **2020**, 10 (3), 356.

[39] Alothman, O. Y.; Jawaid, M.; Senthilkumar, K.; Chandrasekar, M.; Alshammari, B. A.; Fouad, H.; Hashem, M.; Siengchin, S. Thermal characterization of date palm/epoxy composites with fillers from different parts of the tree. *J. Mater. Res. Technol.* **2020**, 9 (6), 15537–15546.

[40] Gheith, M. H.; Aziz, M. A.; Ghori, W.; Saba, N.; Asim, M.; Jawaid, M.; Alothman, O. Y. Flexural, thermal and dynamic mechanical properties of date palm fibres reinforced epoxy composites. *J. Mater. Res. Technol.* **2019**, 8 (1), 853–860.

[41] Özturk, S. Effect of fiber loading on the mechanical properties of kenaf and fiberfrax fiber-reinforced phenol-formaldehyde composites. *J. Compos. Mater.* **2010**, 44 (19), 2265–2288.

[42] Asyraf, M. R.; Ishak, M. R.; Syamsir, A.; Nurazzi, N. M.; Sabaruddin, F. A.; Shazleen, S. S.; Norrrahim, M. N.; Rafidah, M.; Ilyas, R. A.; Abd Rashid, M. Z.; Razman, M. R. Mechanical properties of oil palm fibre-reinforced polymer composites: a review. *J. Mater. Res. Technol.* **2022**, 17, 33–65.

[43] Alshammari, B. A.; Saba, N.; Alotaibi, M. D.; Alotibi, M. F.; Jawaid. M.; Alothman, O. Y. Evaluation of mechanical, physical, and morphological properties of epoxy composites reinforced with different date palm fillers. *Materials.* **2019**, 12 (13), 2145.

[44] Mukhtar, I.; Leman. Z.; Ishak, M. R.; Zainudin, E. S. Sugar palm fibre and its composites: a review of recent developments. *BioResources.* **2016**, 11 (4), 10756–10782.

[45] Costa, R. D.; Rodrigues, A. M.; Silva, L. H. The fruit of peach palm (*Bactris gasipaes*) and its technological potential: an overview. *Food Sci. Technol.* **2022**, 42. https://doi.org/10.1590/fst.82721

[46] Sapuan, S. M.; Ilyas, R. A. Sugar palm: fibers, biopolymers and biocomposites. In *INTROPica*; INTROP: Serdang, Malaysia, **2017**, pp. 5–7.

[47] Huzaifah, M. R. M.; Sapuan, S. M.; Leman, Z.; Ishak, M. R. Comparative study on chemical composition, physical, tensile, and thermal properties of sugar palm fiber (*Arenga pinnata*) obtained from different geographical locations. *BioResources* **2017**, 12, 9366–9382.

[48] Bachtiar, D.; Sapuan, S. M.; Hamdan, M. M. Flexural properties of alkaline treated sugar palm fibre reinforced epoxy composites. *J. Mater. Res. Technol.* **2010**, 1 (1), 79–90.

[49] Rong, M. Z.; Zhang, M. Q.; Liu, Y.; Yang, G. C.; Zeng, H. M. The effect of fiber treatment on the mechanical properties of unidirectional sisal reinforced epoxy composites. *Compos. Sci. Technol.* **2001**, 61, 1437–1447.

[50] Ishak, M. R.; Leman, Z.; Sapuan S. M.; Salleh, M. Y.; Misri, S. The effect of sea water treatment on the impact and flexural strength of sugar palm fibre reinforced epoxy composites. *Int. J. Mech. Mater. Eng.* **2009**, 4 (3), 316–320.

[51] Bachtiar, D.; Sapuan, S. M.; Hamdan, M. M. The influence of alkaline surface fibre treatment on the impact properties of sugar palm fibre-reinforced epoxy composites. *Polym. Plast. Technol. Eng.* **2009**, 48 (4), 379–383.

[52] Wambua, P.; Ivens, J.; Verpoest, I. Natural fibers: can they replace glass in the fibre reinforced plastics? *Compos. Sci. Technol.* **2003**, 63, 1259–1264.

[53] Sahari, J.; Sapuan, S.; Zainudin, E.; Maleque, M. Sugar palm tree: a versatile plant and novel source for biofibres, biomatrices, and biocomposites. *Polym. Renew. Resour.* **2012**, 3 (2), 61.

[54] Monteiro, S. N.; Lopes, F. P. D.; Barbosa, A. P.; Beviroti, A. B.; Silva, I. L. A.; Costa, L. L. Natural lignocellulosic fibers as engineering materials-an overview. *Metall. Mater. Trans A.* **2011**, 42, 2963.

[55] Cordeiro, E. P.; Pita, V. J.; Soares, B. G. Epoxy–fiber of peach palm trees composites: the effect of composition and fiber modification on mechanical and dynamic mechanical properties. *J. Polym. Environ.* **2017**, 25 (3), 913–924.

8 Natural Fibres-Based Bio-Epoxy Composites
Mechanical and Thermal Properties

Carlo Santulli
Università degli Studi di Camerino

Sivasubramanian Palanisamy
Dilkap Research Institute of Engineering
and Management Studies

Shanmugam Dharmalingam
Dr Mahalingam College of Engineering and Technology

CONTENTS

8.1 INTRODUCTION

The use of thermosetting matrix composites appears essential in a number of sectors, such as for nautical, civil, aeronautical, and automotive industry. Even when using mineral fibers, such as basalt, or vegetable ones, such as hemp, flax, etc., as the replacement for glass fibres in composites, thermosetting matrices were prevalently employed. A conspicuous advantage for thermosetting matrices in composites is the ability to be applied with limited pressure and curing at temperatures close to ambient ones, even enhanced by some specific procedures, such as the use of microwaves [1].

Among thermosetting matrices, the most diffuse is definitely epoxy, especially in view of its versatility, which makes it of interest for a number of sectors, ranging from automotive to aerospace, passing through general and construction

engineering and even to nanotechnology [2]. Some of the most significant characteristics that make epoxy a very popular polymer resin are its thermal stability,
effective adhesive strength, and toughness, which allows it to be used in different forms, in particular as a resin for electronic encapsulation, blending, and the
production of composites and nanocomposites, using traditional synthetic fibres,
such as carbon, glass and Kevlar, and natural ones, such as jute, bamboo, coir,
sisal, hemp, flax, etc. [3,4]. In the latter case, some issues may be encountered for
epoxy, which are its relatively low impact strength, low fracture toughness, and
least resistance to crack propagation [5]. A major limitation, which is encountered for epoxy, is in particular its flammability, which requires the application
of flame-retardant additives that often enable obtaining improvements in terms of
other properties, e.g., hardness, impact resistance, wear fatigue, etc. [6]. In this
sense, the use of a number of additional fillers, silica, zinc powder, graphene,
and nanoclay have been proposed [7]. The modification of epoxy would also be
required for some specific applications, which equally extend their profile of use:
in particular, the investigation of tribological properties would be required when
considering the application of bio-wastes with epoxy, among which are different
types of husks or straw, or coconut pith, whose insertion equally belongs to the
wider field of natural fibre composites [8].

Conventional epoxies are based on the presence of at least two epoxide groups
per molecule, to enable crosslinking. Most widely used epoxies in the production
of thermosetting resins are glycidyl ether epoxies, obtained by condensing reactive
hydroxyl-containing compounds, in particular aromatic diols, with epichlorohydrin
[9]. Following this, the reaction with curing agents enables the transformation of
viscous liquid or semi-solid products into solid and rigid thermosetting products.
Originally, the building blocks for epoxy, starting from epichlorohydrin, have been
obtained from petroleum: however, other routes are also possible, which foresee the
use of bio-based building blocks: an example of the relevant respective (from oil and
bio-based) routes are reported in Figure 8.1.

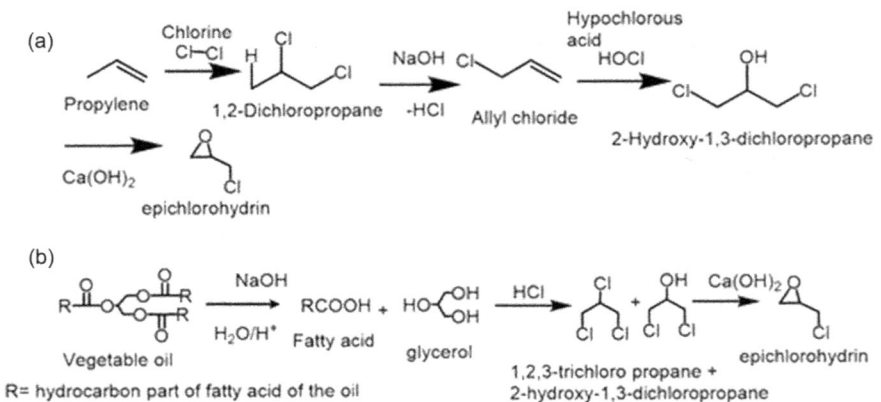

FIGURE 8.1   (a) Petroleum-based route for the synthesis of epichlorohydrin. (b) Vegetable
oil–based route for the synthesis of epichlorohydrin [9].

Two obvious reasons might lead to epoxy formulation of bio-based precursors: first, in a context in which the use of non-renewable resources, such as petroleum, needs to be discouraged, the use of alternatives might be recommended; second, yet confined to some fields of application, such as electronic packaging [10,11] and biomedical ones [12], and in any case for shorter life durations, is the possibility to synthesize epoxy resins that are thermally degradable with facility, hence recyclable [13]. Moreover, the use of bio-epoxy building blocks may lead to further possibilities, such as the reduction of flammability: this is particularly relevant in the case of use of some polyphenols derived from sustainable resources. Lower flammability potential has been observed, for example, for cured epoxidized soybean oil (ESO) with tannic acid (TA) and histidine (His) acting as curing agent and accelerator, respectively [14].

Obtaining some of the building blocks for epoxy with renewable matter led to bio-epoxy. The possible raw materials from the synthesis of bio-epoxies are reported in Figure 8.2.

In practice, bio-epoxy can be summarized as a class of biodegradable resins produced from unsaturated vegetable, saccharides, tannins, cardanol, terpenes, rosins, and lignin [15]. Bio-epoxy is usually combined with organic and inorganic fillers to design composites that can suit varied applications, such as adhesives, resins for bio-composites, coatings, being even adapted to the food sector [16]. Some further details of a number of bio-epoxies and their uses are shown in Table 8.1.

To expand the considerations reported in Table 8.1, it needs to be emphasized that the possibility to obtain various sources of bio-epoxies derived from vegetable and fruit-based materials, such as soybean, castor bean, linseed, rapeseed, sunflower,

Class	Chemical structure	Main sources
Isosorbide-based epoxy	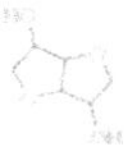	Starch
Furan-based epoxy	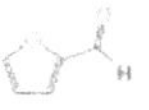	Corn cobs, biomass wastes
Phenolic- and polyphenolic-based epoxy	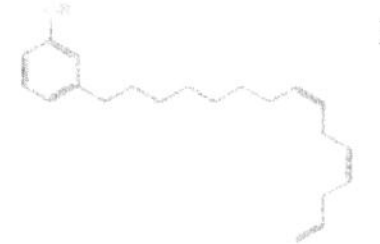	Black Mimosa Bark, Quebracho Wood
Lignin-based epoxy derivatives	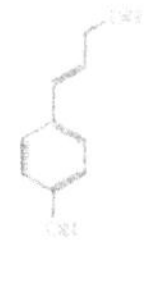	Wood
Rosin-based epoxy	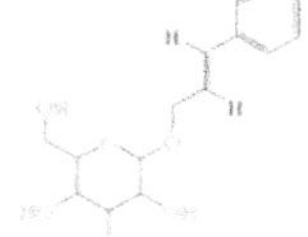	Pine resin

FIGURE 8.2 Chemical structure and main sources of bio-epoxies (original drawing).

TABLE 8.1
Some Bio-Epoxies and Their Uses

Type of Epoxy	Uses
Diglycidyl ethers of isosorbide	epoxy resins in food industry
Epoxidized linseed oil	composites, adhesives, laminates
Furan diepoxy of 2,5-bis(hydroxymethyl)-furan	coatings
Kraft lignin and waste cooking oil (WCO)	asphalt binder
Liquid epoxidized natural rubber	epoxy composites
Terpene-maleic ester type epoxy	composite coatings

cotton, peanut, and palm oils, is well known since a few decades [17]. In particular, the aforementioned conversion of vegetable oils into epoxy substances can be done by oxidizing the fatty residues present in the oils and the oxidization of unsaturated bonds [18]. Miyagawa et al. [19] replaced diglycidyl ether of bisphenol F (DGEBF) in an epoxy resin by epoxidized linseed oil (ELO) amounts from 20 to 100 wt. % by means of an anhydride hardener, after which the thermo-physical and impact properties of such developed bio-based epoxies were investigated. This experimentation proved that the storage modulus, glass transition temperature (T_g), and heat deflection temperature (HDT) decreased with the amount of ELO added. The impact strength remained constant when anhydride curing agent was used, but, in contrast, the impact strength radically increased with an increase in the amount of ELO added if amine hardener was used [20].

8.2 PRODUCTION OF COMPOSITES USING BIO-EPOXY

The production of composites has also been carried out using bio-epoxy, in some cases providing evidence of a performance comparable to the one obtained using conventional epoxies, also exhibiting a significant malleability, so as to allow processing by compression moulding [21]. In the case of using glass fibres, bio-epoxy composites were also successfully tested against seawater resistance, so as to possibly promote their prospective use in ship hulls [22]. Natural fibres are considered from many years as a potential replacement for glass fibres; they have also been tested when introduced in bio-epoxy. The use of jute and hemp has been proposed for instance [23]. Another interesting attempt, to allow enhancing the compatibility of the composite with bio-epoxy, is the use of alternative natural fibres, derived from bio-waste, such as soy stems, treated with various chemicals, e.g., oxalic acid, silane, and sodium hydroxide [24]. In this context, different solutions were adopted to fabricate plant fibre composites. In particular, cashew nutshell liquid blended (the performance of 25% or 30% bio-based were compared) epoxy resins were preferred for use as matrix with jute fibres [25]. A hybrid polymer of epoxy with bio-benzoxazines was used for matrix introducing rice husk and saw dust reinforcement, as two examples of bio-waste to be possibly disposed of in a bio-epoxy [26] (Figure 8.3).

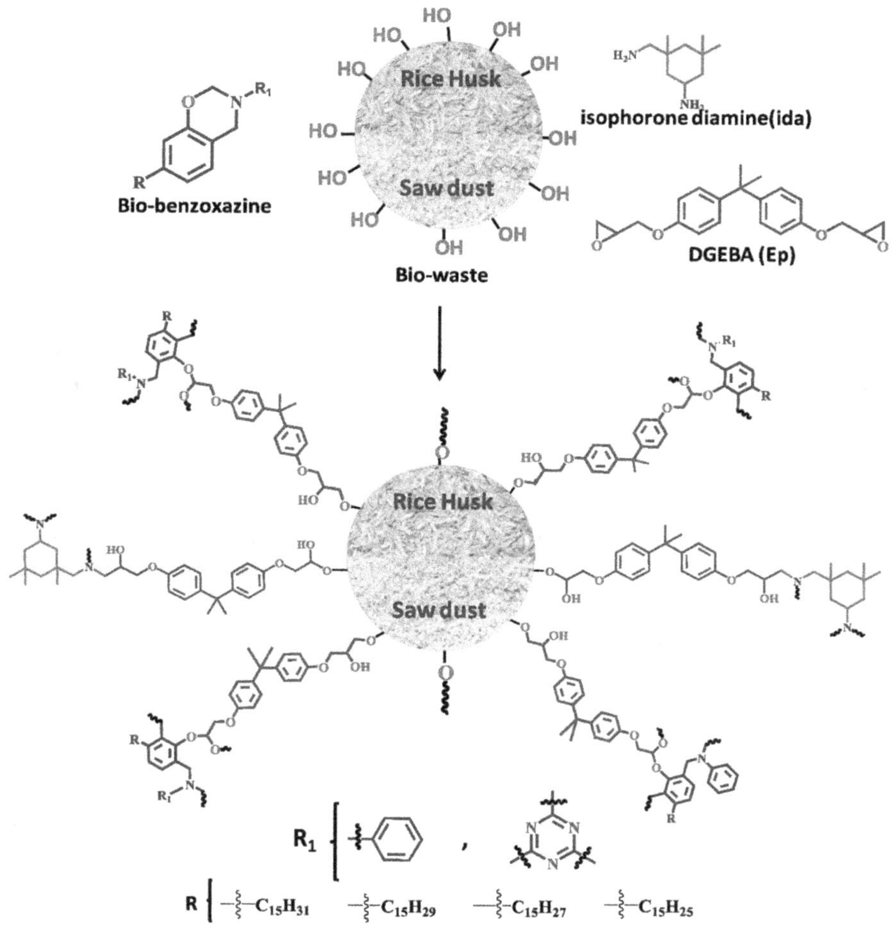

FIGURE 8.3 Rice husk and saw dust reinforced hybrid polybenzoxazine–epoxy composites [26]. (Copyright licence: 54258506410.)

8.3 USE OF NATURAL FIBRES IN BIO-EPOXY COMPOSITES

A large number of natural fibres were used with bio-epoxy matrix for the production of polymer composites: some of these are particularly popular, since they provide reasonably regular fabrics, suitable for obtaining sufficient performance in natural fibre composites. This is the case, e.g., with flax, appropriately improved, when necessary, by the use of fibre treatments, such as with silane [27] or alkali–silane mixtures, the latter proving able to particularly increase the flexural performance of the composites [28]. Another possible treatment, which has been demonstrated to be successful in harsh applications, such as marine ones, is by using sodium bicarbonate [29]. This is a long-term treatment, to be applied over several days' time, which proved effective with *Punica granatum* fibres bio-epoxy resin obtained from cashew nut oil [30]. In other instances, to maintain the regularity of the structure, a possible approach is the fabrication of hybrids, which, despite a higher level of complication, would allow a more thorough control of the composite performance: this has been done, e.g., with bamboo, basalt, and carbon [31], or with roselle and banana and roselle and sisal, in the latter case specifically aiming at automotive application [32].

More specifically, some of the studies that involve the use of commercial bio-epoxy as the replacement of petroleum-based epoxy aimed at the production of lightweight components with optimized fabrication procedures are listed in Table 8.2. In other cases, which are also exposed during the course of this work, bio-epoxy resins are purposely synthesized for the production of composites. In the latter case, it occurs that by-products from other systems, specifically for non-food applications, such as fuel synthesis, e.g., jatropha oil, can be used for the production of bio-epoxies [33].

In this regard, it is widely demonstrated that lignocellulosic fibres have considerable issues in achieving a sufficient compatibility with hydrophobic matrices, such as epoxies, so that they are usually treated for the purpose. This remains true also for bio-epoxies and put considerable limitations on the amount of fibres that can be introduced in the resin to achieve an increase of performance, especially when the fibres have some geometry variations and irregularities. In particular, ELO was produced and hemp fibres were used as the reinforcement along with anhydride as hardener, and it proved not possible to exceed 50 wt.% of fibre content, after which the mechanical performance started to decrease [34].

The use of bio-epoxy adds to the fibres a greater tendency to biodegradation. This in itself would result in a reduction of the properties of the composite overtime. For this reason, the biodegradation is frequently investigated in natural fibre composites with bio-epoxy matrix through water absorption and decomposition tests. Some findings are to be expected in this respect, which deal in mechanical terms with the penetration of water in the resin also during freeze–thaw cycles and, as regards thermal properties, with the variation of degradation temperature, due also to the modification of glass transition temperature of the resin. It was found out that the water

TABLE 8.2
Some Works on Natural Fibre Composites with a Bio-Epoxy Matrix

Fibre	Bio-Epoxy Resin or Component	Reference
Giant reed (*Arundo donax*)	Super Sap® 100	[35]
Hemp (two different grid sizes)	Super Sap® 100	[36]
Hemp sandwich	Super Sap® 100	[37]
Hemp/sisal hybrid	SR Greenpoxy 56®	[38,39]
Flax/basalt/carbon	SR Greenpoxy 56®	[40]
Chicken feathers	SR Greenpoxy 56®	[41]
Kenaf/pineapple	FormuLITE	[42]
Flax, ramie	FormuLITE	[43]
Flax	SuperSap® 300/Recyclamine®301	[44]
Flax/carbon	SuperSap® CLX/Recyclamine® 301	[45]
Jute/basalt	Super Sap® 100	[46]

absorption into flax/bio-epoxy composites, though increasing with higher amounts of reinforcement, did not prove to significantly affect the mechanical performance of the composites under humid and warm environment, where water diffusion did appear to be close to Fick's model [47]. However, it is noteworthy that the real success of bio-epoxy resins in the context of natural fibre composites would only be possibly assessed against the use of conventional epoxy resins. This type of studies is quite rare nonetheless: comparison of two bio-epoxy resins for use in hemp fibre composites was, e.g., performed, yet aimed at the application of a specific manufacturing method, resin transfer moulding (RTM), which requires longer gelling time [48]. A comparative study between the application of an epoxy and a bio-epoxy system on the same natural fibre composite has been realized using 20% and 40% of jute fibres: this demonstrated that, though the bio-epoxy composite showed further water absorption overtime and it got to saturation at higher amounts of water, the moisture behaviour was not drastically modified [49].

8.4 MECHANICAL PROPERTIES OF BIO-EPOXY NATURAL FIBRE COMPOSITES

As it is the case with natural fibre composites, also when the matrix is bio-epoxy, either industrial or purposely synthesized, most studies do concern tensile, flexural, and impact properties of the obtained composites. The use of differently formulated bio-epoxies was investigated for example on 25% volume jute fibre composites to elucidate their effect on tensile strength and modulus, and Charpy impact strength: more specifically, the use of DGEBA-based epoxy resins was compared with vegetable oil synthesized epoxy with three different bio-hardeners [50]. A partial substitution with bio-epoxy was found to be more suitable in terms of mechanical performance.

Another factor that has been proved to be significant for the mechanical properties, especially tensile ones, is the arrangement of the reinforcement, such as unidirectional or random oriented. This was investigated also using bio-based epoxy resin, Supersap by Entropy Resins. In this case, it was possible to find, by using a not very high volume (305) of 3–4 mm long sisal fibres, that stiffness more than doubles the one of the neat resin, from 2.5 to 6 GPa. In contrast, tensile strength was improved by less than the value predictable by the rule of mixtures [51]. A recent trend to provide a tensile performance more suitable for intended applications and at the same time able to develop further the disposal of waste is the possible use as reinforcement of further agrofood refuse. This has been realized, e.g., with the combined introduction of Ceiba pentandra fibres together with keratin-based chicken feathers: for these applications, a bio-epoxy matrix is particularly adapted [52].

In Table 8.3, the tensile strength and stiffness of some bio-epoxy/natural fibre composites is reported. It is apparent from these few data that the main influence is still offered by fibres and their tenor, while the effect of treatment on tensile performance appears to be quite limited.

As regards flexural and impact properties, also in this case the main question, as regards the application of bio-epoxy in natural fibre composites, is their durability under harsh environmental conditions, such as the presence of water or moisture,

TABLE 8.3

Tensile Properties of Some Natural Fibres–Based Bio-Epoxy Composites

Fibres Used	Tensile Strength (MPa)	Tensile Modulus (GPa)	Reference
40 vol.% hemp	63	5.87	[48]
Ficus natalensis bark	33	3	[53]
35 vol.% silanized flax	230	28.8	[28]
35 vol.% alkalized flax	256	28.56	[28]
50 wt.% unidirectional raw flax	222.94	22.30	[54]
50 wt.% non-woven mat flax	76.28	8.04	[54]
Denim/jute (40 wt% total)	43.31	7.45	[55]

as opposed with what is achievable using conventional epoxies. Quite contrary to evidences, in the case of bio-epoxy/flax fibre composites, for an amount of 40 wt.% fibres (768 hours of immersion) flexural strength was improved on an average by 25.5%, while in the composite produced using 55 wt. % flax fibres, a reduction of 20% was observed [56]. This was explained by the possibility that water immersion, together with the relative permeability of bio-epoxy matrix to humidity, would be able to compensate to a point for the relatively ineffective impregnation of the reinforcement by the resin. In another case, the comparison between different hybrids, namely, containing rice husk and coconut shell, rice husk and walnut shell, and coconut shell and walnut shell, all in powder form, in a purposely developed bio-epoxy, led to the observation that the latter combination provided about 10% superior flexural strength than the other two, proving more performing also under impact [57]. It can be noticed that the application of the synthesized bio-epoxy offered some decrease of both properties, though not very significant (normally not exceeding 5%) after moisture absorption in distilled water at 30°C for 5 days.

Another question that appears relevant in the sense of flexural and interlaminar performance of bio-epoxy–natural fibre composites is the stacking sequence adopted to fabricate the laminate, which is directly connected to the adhesion obtained between the matrix and the fibre. This is particularly significant in the case of hybrids: a study on jute (J)/hemp (H) hybrid laminates with SR Greenpoxy 56® bio-epoxy illustrated that the production of hybrids offered some advantage in terms of flexural performance with respect to the pure jute or hemp laminates, with more evidence in the case of H/J/H than it is the case for J/H/J one [23]. This is a demonstration of the fact that also by the use of bio-epoxy it is possible to obtain a positive hybridization effect in flexion, especially considering the higher complication of matrix–fibre interface contact [58]. On the other side, the adoption of bio-epoxy did not impede the achievement of very high performance for flexural strength, as it has been measured on *Mauritia flexuosa*/bio-epoxy composites, when fabricated by vacuum-assisted RTM, in which case a value of 253.7 MPa was obtained [59]. In the same way, and in relation with fibre–matrix interfacial properties, the evaluation of impact properties on bio-epoxy natural composites does mainly concern the weathering effect, which was demonstrated to produce some degradation of performance, as in the case of kenaf/sisal hybrids, where, depending on the configuration, it was never lower than 10% [60].

8.5 THERMAL PROPERTIES OF BIO-EPOXY/NATURAL FIBRE COMPOSITES

The thermal properties of bio-epoxy/natural fibre composites are investigated in the twofold sense of assessing the degradation curves of lignocellulosic materials by thermogravimetric analysis, and to highlight how this would interfere with the properties of the resin with its own glass transition temperature and degradation patterns. It is also important to verify which can be the temperature limit for their application in terms of increase of the mechanical losses, leading to the collapse of the resin, as obtained by the measurement of the complex modulus. It was suggested that the longer aliphatic chains together with the lower crosslinking density would result in inferior thermal properties for bio-epoxy resins, as compared with conventional ones [61].

To try to improve the thermal properties of these composites, namely, to increase the degradation temperature, which would extend the application range, different treatments, even very aggressive ones, were experimented on the fibres, such as in the case for Morinda citrifolia fibres, wherein alkali, silane, and nitric acids were applied, with modest improvements though [62]. In contrast, the inclusion of nanoparticles in the bio-epoxy resin, though a challenging task for the difficulty of avoiding agglomeration, can on the other side result in composites with good load sharing and bearing properties and also improving the thermal behaviour [63]. Also, cellulose-based nanofillers can be used in case where the tribological properties need to be coupled with mechanical properties and delay in thermal degradation [64].

Another aspect that can eventually promote or conversely dissuade users from the application of bio-epoxy are dynamical mechanical properties (DMA), which are linked to the potential employment of the resin in temperature. A limited number of studies on DMA of bio-epoxies does exist, in particular concerning the use of epoxidized hemp oil (EHO) resin, which is of interest as a by-product of the same economical system, which can lead to the production of bio-epoxy/hemp composites: the specific resin showed some improvement in the maintenance of some storage modulus over the temperature with respect to competing bio-epoxies, such as ESO [65]. A thorough recent comparison of five epoxidized bio-resins, some of which are blended with other bio-resins, such as poly(lactic acid) (PLA) or loaded with flax fibres, was also carried out: the latter showed a better matrix–fibre compatibility, indicating from tan δ values a potential use of it up to 148°C [66]. DMA studies were able to clarify the levels of compatibility achieved by different procedures of silane treatment, namely, TEMPO oxidation [67] and pre-hydrolyzed one, indicating the superiority of the former, though with very limited differences [27].

8.6 CONCLUSIONS

The potential that bio-epoxies have into replacing conventional epoxies for the production of natural fibre composites needs to be assessed under a number of aspects, namely, mechanical (e.g., tensile, flexural and impact) and thermal performance (composite degradation, loss of mechanical performance as the effect of temperature). Studies comparing bio-epoxies with conventional ones are limited, though the number of natural fibres used for the production of natural fibre composites is countless, and some of these, not necessarily amongst the most diffuse, have also been applied in combination with bio-epoxies (industrial ones or synthesized for the purpose). Despite this, bio-epoxies, in general terms, appear to be applicable in the production of biocomposites, though a number of aspects need to be taken care of. These include stacking sequence, fibre treatment, possible production of hybrids with two different fibres, or with other fillers (e.g., cellulose materials, eggshell fragments, chicken feathers). The matter needs further investigation, especially in view of the availability of more comparative studies on biocomposites obtained with different bio-epoxy resins, maintaining equal all other conditions regarding the reinforcement and processing.

REFERENCES

1. C. O. Mgbemena, D. Li, M. F. Lin, P. D. Liddel, K. B. Katnam, V. K. Thakur, H. Y. Nezhad (2018). Accelerated microwave curing of fibre-reinforced thermoset polymer composites for structural applications: A review of scientific challenges. *Composites Part A: Applied Science and Manufacturing*, 115, 88–103.
2. Z. Ahmadi (2019). Epoxy in nanotechnology: A short review. *Progress in Organic Coatings*, 132, 445–448.
3. P. Mohan (2013). A critical review: The modification, properties, and applications of epoxy resins. *Polymer-Plastics Technology and Engineering*, 52(2), 107–125.
4. S.K. Mazumdar (2002), *Composites Manufacturing*, CRC Press LLC, Boca Raton.
5. S.M. Rangappa, S. Siengchin, H.N. Dhakal (2020). Green-composites: Eco friendly and sustainability. *Applied Science and Engineering Progress*, 13, 183–184, doi: 10.14416/j. asep.2020.06.001.
6. S.V. Levchik, E.D. Weil (2004). Thermal decomposition, combustion and flame-retardancy of epoxy resins – a review of the recent literature. *Polymer International*, 53(12), 1901–1929.
7. J. Gao, J. Li, B. C. Benicewicz, S. Zhao, H. Hillborg, L. S. Schadler (2012). The mechanical properties of epoxy composites filled with rubbery copolymer grafted SiO_2. *Polymers*, 4, 187–210.
8. N. Saba, M. Jawaid, O. Y. Alothman, M. T. Paridah, A. Hassan (2016). Recent advances in epoxy resin, natural fiber-reinforced epoxy composites and their applications. *Journal of Reinforced Plastics and Composites*, 35(6), 447–470.
9. N. Karak (2021). *Overview of Epoxies and Their Thermosets. In Sustainable Epoxy Thermosets and Nanocomposites* (pp. 1–36). American Chemical Society, Washington, DC.
10. M. L. Sham, J. K. Kim (2004). Evolution of residual stresses in modified epoxy resins for electronic packaging applications. *Composites Part A: Applied Science and Manufacturing*, 35(5), 537–546.

11. X. Chen, S. Chen, Z. Xu, J. Zhang, M. Miao, D. Zhang (2020). Degradable and recyclable bio-based thermoset epoxy resins. *Green Chemistry*, 22(13), 4187–4198.
12. S. Bobby, M. A. Samad (2019). Epoxy composites in biomedical engineering. In *Materials for Biomedical Engineering* (pp. 145–174). Elsevier, Amsterdam.
13. A. Takahashi, T. Ohishi, R. Goseki, H. Otsuka (2016). Degradable epoxy resins prepared from diepoxide monomer with dynamic covalent disulfide linkage. *Polymer*, 82, 319–326.
14. M. Qi, Y.-J. Xu, W.-H. Rao, X. Luo, L. Chen, Y.-Z. Wang (2018), Epoxidized soybean oil cured with tannic acid for fully bio-based epoxy resin. *RSC Advances* 8, 26948–26958.
15. E. Ramon, C. Sguazzo, P. Moreira (2018), A review of recent research on bio-based epoxy systems for engineering applications and potentialities in the aviation sector. *Aerospace*, 5, 110.
16. E.A. Baroncini, S.K. Yadav, G.R. Palmese, J.F. Stanzione (2016), Recent advances in bio-based epoxy resins and bio-based epoxy curing agents. *Journal of Applied Polymer Science*, 133, 44103.
17. V. Sharma, P. P. Kundu (2006). Addition polymers from natural oils—A review. *Progress in Polymer Science*, 31(11), 983–1008.
18. A. Köckritz, A., A. Martin (2008). Oxidation of unsaturated fatty acid derivatives and vegetable oils. *European Journal of Lipid Science and Technology*, 110(9), 812–824.
19. H. Miyagawa, A. K. Mohanty, M. Misra, L. T. Drzal (2004). Thermo-physical and impact properties of epoxy containing epoxidized linseed oil 1. Anhydride cured epoxy. *Macromolecular Materials and Engineering*, 289(7), 629–635.
20. H. Miyagawa, A. K. Mohanty, M. Misra, L. T. Drzal (2004). Thermo-physical and impact properties of epoxy containing epoxidized linseed oil, 2. Amine cured epoxy. *Macromolecular Materials and Engineering*, 289(7), 636–641.
21. Y. Liu, B. Wang, S. Ma, T. Yu, X. Xu, Q. Li, S. Wang, Y. Han, Z. Yu, J. Zhu (2021). Catalyst-free malleable, degradable, bio-based epoxy thermosets and its application in recyclable carbon fiber composites. *Composites Part B: Engineering*, 211, 108654.
22. J. A. Velasco-Parra, B. A. Ramon-Valencia, A. Lopez-Arraiza (2022). Effects of seawater immersion on a glass fibre reinforced bioepoxy mechanical properties and its application in the ship hull finite-element analysis. *Proceedings of the Institution of Mechanical Engineers, Part L: Journal of Materials: Design and Applications*, 236(1), 147–154.
23. A. Vinod, Jiratti Tengsuthiwat, Yashas Gowda, R. Vijay, M.R. Sanjay, S. Siengchin, H. N. Dhakal (2022). Jute/Hemp bio-epoxy hybrid bio-composites: Influence of stacking sequence on adhesion of fiber-matrix. *International Journal of Adhesion and Adhesives*, 113, 103050.
24. A. Vinod, M.R. Sanjay, S. Siengchin, S. Fischer (2021). Fully bio-based agro-waste soy stem fiber reinforced bio-epoxy composites for lightweight structural applications: Influence of surface modification techniques. *Construction and Building Materials*, 303, 124509.
25. B. Shivamurthy, N. Naik, B.H.S. Thimappa, R. Bhat (2020). Mechanical property evaluation of alkali-treated jute fiber reinforced bio-epoxy composite materials, *Materials Today: Proceedings*, 28, 2116–2120.
26. H. Arumugam, B. Krishnasamy, G. Perumal, A. Anto Dilip, M.I. Abdul Aleem, A. Muthukaruppan (2021), Bio-composites of rice husk and saw dust reinforced bio-benzoxazine/epoxy hybridized matrices: Thermal, mechanical, electrical resistance and acoustic absorption properties. *Construction and Building Materials*, 312, 125381.
27. B. Fathi, M. Foruzanmehr, S. Elkoun, M. Robert (2019). Novel approach for silane treatment of flax fiber to improve the interfacial adhesion in flax/bio epoxy composites. *Journal of Composite Materials*, 53(16), 2229–2238.

28. D. Perremans, Y. Guo, J. Baets, A. W. Van Vuure, I. Verpoest (2014). Improvement of the interphase strength and the moisture sensitivity of flax fibre reinforced bio-epoxies: Effect of various fibre treatments. In *Proceedings of the ECCM-16 European Conference on Composite Materials*, Seville, Spain (pp. 22–26).

29. V. Fiore, T. Scalici, F. Nicoletti, G. Vitale, M. Prestipino, A. Valenza (2016). A new eco-friendly chemical treatment of natural fibres: Effect of sodium bicarbonate on properties of sisal fibre and its epoxy composites. *Composites Part B: Engineering*, 85, 150–160.

30. D. Zindani, S. Kumar, S. R. Maity, S. Bhowmik (2021). Mechanical characterization of bio-epoxy green composites derived from sodium bicarbonate treated Punica granatum short fiber agro-waste. *Journal of Polymers and the Environment*, 29(1), 143–155.

31. K. Yorseng, S. M. Rangappa, J. Parameswaranpillai, S. Siengchin (2022). Towards green composites: Bioepoxy composites reinforced with bamboo/basalt/carbon fabrics. *Journal of Cleaner Production*, 363, 132314.

32. D. Chandramohan, J. Bharanichandar (2013). Natural fiber reinforced polymer composites for automobile accessories. *American Journal of Environmental Sciences*, 9(6), 494.

33. S.K. Sahoo, V. Khandelwal, G. Manik (2019). Sisal fibers reinforced epoxidized nonedible oils based epoxy green composites and its potential applications. In S. S. Muthu (ed.), *Green Composites* (pp. 73–102). Springer, Singapore.

34. N. Boquillon (2006). Use of an epoxidized oil-based resin as matrix in vegetable fibers-reinforced composites. *Journal of Applied Polymer Science*, 101, 4037–4043.

35. T. Scalici, V. Fiore, A. Valenza (2016). Effect of plasma treatment on the properties of *Arundo donax* L. leaf fibres and its bio-based epoxy composites: A preliminary study. *Composites Part B: Engineering*, 94, 167–175.

36. L. Boccarusso, M. Durante, A. Langella (2018). Lightweight hemp/bio-epoxy grid structure manufactured by a new continuous process. *Composites Part B: Engineering*, 146, 165–175.

37. R. Dragonetti, M. Napolitano, L. Boccarusso, M. Durante (2020). A study on the sound transmission loss of a new lightweight hemp/bio-epoxy sandwich structure. *Applied Acoustics*, 167, 107379.

38. S. M. K. Thiagamani, S. Krishnasamy, C. Muthukumar, J. Tengsuthiwat, R. Nagarajan, S. Siengchin, S. O. Ismail (2019). Investigation into mechanical, absorption and swelling behaviour of hemp/sisal fibre reinforced bioepoxy hybrid composites: Effects of stacking sequences. *International Journal of Biological Macromolecules*, 140, 637–646.

39. K. Senthilkumar, T. Ungtrakul, M. Chandrasekar, T. Senthil Muthu Kumar, N. Rajini, S. Siengchin, H. Pulikkalparambil, J. Parameswaranpillai, N. Ayrilmis (2021). Performance of sisal/hemp bio-based epoxy composites under accelerated weathering. *Journal of Polymers and the Environment*, 29, 624–636.

40. T. G. Yashas Gowda, A. Vinod, P. Madhu, V. Kushvaha, M. R. Sanjay, S. Siengchin (2021). A new study on flax-basalt-carbon fiber reinforced epoxy/bioepoxy hybrid composites. *Polymer Composites*, 42(4), 1891–1900.

41. J. Bessa, J. Souza, J. B. Lopes, J. Sampaio, C. Mota, F. Cunha, R. Fangueiro (2017). Characterization of thermal and acoustic insulation of chicken feather reinforced composites. *Procedia Engineering*, 200, 472–479.

42. S. Kumar, A. Saha, S. Bhowmik (2022). Accelerated weathering effects on mechanical, thermal and viscoelastic properties of kenaf/pineapple biocomposite laminates for load bearing structural applications. *Journal of Applied Polymer Science*, 139(2), 51465.

43. S. Kumar, D. Zindani, S. Bhowmik (2020). Investigation of mechanical and viscoelastic properties of flax-and ramie-reinforced green composites for orthopedic implants. *Journal of Materials Engineering and Performance*, 29(5), 3161–3171.

44. G. Cicala, E. Pergolizzi, F. Piscopo, D. Carbone, G. Recca (2018), Hybrid composites manufactured by resin infusion with a fully recyclable bioepoxy resin. *Composites Part B: Engineering*, 132, 69–76.
45. G. Cicala, A. D. La Rosa, A. Latteri, R. Banatao, S. Pastine (2016). The use of recyclable epoxy and hybrid lay up for biocomposites: Technical and LCA evaluation. In *Proceedings of the CAMX*.
46. V. Fiore, T. Scalici, D. Badagliacco, D. Enea, G. Alaimo, A. Valenza (2017). Aging resistance of bio-epoxy jute-basalt hybrid composites as novel multilayer structures for cladding. *Composite Structures*, 160, 1319–1328.
47. A. Moudood, A. Rahman, H. Mohammad Khanlou, W. Hall, A. Öchsner, G. Francucci (2019). Environmental effects on the durability and the mechanical performance of flax fiber/bio-epoxy composites. *Composites Part B: Engineering*, 171, 284–293.
48. L. Di Landro, G. Janszen (2014). Composites with hemp reinforcement and bio-based epoxy matrix. *Composites Part B: Engineering*, 67, 220–226.
49. R. Masoodi, K. M. Pillai (2012). A study on moisture absorption and swelling in bio-based jute-epoxy composites. *Journal of Reinforced Plastics and Composites*, 31(5), 285–294.
50. A.K. Bledzki, M. Urbaniak, A. Boettcher, C. Berger, R. Pilawka (2013). Bio-based epoxies and composites for technical applications. *Key Engineering Materials*, 559, 1–6.
51. A. Mancino, G. Marannano, B. Zuccarello (2018). Implementation of eco-sustainable bio composite materials reinforced by optimized agave fibers. *Procedia Structural Integrity*, 8, 526–538.
52. S. M. Rangappa, J. Parameswaranpillai, S. Siengchin, M. Jawaid, T. Ozbakkaloglu (2022). Bioepoxy based hybrid composites from nano-fillers of chicken feather and lignocellulose Ceiba Pentandra. *Scientific Reports*, 12(1), 1–18.
53. S. Rwawiire, B. Tomkova, J. Militky, A. Jabbar, B. M. Kale (2015). Development of a biocomposite based on green epoxy polymer and natural cellulose fabric (bark cloth) for automotive instrument panel applications. *Composites Part B: Engineering*, 81, 149–157.
54. C. Avril, P. A. Bailly, J. Njuguna, E. Nassiopoulos, A. De Larminat (2012). Development of flax-reinforced bio-composites for high-load bearing automotive parts. In *Proceeding of European Conference on Composite Materials (ECCM)*, Venice, Italy (vol. 2428).
55. R. Temmink, B. Baghaei, M. Skrifvars (2018). Development of biocomposites from denim waste and thermoset bio-resins for structural applications. *Composites Part A: Applied Science and Manufacturing*, 106, 59–69.
56. E. Muñoz, J. A. García-Manrique (2015). Water absorption behaviour and its effect on the mechanical properties of flax fibre reinforced bioepoxy composites. *International Journal of Polymer Science*, 2015, 390275.
57. D. Chandramohan, A. J. P. Kumar (2017). Experimental data on the properties of natural fiber particle reinforced polymer composite material. *Data in Brief*, 13, 460–468.
58. C. Santulli (2019). Mechanical and impact damage analysis on carbon/natural fibers hybrid composites: A review. *Materials*, 12(3), 517.
59. W. J. M. Espinosa, B. A. R. Valencia, G. G. M. Contreras (2019). Physical-mechanical characterization of moriche natural fibre (*Mauritia flexuosa*) and composite with bioepoxy resin. *Strojniski Vestnik/Journal of Mechanical Engineering*, 65(3), 181–188.
60. K. Yorseng, S. M. Rangappa, H. Pulikkalparambil, S. Siengchin, J. Parameswaranpillai (2020). Accelerated weathering studies of kenaf/sisal fiber fabric reinforced fully bio-based hybrid bioepoxy composites for semi-structural applications: Morphology, thermomechanical, water absorption behavior and surface hydrophobicity. *Construction and Building Materials*, 235, 117464.

61. X. Q. Liu, W. Huang, Y. H. Jiang, J. Zhu, C. Z. Zhang (2012), Preparation of a bio-based epoxy with comparable properties to those of petroleum-based counterparts. *Express Polymer Letters*, 6(4), 293–298.

62. A. Vinod, M. R. Sanjay, S. Siengchin (2021). Fatigue and thermo-mechanical properties of chemically treated *Morinda citrifolia* fiber-reinforced bio-epoxy composite: A sustainable green material for cleaner production. *Journal of Cleaner Production*, 326, 129411.

63. S. K. Bobade, N. R. Paluvai, S. Mohanty, S. K. Nayak (2016). Bio-based thermosetting resins for future generation: A review. *Polymer-Plastics Technology and Engineering*, 55(17), 1863–1896.

64. B. Barari, E. Omrani, A. D. Moghadam, P. L. Menezes, K. M. Pillai, P. K. Rohatgi, P. K. (2016). Mechanical, physical and tribological characterization of nano-cellulose fibers reinforced bio-epoxy composites: An attempt to fabricate and scale the 'Green' composite. *Carbohydrate Polymers*, 147, 282–293.

65. N. W. Manthey, F. Cardona, G. Francucci, T. Aravinthan, T. (2013). Thermo-mechanical properties of epoxidized hemp oil-based bioresins and biocomposites. *Journal of Reinforced Plastics and Composites*, 32(19), 1444–1456.

66. C. Di Mauro, A. Genua, M. Rymarczyk, C. Dobbels, S. Malburet, A. Graillot, A. Mija (2021). Chemical and mechanical reprocessed resins and bio-composites based on five epoxidized vegetable oils thermosets reinforced with flax fibers or PLA woven. *Composites Science and Technology*, 205, 108678.

67. B. Fathi, M. Harirforoush, M. Foruzanmehr, S. Elkoun, M. Robert (2017). Effect of TEMPO oxidation of flax fibers on the grafting efficiency of silane coupling agents. *Journal of Materials Science*, 52(17), 10624–10636.

9 Natural Fiber/ Epoxy-Based Hybrid Composites
Thermal and Mechanical Properties

Sabarish Radoor
Jeonbuk National University
King Mongkut's University of Technology

Amritha Bemplassery
National Institute of Technology

Aswathy Jayakumar
King Mongkut's University of Technology
Kyung Hee University

Jasila Karayil
Government Engineering College

Jyothi Mannekote Shivanna
AMC Engineering College

Jun Tae Kim
Kyung Hee University

Jaewoo Lee
Jeonbuk National University

Jyotishkumar Parameswaranpillai
Alliance University

Suchart Siengchin
King Mongkut's University of Technology

DOI: 10.1201/9781003271017-9

CONTENTS

9.1   INTRODUCTION

In recent years, there is an increase in the demand for natural fibers for applications in different sectors such as households, automotive, packaging, building, construction, etc. The chemical composition of natural fibers includes cellulose, pectin, lignin, hemicellulose, fat and water-soluble substances. They are cheap and less toxic than synthetic fibers. Moreover, they are lightweight, less abrasive, non-toxic, biodegradable, easily available, recyclable, and less dense [1,2]. Therefore, in recent years, the replacement of synthetic fibers (glass, aramid, and carbon) with natural fibers is gaining interest and profound research has been carried out on the development of natural fiber (hemp, abaca, sisal, coir, pineapple, bamboo, banana, cotton, jute, flax, etc.)-reinforced polymer matrix for industrial applications [3,4]. Unfortunately, natural fibers have several drawbacks such as low thermal stability, poor mechanical properties, high moisture uptake and enmity with hydrophobic polymer matrix [5–7]. The physical/mechanical properties of natural and synthetic fiber are shown in Table 9.1 [8,9]. It is clearly observed from Table 9.1 that the mechanical properties

TABLE 9.1
Mechanical Properties of Natural Fibers

Fiber	Density (g/cm^3)	Young's Modulus (GPa)	Tensile Strength (MPa)	Elongation at Break (%)
Flax	1.43–1.52	27.5–85	345–2000	1–4
Ramie	1.48	61.4–128	925	1.2–3.8
Hemp	1.4–1.5	58–70	530	1.6
Jute	1.46	13–26.5	393–800	1.16–1.5
Sisal	1.16–1.5	9.4–22	350–700	3–7
Coir	1.2	4–6	175–220	15–30
Cotton	1.5–1.6	5.5–12.6	330–585	7–8
Curaca	1–1.3	11.8	825	3.7–4.3
Kenaf	1.5	53	743	1.6
Pineapple	1.32	60–82	1020	2.4
Bamboo	0.6–1.1	1–17	140–230	-
Banana	1.3	29	500–700	3
Kevlar	1–1.4	60	2000–3000	3.8
Carbon	1.4	240–425	4000	1.4–1.8

of natural fiber are comparatively lesser than synthetic fiber. This complication could be overthrown by surface treatment of natural fibers. As the chemical treatment enhances the surface roughness and hydrophobicity of the fiber, the surface-treated natural fiber has excellent bonding with the hydrophobic polymer matrix [10–12].

9.2 MECHANICAL PROPERTIES OF NATURAL FIBER/ EPOXY-BASED HYBRID COMPOSITES

Alkali treatment, acetylation, benzoylation, peroxide treatment, etc. are the commonly used chemical treatments for natural fibers [13,14]. Bledzki et al. [15] modified flax fiber using acetylation treatment. The modified fiber is superior to unmodified fiber in terms of properties such as thermal and mechanical. During acetylation, the wax, lignin and hemicellulose present on the surface of the fiber was removed. This improved the interfacial adhesion and thereby improved the mechanical properties such as tensile and flexural properties and thermal behavior of the material. However, high degree of acetylation (18%) results in the degradation of cellulose and leads to the formation of cracks in the fiber. Therefore, the tensile strength decreases at high degree of acetylation. Seki [16] investigated the effect of both alkali and siloxane treatment on the tensile and flexure strength of jute/polyester composite. They observed that surface-treated polyester composites displayed high tensile and flexural strength. The interlaminar shear strength (ILSS) value for untreated fiber was 1.5 MPa. Meanwhile, the alkali- and siloxane-treated fibers show a high ILSS value of 14.2 and 24 MPa, respectively (Figure 9.1). As a result, surface treatment improves the interfacial bonding between the fiber and the polymer matrix. The surface treatment also enhances the effective surface area of the fiber and therefore a better mechanical interlocking is achieved for surface-treated fiber. The excellent interfacial bonding between fiber and matrix was confirmed through scanning electron microscopy (SEM) analysis. Less fiber pull-out and good adherence of the fiber with the matrix were observed in the SEM image.

Singh and co-workers [17] investigated the influence of chemical modification to strengthen the adhesion between sisal fiber and polyester by chemical treatments such as silane, organ titanate, N-substituted methacrylamide and zirconate. These coupling agents can deposit on the surface as well as on the interfibrillar regions and thereby prevent the moisture uptake of the fiber. Hence, the surface modification enhances the hydrophobicity of the fiber. Consequently, surface-treated composite exhibit improved physicomechanical properties such as density, tensile strength, elongation, energy to break, flexural strength and flexural modulus. The authors also monitored the consequence of humidity on the flexural strength of both untreated and surface-treated composites. On exposure to wet environment, the flexural strength of the composite declined by 50%–70%. Abbas et al. [18] employed organic waste (cascara/testa) and E-glass fiber to reinforce the epoxy matrix and investigate its effect on tensile, flexural, impact and water absorption properties of the composite. The result point out that the hybrid composite is a promising candidate for industrial applications as it possesses desirable tensile and flexural strength. Meanwhile, Neuba et al. [19] chose natural lignocellulosic fiber (NLF) *Cyperus malaccensis* (CM) sedge fiber

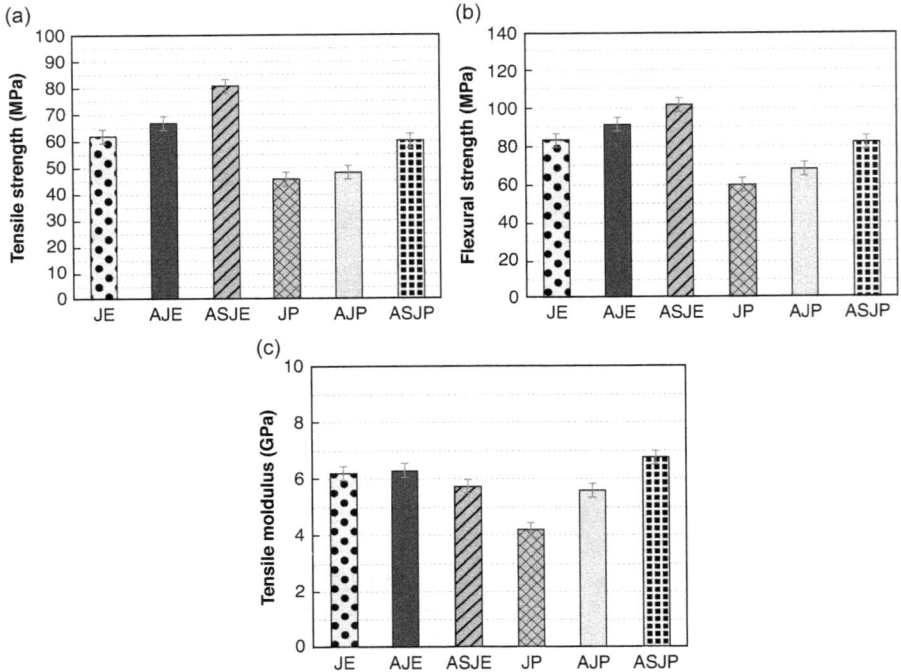

FIGURE 9.1 The influence of surface modification on of jute/polyester composite: (a) tensile strength, (b) flexural strength and (c) flexural modulus. (Reproduced with permission from Elsevier, License Number: 5217940236235.)

as reinforcing agent for epoxy composites. They observed that incorporating the CM fibers enhances the toughness and Izod impact strength of the composite. The good thermal stability along with its desirable mechanical property makes CM fiber a suitable substitute for synthetic fiber.

Yorseng et al. [20] revealed the influence of accelerated weathering on the physical properties of neat epoxy, kenaf/sisal and hybrid kenaf/sisal composite. The experimental studies show that the hybrid composite retained its properties even after subjected to accelerated weathering. Tripathy and co-workers [21] studied the influence of surface modification on the physical and mechanical properties of date palm stem fiber-reinforced epoxy composite. They observed that with enhancement in the fiber content from 0 to 10 wt%, the tensile strength also increases and reaches a maximum value of 29.03 MPa. As a result, a favorable interfacial bonding between the fiber and matrix was achieved. However, high fiber content causes destitute compatibility between the fiber and matrix, and consequently the tensile strength decreases. A similar trend was noticed for flexural strength. Their studies further show that hybrid composite treated with 5% NaOH exhibit good mechanical property. Sumesh et al. [22] investigated the effect of bio-fillers such as pineapple fly ash (PFA), banana fly ash (BFA) and coir fly ash (CFA) on the mechanical properties of sisal (S)/pineapple (P) hybrid fiber composites. The tensile strength value increases from 23 to 33 MPa with the incorporation of BFA, PFA and CFA filler as compared with neat natural fiber composites. As a result, a better adhesion between the fiber and matrix is achieved. Besides this, the impact property, hardness and density of the composite also increase with the addition of filler. The same group improved the mechanical properties of natural fiber s by incorporating it with nanofiller, TiO_2. The fiber loaded with 3% of TiO_2 possesses high tensile, impact and flexural strength. The tensile property values of nanofiller-loaded sisal/banana, banana/pineapple and pineapple/sisal fiber are 42.94, 45.25 and 48.10 MPa, respectively. This enhancement could be due to better dispersion of nano-TiO_2 filler as well as due to the excellent interaction between fiber /filler and matrix. Beyond 3% filler, a reduction in mechanical property was noted, probably due to irregular dispersion or agglomeration of nanofillers [23]. Prasad et al. [24] developed a hybrid composite of hemp and nettle fiber-reinforced polyester composites. The hybrid composites exhibit high tensile strength value than neat natural fiber composites. The tensile strength, bending strength and impact energy of hybrid polyester composite are 72.67 MPa, 16.89 MPa and 3.2 J, respectively. The wear resistance of the composite could be tuned by fiber loading. Composite loaded with high fiber content has less wear resistance. Kavya et al. [25] thoroughly investigated the effect of TiC and fly ash filler on the tensile, flexural, impact strength and inter-laminar shear of coir epoxy hybrid composite. From the results, it is clear that the introduction of TiC and fly ash into coir/epoxy composite significantly improves its mechanical property.

9.3 THERMOGRAVIMETRIC ANALYSIS (TGA) OF NATURAL FIBER/EPOXY-BASED HYBRID COMPOSITES

TGA is a highly potent analytical method used for the measurement of thermal stability of materials including the polymers. In this technique, the specimen mass was changed with rise in temperature [26–28]. Thermogram is plotted with percentage weight loss as a function of temperature. TGA detects the moisture and volatile contents of a sample [26,29]. This method is beneficial in extracting information about physical phenomena such as absorption, adsorption and phase transitions, as well as chemical phenomena like oxidation, reduction, chemisorption, thermal desorption, etc.[30]. Another important application of TGA would be to determine the decomposition temperature, purity/composition of samples and also drying temperatures. In differential thermogravimetry (DTG) or differential thermal analysis, the temperature difference between specimen and inert reference is calculated as a function of temperature [26]. This method is useful in measuring the melting point of substances and temperatures of transitions. The maximum temperature limit for TGA is 1000°C, while that for DTG is as high as 1600°C.

The demand for the use of greener environmentally friendly natural products or fibers has enhanced nowadays. Reinforcement of natural fibers using coir, coconut shell, sisal, jute fibers, etc. has been well documented [29,31,32]. Renewability, less cost and biodegradability are some of the numerous advantages of natural fibers (Singh et al., 2020). Increased thermal, mechanical and flexibility properties of design are guaranteed for hybrid fiber-reinforced composites compared to individual fibers [33]. Coir fibers possess many attractive properties like renewability, biodegradability, highest elongation, lowest thermal conductivity, etc. [32]. These properties enable to use coir fibers as building material composites and for low energy-consuming applications. Carbon fibers enhance the power of composites, as they are stiff, strong and have higher stiffness-to-weight ratio compared to steel or plastic [32,34]. Although numerous studies are focused on the coir fiber epoxy resin-reinforced polymer composites, the influence of carbon fibers on the characteristics of hybrid composites is still less explored. Singh et al. studied the development and characterizations of coir/carbon-reinforced epoxy resin hybrid composites and investigated the possibility of fabrication of helmet shells according to their thermal and mechanical properties [32,34].

As a part of thermal characterization, thermogravimetric analyzer was employed to evaluate the thermal behavior of coir fiber/carbon fiber/epoxy resin hybrid composites with respect to temperature. The samples were heated at 20°C/min to a maximum of 800°C under nitrogen atmosphere. Corresponding weight losses were recorded and plotted against temperature. The results were presented with 30%, 20% and 10% fiber loading. The hemicelluloses content of alkali-treated carbon/coir/epoxy resin hybrid composite was found to be lower as evident from less weight loss due to low moisture absorption. The presence of carbon fiber imparts higher thermal stability [35]. This thermal stability was accountable for the thermal expansion contraction and moisture absorption rate. Owing to the higher coir fiber weight percentage, 30% fiber loading was observed to have maximum weight loss.

The worldwide annual production of date palm was estimated to be about 40% more than that of coir. Date palm fiber (DPF) is multicellular and resembles coir fiber in shape and structure. It contains lignin, cellulose, fat, etc. [36,37]. These fibers find wide applications as filler in thermosetting and thermoplastic materials, in industries, automobile fields, etc. [38,39]. Gheith et al. investigate the influence of DPF loading with different weight percentages such as 40%, 50% and 60% on mechanical properties, i.e., modular strength and dynamic mechanical properties. In these studies, they used DPF as a filler for the development of DPF/epoxy composites with varying concentrations. TGA was conducted to estimate the thermal stability of epoxy and DPF/epoxy composites. The experiment was carried out at 20°C/min in the range of 30°C–700°C under room temperature.

TGA of pure epoxy composites was recorded and compared with that of reinforced samples. The results revealed that at 305.02°C, there is a weight loss of 65.11% observed for pure epoxy composites and the remaining residue was only 9.58%. The mass loss at high temperature (360 and 500 °C) corresponds to the scission of the aliphatic bonds of bisphenol A [39,40]. However, the thermogram of 40% DPF-filled samples displayed higher mass loss around 70.99% and final residue of 12.51% at a temperature of 299.72°C. The third mass loss was higher compared to pure epoxy composites on account of lignin content of DPF. 50% loaded palm fiber-reinforced composite showed higher thermal stability at temperature 316.9°C. Though 60% loaded palm fiber composites also exhibit higher mass content around 15.2%, 50% loading was the highest. It could be concluded that all DPF-incorporated samples show maximum mass loss at higher temperatures, which could be accounted for the thermal degradation of cellulose, pectin, lignin and glycosidic linkages of natural fibers (Figure 9.2) [41]. Addition of palm fiber into epoxy composites enhanced the thermal stability and 50% DPF/epoxy composites showed better residual content.

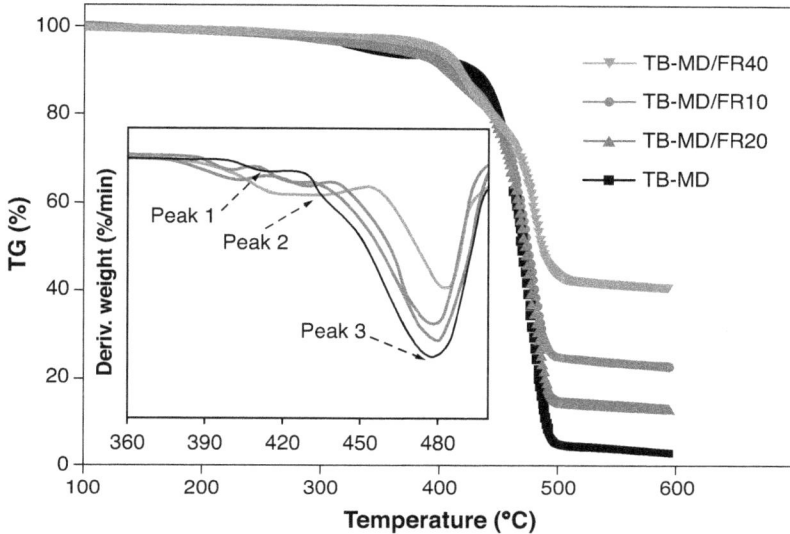

FIGURE 9.2 Thermogravimetric curve of composite with various percentages of flame retardant (inset: DSC). (Reproduced with permission from Elsevier, License Number: 5217970329776.)

Differential TGA was also employed to examine the derivative weight loss of neat and DPF-reinforced epoxy samples. DTG of pure epoxy composites revealed only one peak at 330°C with higher mass degradation around 19%/min. The initial peak corresponds to water molecules present in the hemicelluloses of DPF-reinforced composites or voids. Second peak was higher, and this indicates that at higher temperature the palm fiber-reinforced composite degrades as compared to neat epoxy composites with DPF/epoxy (50%) being the highest. With regard to differential thermal behavior, the neat epoxy sample show higher derivate mass loss as compared to DPF/epoxy composites.

The morphological features of natural fibers extracted from various components of trees, commonly coconut, date palm, sugar palm, etc., are quite different. These differences cause variations in thermal and mechanical properties [42,43]. Alothman et al. investigated the characterization of certain DPF/epoxy composites on the basis of thermal degradation and elevated temperature studies using TGA, thermomechanical analysis (TMA) and dynamic mechanical analysis (DMA) [44]. Different DPF/epoxy composites were fabricated using various components of DPF mainly from leaf sheath, fruit bunch stalk tree trunk and leaf stalk. TGA was used to study the thermal degradation behavior of the composites. The samples were analyzed between 25°C and 670°C at a heating rate of 10°C/min. Thermograms of pure and DPF/epoxy composites on comparison revealed that initial weight loss due to water loss occurred between 80°C and 130°C. The final residue after all thermal degradations was 6%. The onset and inflection temperature as well as initial and final major weight losses were lower for DPF/epoxy composites compared to pure epoxy fibers. These may be attributed to the decomposition of fiber constituents such as lignin, cellulose and hemicelluloses. DTG curves also revealed smaller peaks between 180°C and 210°C and also between 210°C and 250°C corresponding to the degradation of hemicelluloses and pectin, respectively. The disintegration of lignin and cellulose is represented by peaks at higher temperatures at 250°C and 400°C [45]. It was estimated that different DPFs from various parts of tree differ in morphology, fiber aspect ratio and constituents which play a vital function in determining thermal stability and decomposition characteristic of samples. On analysis, the final residue obtained after decomposition of DPF fiber with tree trunk as filler possesses better thermal stability. This was substantiated by the 22% less weight loss, larger onset and inflection temperatures. However, the result showed that the pure epoxy fiber has lower residual percentage as compared to DPF fiber. This justifies the superiority of DPF fiber composites over pure epoxy fibers.

Piassava fiber, extracted from palm tree with rigid fiber structure and commonly seen in Brazilian Atlantic rain forest, is a unique NLF which possesses similar potential like coir or bagasse fibers. The pioneering studies regarding the thermal properties of piassava fibers-reinforced epoxy composites were attempted by Garcia Filho et al. [46]. The adhesion of hydrophilic fibers to hydrophobic epoxy matrix was poor [47]. Hence, they investigated the possibility of functionalization with graphene oxide (GO) to enhance the adhesion with an epoxy composite. The TGA/DTG and differential scanning calorimetry (DSC) analyses were utilized to analyze the thermal stability of GO-incorporated piassava fiber composites. TGA studies were performed at a heating rate of 10°C/min to a maximum temperature of 500°C under nitrogen atmosphere. Three main temperature stages at around 150°C, 288°C and 359°C indicating mass losses of 7.8%, 26.2% and 32.4%, respectively, were identified in the TGA curves of neat piassava fiber curves. These results were similar to previously reported data at peak temperatures of 74°C, 276.4°C and 347.8°C with corresponding weight losses (5.18%, 18.8% and 34.33%), respectively (Figure 9.3) [48]. Slight differences in values could be attributed to the difference in heterogeneity of fiber. In the case of GO-coated piassava fibers, similar mass loss percentages as neat piassava fibers were observed at 7.3%, 22.1% and 32.5%. However, the hemicellular and lignin decomposition occurred at higher temperatures at around 317°C and 479°C, respectively. This clearly suggests the protective effect of GO coating. Hence, TGA validates the improved thermal stability of fibers imparted by GO coating.

9.4 DSC OF NATURAL FIBER/EPOXY-BASED HYBRID COMPOSITES

DSC is a thermal analyzer used in the characterization of materials. It is a widely used method owing to its speed, availability and simplicity. It is considered to be a quantitative method. After weighing an empty sample pan, sample and reference are placed in holder of DSC instrument [49]. Temperature is either varied at a specified rate or calorimeter is held at a particular temperature. The differences in heat flow between specimen and reference as a function of temperature are measured. DSC records the difference in the amount of heat needed to lift the temperatures of a specimen and reference as a function of temperature. Quantitative applications of DSC include the determination of extent of crystallization, heat of fusion, etc. for crystalline substances [49]. Purity of samples is excellently established from melting point

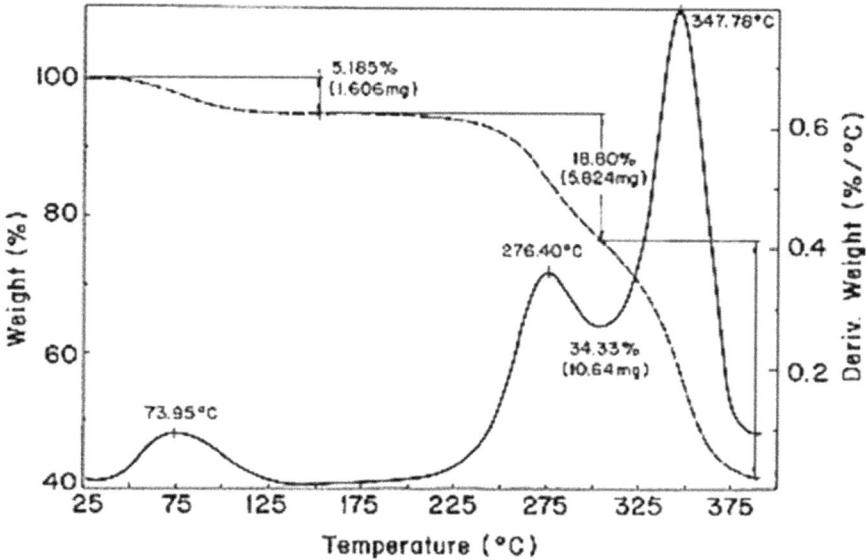

FIGURE 9.3   Thermogravimetric plot of piassava fibers. (Reproduced with permission from Elsevier, License Number: 5217970865376.)

determinations. One of the most prominent applications of DSC would be determining glass transition temperature (T_g). DSC effectively and readily measures the heat capacity changes during transition of polymers from glassy state to rubbery state [49]. Hence, DSC plays a key role in characterization of polymers.

Garcia Filho et al. extended the thermal characterizations of GO-coated piassava fibers to DSC analysis [46]. For neat piassava fiber, an endothermic peak was observed 125°C with an enthalpy of 71.3% J/g associated with this peak. These results indicated that the inter and intra bonds rupture and with the same magnitude of enthalpy. This bond breaking could be considered as the primary stage of decomposition of fiber. The piassava-coated GO composite identified an endothermic peak at 61°C with enthalpy (138.8 J/g). This peak corresponds to water desorption. Moreover, the disappearance of peak at 125°C in GO-coated piassava fiber points out that it might appear at higher temperatures. These findings justify that the thermal stability of piassava fiber could be enhanced by GOcoating.

Composite materials fabricated with synthetic fiber naturals are popular nowadays owing to their unique properties and excellent applications. Among the synthetic fibers, polyaramid fibers (Kevlar fibers) possess exceptional properties [50]. It is a highly stiff material and has applications in industries and ballistic, armor technology, etc. High tensile strength, long elongation, etc. are some of the distinctive features of Kevlar fibers [51]. Chinnasamy et al. attempted to analyze the influence of nanoclay dosage on thermal properties of composites (glass fiber and Kevlar fiber) [52]. They reinforced epoxy samples with different fractions of nanoclay containing glass and Kevlar fiber and the resulting samples were further analyzed. All the

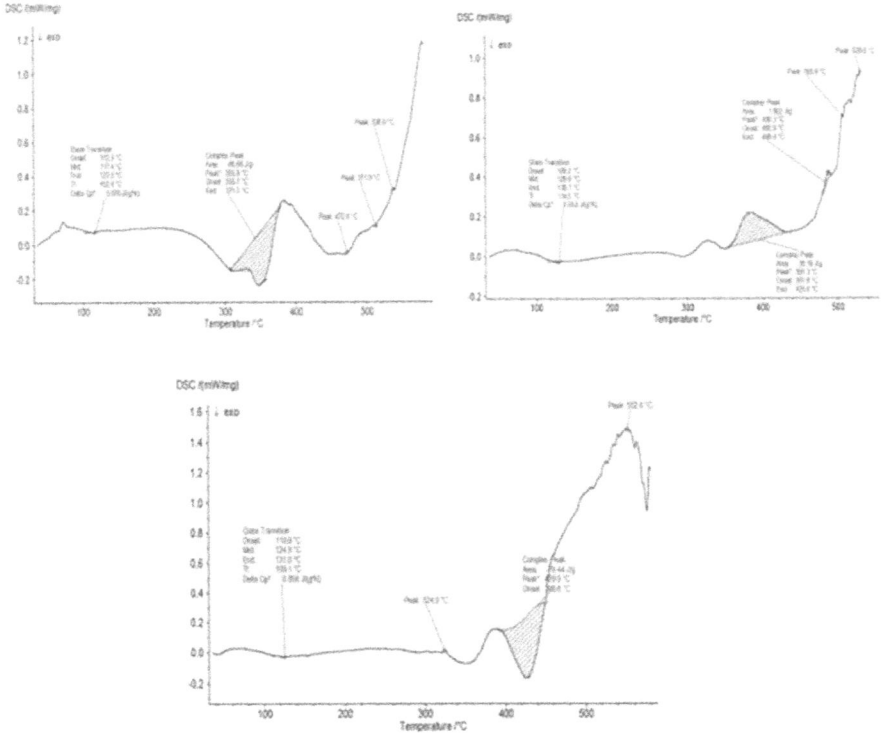

FIGURE 9.4 Thermal transition DSC of Kevlar/glass fiber composites with different ratios. (Reproduced with permission from Elsevier, License Number: 5217980222081.)

synthesized Kevlar/glass composite were analyzed by DSC under N_2 atmosphere with a flow rate of 50 mL/min. About 5–6 mg of each specimen was taken for the analysis. Heating rate was maintained at 10°C from 25°C to 580°C. In the hybrid curves of epoxy composites, at the temperature of 122.5°C, endothermic peak was observed for Kevlar. The T_g value for textured woven Kevlar was observed at 112.3°C. Around 122.5°C, 139.7°C, 131.1°C and 127.7°C a sharp endothermic peak was recorded for Kevlar woven. Endothermic peak was observed at three different points 472.0°C, 511.9°C and 536.0°C, respectively, for DSC curves of Kevlar fibers which were much higher than the melting point of Kevlar fibers. This leads to the conclusion that the Kevlar and glass fiber undergo decomposition while melting, and around 122.5°C Kevlar fiber acquires glass transition temperature (Figure 9.4).

9.5   CONCLUSION

Owing to its eco-friendly and less expensive nature, natural fibers are extensively explored as filler in polymer composites. The addition of natural fiber highly influences the properties of composite such as mechanical, thermal and microstructural. The physical and chemical properties of natural fibers could be further enhanced by various types of chemicals such as alkaline, benzoyl chloride, silane, ultrasonic and permanganate. The chemical treatment reduces the hydrophilic nature and thereby enhances the interaction between the fiber and polymers.

ACKNOWLEDGMENTS

The authors gratefully thank the King Mongkut's University of Technology North Bangkok (KMUTNB), Thailand for the support through the Post-Doctoral Program (Grant No. KMUTNB-64-Post-03 and KMUTNB-63-Post-03 to SR).

REFERENCES

[1] M. Chandrasekar, K. Senthilkumar, T. Senthil Muthu Kumar, R. Sabarish, and S. Siengchin, Morphological characterization of the wood polymer composites, in Sanjay Mavinkere Rangappa, Jyotishkumar Parameswaranpillai, Mohit Hemanth Kumar, Suchart Siengchin (eds.) *Wood Polymer Composites*. Singapore: Springer. 2021. pp. 93–112.

[2] S. Radoor, J. Karayil, J.M. Shivanna, and S. Siengchin, Water absorption and swelling behaviour of wood plastic composites, in Sanjay Mavinkere Rangappa, Jyotishkumar Parameswaranpillai, Mohit Hemanth Kumar, Suchart Siengchin (eds.) *Wood Polymer Composites*. Singapore: Springer. 2021. pp. 195–212.

[3] S. Krishnasamy, S.M.K. Thiagamani, C. Muthukumar, R. Nagarajan, and S. Siengchin, *Natural Fiber-Reinforced Composites*. John Wiley & Sons. 2022.

[4] S. Radoor, J. Karayil, R. Soman, A. Jayakumar, E.K. Radhakrishnan, J. Parameswaranpillai, and S. Siengchin, Influence of nanoclay on the thermal properties of the natural fiber-based hybrid composites, in Sanjay Senthilkumar Krishnasamy, Senthil Muthu Kumar Thiagamani, Chandrasekar Muthukumar, Rajini Nagarajan, Suchart Siengchin (eds.) *Natural Fiber-Reinforced Composites*. John Wiley & Sons. 2022. pp. 239–254.

[5] O.Y. Alothman, M. Jawaid, K. Senthilkumar, M. Chandrasekar, B.A. Alshammari, H. Fouad et al., Thermal characterization of date palm/epoxy composites with fillers from different parts of the tree, *Journal of Materials Research and Technology* 9 (2020) 15537–15546.

[6] M. Chandrasekar, I. Siva, T.S.M. Kumar, K. Senthilkumar, S. Siengchin, and N. Rajini, Influence of fibre inter-ply orientation on the mechanical and free vibration properties of banana fibre reinforced polyester composite laminates, *Journal of Polymers and Environment* 28 (2020) 2789–2800.

[7] S.K. Thomas, J. Parameswaranpillai, S. Krishnasamy, P.M.S. Begum, D. Nandi, S. Siengchin et al., A comprehensive review on cellulose, chitin, and starch as fillers in natural rubber biocomposites, *Carbohydrate Polymer Technologies and Applications* 2 (2021) 100095.

[8] K. Senthilkumar, N. Saba, M. Chandrasekar, M. Jawaid, N. Rajini, S. Siengchin et al., Compressive, dynamic and thermo-mechanical properties of cellulosic pineapple leaf fibre/polyester composites: Influence of alkali treatment on adhesion, *International Journal of Adhesion and Adhesives* 106 (2021) 102823.

[9] K. Senthilkumar, I. Siva, N. Rajini, and P. Jeyaraj, Effect of fibre length and weight percentage on mechanical properties of short sisal/polyester composite, *International Journal of Computer Aided Engineering and Technology* 7 (2015) 60.

[10] S. Radoor, J. Karayil, A. Jayakumar, E.K. Radhakrishnan, L. Muthulakshmi, S.M. Rangappa, S. Siengchin, and J. Parameswaranpillai, Structure and surface morphology techniques for biopolymers, in Sanjay Anish Khan, Sanjay Mavinkere Rangappa, Suchart Siengchin, Abdullah M. Asiri (eds.) *Biofibers and Biopolymers for Biocomposites*. Cham: Springer. 2020. pp. 35–70.

[11] S. Radoor, J. Karayil, S.M. Rangappa, S. Siengchin, and J. Parameswaranpillai, A review on the extraction of pineapple, sisal and abaca fibers and their use as reinforcement in polymer matrix, *Express Polymer Letters* 14(4) (2020) 309–335.

[12] K. Senthilkumar, S. Subramaniam, T. Ungtrakul, T.S.M. Kumar, M. Chandrasekar, N. Rajini et al., Dual cantilever creep and recovery behavior of sisal/hemp fibre reinforced hybrid biocomposites: Effects of layering sequence, accelerated weathering and temperature, *Journal of Industrial Textiles* 51 (2022) 2372S–2390S.

[13] S. Radoor, J. Karayil, A. Jayakumar, and S. Siengchin, Investigation on mechanical properties of surface-treated natural fibers-reinforced polymer composites, in Sanjay Rajini Nagarajan, Senthil Muthu Kumar Thiagamani, Senthilkumar Krishnasamy, Suchart Siengchin (eds.) *Mechanical and Dynamic Properties of Biocomposites*. 2021. Wiley. pp. 135–162.

[14] R.M. Shahroze, M. Chandrasekar, K. Senthilkumar, T. Senthil Muthu Kumar, M.R. Ishak, N. Rajini et al., Mechanical, interfacial and thermal properties of silica aerogel-infused flax/epoxy composites, *International Polymer Processing* 36 (2021) 53–59.

[15] A.K. Bledzki, A.A. Mamun, M. Lucka-Gabor, and V.S. Gutowski, The effects of acetylation on properties of flax fibre and its polypropylene composites, *Express Polymer Letters* 2(6) (2008) 413–422.

[16] Y. Seki, Innovative multifunctional siloxane treatment of jute fiber surface and its effect on the mechanical properties of jute/thermoset composites, *Materials Science and Engineering: A* 508(1–2) (2009) 247–252.

[17] B. Singh, M. Gupta, and A. Verma, Influence of fiber surface treatment on the properties of sisal-polyester composites, *Polymer Composites* 17(6) (1996) 910–918.

[18] S. Mohamed Abbas, R. Manikandan, G. Suresh, S. Suberiya Begum, and S. Selvi, Effect of cascara/testa natural fiber reinforced (epoxy based) hybrid composites, *Journal of Physics: Conference Series* 1921(1) (2021) 012093.

[19] L. de Mendonça Neuba, R.F. Pereira Junio, M.P. Ribeiro, A.T. Souza, E. de Sousa Lima, F.d.C. Garcia Filho, A.B.-H.d.S. Figueiredo, F.d.O. Braga, A.R.G.d. Azevedo, and S.N. Monteiro, Promising mechanical, thermal, and ballistic properties of novel epoxy composites reinforced with *Cyperus malaccensis* sedge fiber, *Polymers* 12(8) (2020) 1776.

[20] K. Yorseng, S. Mavinkere Rangappa, J. Parameswaranpillai, and S. Siengchin, Influence of accelerated weathering on the mechanical, fracture morphology, thermal stability, contact angle, and water absorption properties of natural fiber fabric-based epoxy hybrid composites, *Polymers* 12(10) (2020) 2254.

[21] S. Tripathy, J. Dehury, and D. Mishra, A study on the effect of surface treatment on the physical and mechanical properties of date-palm stem liber embedded epoxy composites, *IOP Conference Series: Materials Science and Engineering* 115 (2016) 012036.

[22] K.R. Sumesh, V. Kavimani, G. Rajeshkumar, S. Indran, and G. Saikrishnan, Effect of banana, pineapple and coir fly ash filled with hybrid fiber epoxy based composites for mechanical and morphological study, *Journal of Material Cycles and Waste Management* 23(4) (2021) 1277–1288.

[23] K.R. Sumesh and K. Kanthavel, Effect of TiO_2 nano-filler in mechanical and free vibration damping behavior of hybrid natural fiber composites, *Journal of the Brazilian Society of Mechanical Sciences and Engineering* 42(4) (2020) 1–12.

[24] L. Prasad, A. Kumain, R.V. Patel, A. Yadav, and J. Winczek, Physical and mechanical behavior of hemp and nettle fiber-reinforced polyester resin-based hybrid composites, *Journal of Natural Fibers* 19 (2020) 2632–2647.

[25] K. H. M, S. Bavan, Y. B, S. M. R, S. Siengchin, and S. Gorbatyuk, Effect of coir fiber and inorganic filler on physical and mechanical properties of epoxy based hybrid composites, *Polymer Composites* 42(8) (2021) 3911–3921.

[26] S. Ebnesajjad, Surface and material characterization techniques, in Sina Ebnesajja (ed.) *Handbook of Adhesives and Surface Preparation*. Elsevier. 2011. pp. 31–48.

[27] N.M. Nurazzi, M.R.M. Asyraf, M. Rayung, M.N.F. Norrrahim, S.S. Shazleen, M.S.A. Rani, A.R. Shafi, H.A. Aisyah, M.H.M. Radzi, F.A. Sabaruddin, R.A. Ilyas, E.S. Zainudin, and K. Abdan, Thermogravimetric analysis properties of cellulosic natural fiber polymer composites: A review on influence of chemical treatments, *Polymers* 13(16) (2021) 2710.

[28] S.N. Monteiro, V. Calado, R.J.S. Rodriguez, and F.M. Margem, Thermogravimetric behavior of natural fibers reinforced polymer composites—An overview, *Materials Science and Engineering: A* 557 (2012) 17–28.

[29] N.A. Nasimudeen, S. Karounamourthy, J. Selvarathinam, S.M. Kumar Thiagamani, H. Pulikkalparambil, S. Krishnasamy et al., Mechanical, absorption and swelling properties of vinyl ester based natural fibre hybrid composites, *Applied Science and Engineering Progress* 14 (2021) 680–688.

[30] A.W. Coats and J.P. Redfern, Thermogravimetric analysis. A review, *The Analyst* 88(1053) (1963) 906–924.

[31] T. Senthil Muthu Kumar, K. Senthilkumar, M. Chandrasekar, S. Karthikeyan, N. Ayrilmis, N. Rajini et al., Mechanical, thermal, tribological, and dielectric properties of biobased composites, in Anish Khan, Sanjay M. Rangappa, Suchart Siengchin, Abdullah M. Asiri (eds.) *Biobased Composites: Processing, Characterization, Properties, and Applications*. Wiley. 2021. pp. 53–73.

[32] K. Senthilkumar, N. Saba, N. Rajini, M. Chandrasekar, M. Jawaid, S. Siengchin et al., Mechanical properties evaluation of sisal fibre reinforced polymer composites: A review, *Construction and Building Materials* 174 (2018) 713–729.

[33] T.P. Sathishkumar, J. Naveen, and S. Satheeshkumar, Hybrid fiber reinforced polymer composites – A review, *Journal of Reinforced Plastics and Composites* 33(5) (2014) 454–471.

[34] Y. Singh, J. Singh, S. Sharma, T.-D. Lam, and D.-N. Nguyen, Fabrication and characterization of coir/carbon-fiber reinforced epoxy based hybrid composite for helmet shells and sports-good applications: Influence of fiber surface modifications on the mechanical, thermal and morphological properties, *Journal of Materials Research and Technology* 9(6) (2020) 15593–15603.

[35] M.H. Gheith, M.A. Aziz, W. Ghori, N. Saba, M. Asim, M. Jawaid, and O.Y. Alothman, Flexural, thermal and dynamic mechanical properties of date palm fibres reinforced epoxy composites, *Journal of Materials Research and Technology* 8(1) (2019) 853–860.

[36] A. Bourmaud, H. Dhakal, A. Habrant, J. Padovani, D. Siniscalco, M.H. Ramage, J. Beaugrand, and D.U. Shah, Exploring the potential of waste leaf sheath date palm fibres for composite reinforcement through a structural and mechanical analysis, *Composites Part A: Applied Science and Manufacturing* 103 (2017) 292–303.

[37] F.M. Al-Oqla, O.Y. Alothman, M. Jawaid, S.M. Sapuan, and M.H. Es-Saheb, Processing and properties of date palm fibers and its composites, in Khalid Rehman Hakeem, Mohammad Jawaid, Umer Rashid (eds.) *Biomass and Bioenergy*. Springer. 2014. pp. 1–25.

[38] B. Agoudjil, A. Benchabane, A. Boudenne, L. Ibos, and M. Fois, Renewable materials to reduce building heat loss: Characterization of date palm wood, *Energy and Buildings* 43(2–3) (2011) 491–497.

[39] A. Abdal-hay, N.P.G. Suardana, D.Y. Jung, K.-S. Choi, and J.K. Lim, Effect of diameters and alkali treatment on the tensile properties of date palm fiber reinforced epoxy composites, *International Journal of Precision Engineering and Manufacturing* 13(7) (2012) 1199–1206.

[40] N. Saba, M.T. Paridah, M. Jawaid, and O.Y. Alothman, Thermal and flame retardancy behavior of oil palm based epoxy nanocomposites, *Journal of Polymers and the Environment* 26(5) (2017) 1844–1853.

[41] K.M. Zadeh, D. Ponnamma, and M. Al Ali Al-Maadeed, Date palm fibre filled recycled ternary polymer blend composites with enhanced flame retardancy, *Polymer Testing* 61 (2017) 341–348.

[42] N. Saba, M. Jawaid, and M.T.H. Sultan, Thermal properties of oil palm biomass based composites, in Mohammad Jawaid, Paridah Md Tahir, Naheed Saba (eds.) *Lignocellulosic Fibre and Biomass-Based Composite Materials.* Elsevier. 2017. pp. 95–122.

[43] J. Sahari, S.M. Sapuan, E.S. Zainudin, and M.A. Maleque, Mechanical and thermal properties of environmentally friendly composites derived from sugar palm tree, *Materials & Design* 49(2013) 285–289.

[44] O.Y. Alothman, M. Jawaid, K. Senthilkumar, M. Chandrasekar, B.A. Alshammari, H. Fouad, M. Hashem, and S. Siengchin, Thermal characterization of date palm/epoxy composites with fillers from different parts of the tree, *Journal of Materials Research and Technology* 9(6) (2020) 15537–15546.

[45] B. Neher, M.M.R. Bhuiyan, H. Kabir, M.A. Gafur, M.R. Qadir, and F. Ahmed, Thermal properties of palm fiber and palm fiber-reinforced ABS composite, *Journal of Thermal Analysis and Calorimetry* 124(3) (2016) 1281–1289.

[46] F.d.C. Garcia Filho, F.S.d. Luz, M.S. Oliveira, A.C. Pereira, U.O. Costa, and S.N. Monteiro, Thermal behavior of graphene oxide-coated piassava fiber and their epoxy composites, *Journal of Materials Research and Technology* 9(3) (2020) 5343–5351.

[47] O. Güven, S.N. Monteiro, E.A.B. Moura, and J.W. Drelich, Re-emerging field of lignocellulosic fiber – polymer composites and ionizing radiation technology in their formulation, *Polymer Reviews* 56(4) (2016) 702–736.

[48] J.R.M. d'Almeida, R.C.M.P. Aquino, and S.N. Monteiro, Tensile mechanical properties, morphological aspects and chemical characterization of piassava (*Attalea funifera*) fibers, *Composites Part A: Applied Science and Manufacturing* 37(9) (2006) 1473–1479.

[49] O. Koshy, L. Subramanian, and S. Thomas, Differential scanning calorimetry in nanoscience and nanotechnology, in Sabu Thomas, Raju Thomas, Ajesh K. Zachariah, Raghvendra Kumar Mishra (eds.) *Thermal and Rheological Measurement Techniques for Nanomaterials Characterization.* Elsevier. 2017. pp. 109–122.

[50] H. Ghouti, A. Zegaoui, M. Derradji, W.-A. Cai, J. Wang, W.-B. Liu, and A. Dayo, Multifunctional hybrid composites with enhanced mechanical and thermal properties based on polybenzoxazine and chopped kevlar/carbon hybrid fibers, *Polymers* 10(12) (2018) 1308.

[51] R.J. Young, D. Lu, R.J. Day, W.F. Knoff, and H.A. Davis, Relationship between structure and mechanical properties for aramid fibres, *Journal of Materials Science* 27(20) (1992) 5431–5440.

[52] V. Chinnasamy, S. Pavayee Subramani, S.K. Palaniappan, B. Mylsamy, and K. Aruchamy, Characterization on thermal properties of glass fiber and kevlar fiber with modified epoxy hybrid composites, *Journal of Materials Research and Technology* 9(3) (2020) 3158–3167.

10 Natural Fiber-Based Bionanocomposites
Thermal, Morphological and Mechanical Properties

Aswathy Jayakumar
King Mongkut's University of Technology
Kyung Hee University

Reshma Soman
Mahatma Gandhi University

Sabarish Radoor
King Mongkut's University of Technology

Jyotishkumar Parameswaranpillai
Alliance University

Jun Tae Kim
Kyung Hee University

Jong Whan Rhim
Kyung Hee University

Suchart Siengchin
King Mongkut's University of Technology

CONTENTS

DOI: 10.1201/9781003271017-10

10.1   INTRODUCTION

The major concerns related to synthetic fibers like non-biodegradability, high cost and energy consumption paved the way to explore the use of natural fibers in environmental as well as other applications (Karimah et al., 2021). Natural fibers are considered as biodegradable materials with renewable nature, abundant availability, higher mechanical properties, and low energy consumption and are low cost in nature. They are mainly classified into four categories and include seed fiber, leaf fibers, bast fibers and stalk fibers (Alothman et al., 2020; Chandrasekar et al., 2020). Cotton, coir and kapok are considered as seed fibers, whereas sisal, agave, pineapple and abaca are leaf fibers. Bast fibers consist of kenaf, ramie, hemp, jute and flax. And stalk fibers include wood, straw and bamboo (Hasan et al., 2020). Among them, cotton, bamboo, rice straw, kenaf, hemp and flax fibers are the most commonly used natural fibers for reinforcements.

The overall performance of the natural fibers will depend on the crystallinity, size of fibers, structural composition, hydrophobicity or hydrophilicity depending upon the applications and the surface functionality (Krishnasamy et al., 2021). Moreover, the quality of the fibers will depend upon the environmental factors, transport method, storing time and conditions and the extraction process (Karimah et al., 2021). But most natural fibers can absorb moisture which in turn affects its applications. For absorption-related applications, natural fiber-based composites are most suitable. The major disadvantages associated with natural fiber-based composites are its low thermal stability, hydrophilicity and variations that occur during the extraction or seasonal changes (Karimah et al., 2021). Due to this, most natural fibers are modified before its use in various applications.

During the last 20 years, the innovations in nanotechnology continue to grow to meet the demands generated by various industries. Natural materials-incorporated products are making good progress in several applications. Moreover, most of the industries are also welcoming the entry of nanocomposites instead of microcomposites or traditional composites. In the research itself, we can see the continuous emergence of bionanocomposite-related articles. Several studies have reported the use of organic and inorganic nanomaterials for the development of bionanocomposites (Nasimudeen et al., 2021; Senthil Muthu Kumar et al., 2021). Among them, cellulose nanomaterials obtained from natural fibers are the most studied. In this chapter, properties of natural fibers, treatments for enhancing the properties of natural fibers, nanomodification strategies and bionanocomposites based on natural fibers and their thermal, morphological and mechanical properties will be discussed.

10.2 PROPERTIES OF NATURAL FIBERS

Natural fibers are considered to be renewable and have low density, renewable nature, less cost, better availability, high strength and biodegradable in nature (Ammar et al., 2018). Due to these properties, they are highly preferred in environment-related applications. The properties of natural fibers are highly dependent on the shape, size and the chemical constituents, and the major part of most fibers is composed of cellulose (Djafari Petroudy, 2017). These properties are also dependent upon the harvesting place, time, soil and other physical as well as chemical environments. But the high moisture absorption, poor mechanical barrier and their quality variations due to seasons may limit their applications. The hydrophilicity of the fibers could result in the decrease in mechanical properties and delamination (Senthilkumar et al., 2018, 2021, 2022). The mechanical properties of natural fibers are dependent upon the cellulose content and crystallinity. The strength of the fibers increases with increase in cellulose content and crystallinity. And also, the components like cellulose, hemicellulose and lignin have been reported to degrade at higher temperatures (Asim et al., 2020). Hence, the thermal stability of natural fibers is of great concern owing to its applications. Due to the presence of hydroxyl groups, polar groups, wax, oil and dead cells, they are not compatible with polymers (Karimah et al., 2021a; Sanadi et al., 2002). And also, the compatibility of natural fibers is also a great concern as it can result in the aggregation in the polymer matrix. But they are used as reinforcing agents due to their higher strength, stiffness and low density. And also, the inefficiency to provide antimicrobial barriers may contribute to microbial contamination and will spoil the entire product.

10.3 TREATMENTS FOR ENHANCING THE PROPERTIES OF NATURAL FIBERS

In order to improve the properties of natural fibers, several chemical and surface treatments are used. Chemical treatments include alkaline treatment, acetylation, benzoylation, stearic acid treatment, silane treatment and the treatment with potassium permanganate (Al Rashid et al., 2020; Lu et al., 2015; Thomas et al., 2021). These kinds of treatments have a determining role in enhancing the surface morphology, mechanical properties and chemical groups of natural fibers. During this chemical treatment, the interfacial adhesion between the surface of natural fibers and the polymer matrix will be enhanced and thus in turn enhances the properties of that material (Zhou et al., 2016). The changes in structural and morphological properties of natural fibers are due to the removal of non-cellulosic materials. Moreover, the reduction in hydrophilic nature enhances the better adhesion of natural fibers with their matrix and this can be achieved by treatments like benzoylation (Thyavihalli Girijappa et al., 2019).

Benzoylation treatments can improve the bonding between the fiber and the matrix that might result in enhancement in mechanical strength of polymer composites (Thiruchitrambalam & Shanmugam, 2012). Similarly, the interfacial adhesion between the natural fiber and the polymer matrix can be enhanced by treating with potassium permanganate also. Usually, the treatments with peroxides are done after

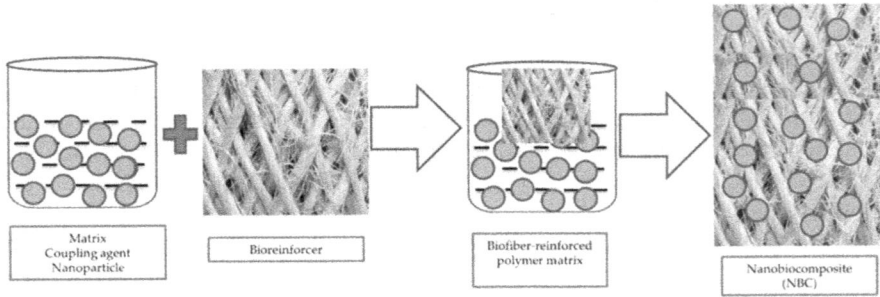

FIGURE 10.1   Development of bionanocomposites. (Hasan et al., 2020.)

alkali modification (Senthilkumar et al., 2015; Shahroze et al., 2021). Peroxide treatment can result in the generation of free radicals and in turn react with hydrogen groups of both the fibers and the polymers. Due to this, there is a huge improvement in mechanical properties of fiber-based composites (Kabir et al., 2012). The larger interstitial voids developed in biocomposites during the reinforcement of natural fibers may hinder the properties' enhancement. Due to this, premature breakage may occur during the mechanical testing. Such cases can only be handled by the intercalation or exfoliation of layered silicates.

10.4   NATURAL FIBER-BASED BIONANOCOMPOSITES

The innovations in nanotechnology-based technologies can be applied to improve the properties of natural fiber-based polymer composites. The drawbacks of natural fiber like its hydrophilic nature, non-antimicrobial nature and low thermal resistance can be enhanced by the treatments with nanoparticles. Extreme research is going on in the field of natural fiber-based composites and nanocomposites. The integration of nanoparticles into the biofiber-reinforced polymeric matrix is shown in Figure 10.1. The uniform and the homogenous distribution of nanofillers in the polymer matrix will result in the improved mechanical, thermal, water barrier, antimicrobial and antioxidant properties of natural fiber-based bionanocomposites. Normally, inorganic or organic nanofillers like clay nanoparticles, carbon nanotubes, titanium nanoparticles, cellulose nanoparticles, chitosan nanoparticles, silicon dioxide nanoparticles and zinc oxide nanoparticles are used to reinforce natural fiber-based biocomposites (Hasan et al., 2020).

10.4.1   Kenaf Fiber-Based Bionanocomposites

Kenaf fibers are obtained from *Hibiscus cannabinus* plant of the Malvaceae family (Hao et al., 2018). The length of individual fiber is about 2–6 mm long and is slender in nature. Lignin and hemicellulose are rich in lighter and porous kenaf core fiber. The incorporation of kenaf fibers into polymer composites has shown significant improvement in mechanical properties, as well as results in lightweight polymer composites with environment-friendly nature (Alam et al., 2017;

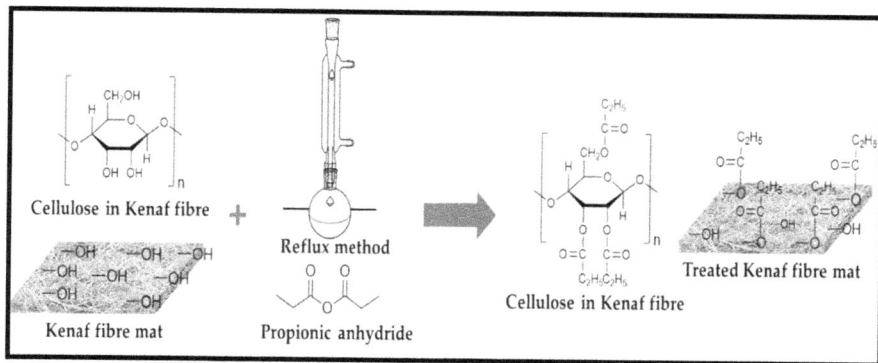

FIGURE 10.2 Propionic anhydride modification of kenaf fiber. (Rizal et al., 2021.)

Rahman et al., 2019). The orientation and the percentage of fiber have a great impact in strengthening the tensile properties as well as other barrier performances (Fairuz et al., 2016).

The major modifications that are used for improving the properties of kenaf fibers are alkaline treatment, treating with isocyanate and acetylation. Rizal and his coworkers employed propionic anhydride-related chemical modification to improve the water barrier properties and compatibility between the kenaf fiber and polymeric matrix (Rizal et al., 2021a). The modification of kenaf fiber with propionic anhydride is represented in Figure 10.2. For the development of bionanocomposite, they utilized different weight percentages of 0, 1, 3 and 5 weight percentages of bio-nanocarbon. The tensile, flexural and impact strength were enhanced to 63.91%, 49.61% and 54.82%, respectively. Their results suggest that the loading of 3% bio-nanocarbon resulted in the improved structural, morphological and functional properties as well as excellent fiber–matrix interaction. The poor interfacial interaction between the fiber and polymeric matrix is represented in Figure 10.3. The nanofiller incorporation has led to the improvement in fiber–matrix interactions. The thermal analysis confirmed the enhanced thermal stability of modified kenaf fiber-based bionanocomposites. The enhanced functionalities of the bionanocomposites could be due to the incorporation of nanofillers which promoted the better fiber–matrix interactions.

Bio-nanocarbon developed from oil palm shell by single-step activation process was found to enhance the non-woven kenaf fiber-based vinyl ester composites (Rizal et al., 2021). As mentioned earlier, different weight percentages (0,1, 3 and 5) of nanomaterial significantly improved the fiber–matrix interaction and other mechanical as well as barrier properties of the bionanocomposites. The mechanism of interaction of composites is represented in Figure 10.4. The loading of 1% bio-nanocarbon resulted in weak interfacial bonding, whereas 3% enhanced the interfacial bonding. But the 5% loading decreased the interfacial bonding which could be due to the agglomeration of bio-nanocarbon.

Sabaruddin and his coworkers studied the effect of bleached (NCC-B) and unbleached (NCC-UB) nanocellulose on thermal and mechanical properties of kenaf fiber-based bionanocomposites (Sabaruddin et al., 2020). They incorporated

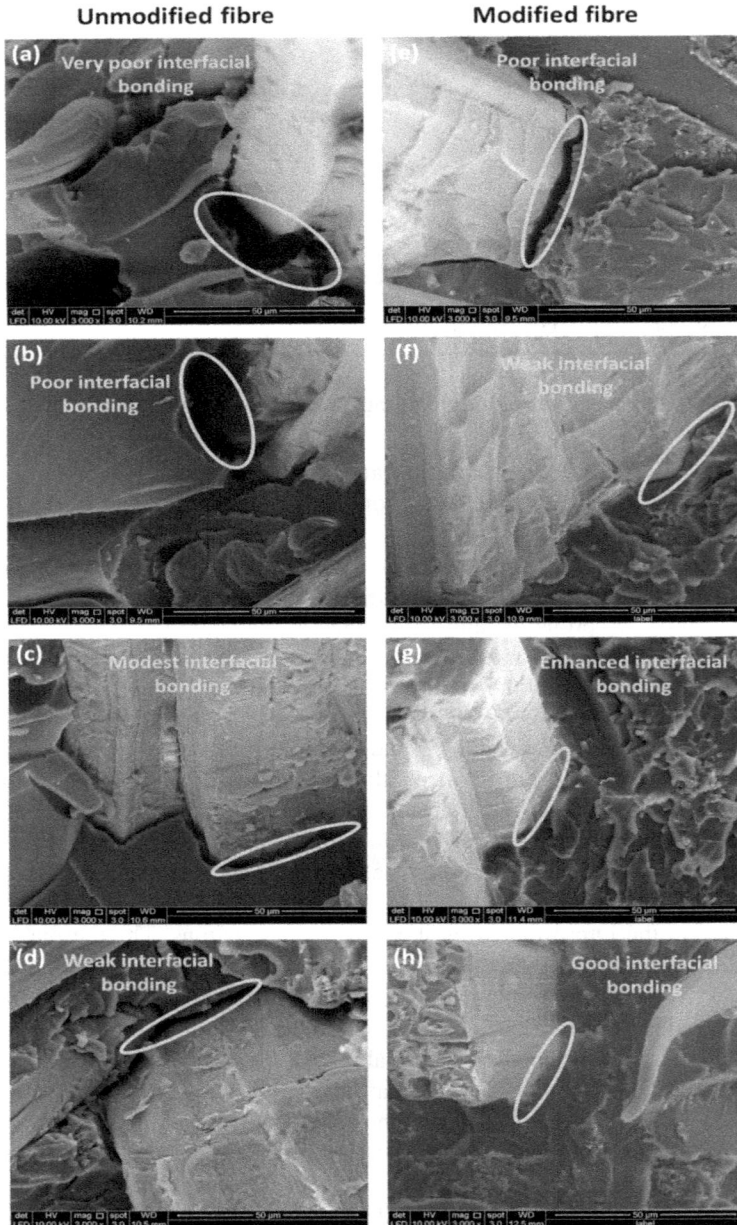

FIGURE 10.3 Field emission scanning electron microscopy (FESEM) micrographic image of unmodified and modified tensile fracture sample: (a) VE/UK/NC0, (b) VE/UK/NC1, (c) VE/UK/NC3, (d) VE/UK/NC5, (e) VE/MK/NC0, (f) VE/MK/NC1, (g) VE/MK/NC3 and (h) VE/MK/NC5. MK, modified kenaf fiber; NC, bio-nanocarbon; VE, vinyl esters; UK, unmodified kenaf fiber; all these three have 0, 1, 3 and 5 as weight percentage, respectively). (Rizal et al., 2021.)

(a) Nonwoven kenaf bio-nanocomposites without bio-nanocarbon

(b) Nonwoven kenaf bio-nanocomposites with bio-nanocarbon

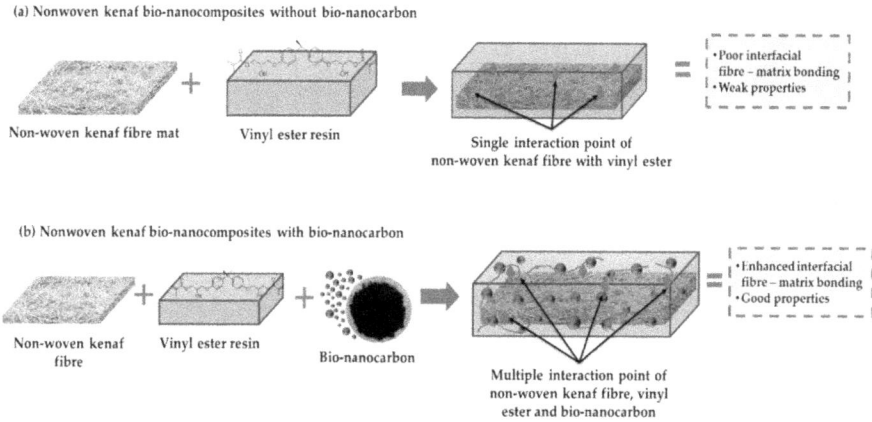

FIGURE 10.4 Mechanism of interaction of kenaf fiber-based composite with and without bio-nanocarbon. (a) The development of non-woven kenaf composites without bio-nanocarbon; (b) non-woven kenaf fiber-based bionanocomposites with bio-nanocarbon. (Rizal et al., 2021.)

nanocellulose in polypropylene matrix along with kenaf core as reinforcing agent, and found that the presence of higher lignin content significantly affected the thermal properties of the bionanocomposites. The unbleached nanocomposites have shown improved thermal properties due to the presence of residual lignin content that could act as compatibilizer between the matrix and NCC-UB. The morphology studies of the nanocomposites showed a clear surface without any cracks, thus suggesting its low brittleness nature. Mechanical properties of the nanocomposites were also suggested to get improved by the addition of nanocellulose.

10.4.2 BANANA FIBER-BASED BIONANOCOMPOSITES

Banana fibers are extracted from the pseudo-stem of the Musaceae family. Cellulose, hemicellulose and lignin are the major chemical components of banana fiber. They are lightweight fibers with higher strength, thermal stability, resistant to fire and biodegradable in nature (Vidya Bharathi et al., 2021). Banana fibers can absorb and release water with great efficiency. The tensile strength of banana fibers is generally between 529 and 914 MPa, whereas specific tensile strength is between 392 and 677 (Waghmare, 2017). The length and the loading of banana fiber have a significant impact on the mechanical properties of the banana fiber-based composites. The tensile strength, hardness and impact energy might increase with an increase in fiber loading and length. Due to these, they are used for reinforcing polymer-based composites.

V.P et al and his coworkers employed mercerization and silane treatment for the development of banana fiber and C30B nanoclay-incorporated polylactic acid (PLA) bionanocomposite by melt blending and injection molding. Their report suggests that the incorporation of 3% nanoclay significantly improved the mechanical as well as thermal properties of the bionanocomposites. And also, the treatment has led to the improvement in interfacial adhesion between the fibers and the polymeric matrix (V P et al., 2014). Other studies also suggested the same result for

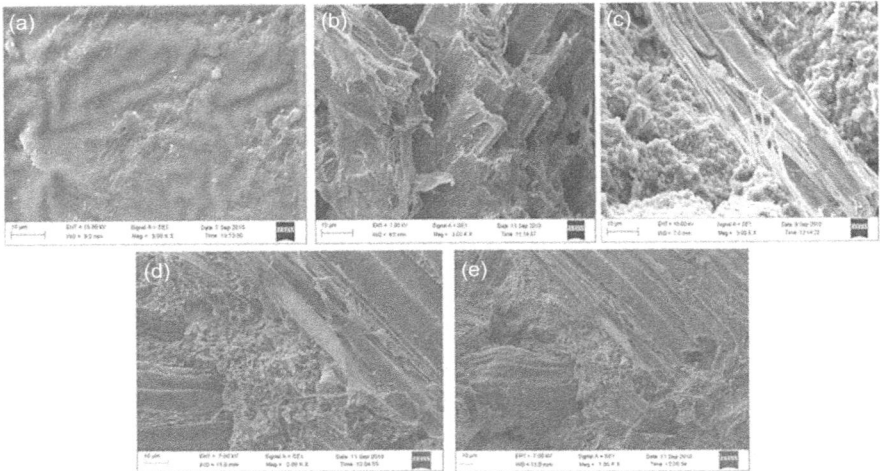

FIGURE 10.5 Scanning electron microscopic images of biocomposites based on PLA and silane-treated and non-treated banana fibers. (a) Virgin PLA, (b) PLA/untreated banana fibers, (c) PLA/sodium hydroxide treated banana fibers, (d) PLA/3-amino propyltriethoxysilane (APS) treated banana fibers, (e) PLA/silane 69 treated banana fibers. [Reproduced with permission from Elsevier, License Number: 5275100076615] (Jandas et al., 2013.)

FIGURE 10.6 Biodegradable cutlery based on polylactic acid, banana fiber and nanoclay-based bionanocomposites. [Reproduced with permission from Elsevier, License Number: 5275100076615] (Jandas et al., 2013.)

hybrid bionanocomposites based on banana fiber, nanoclay and PLA (Sajna et al., 2014; 2017). Further, Jandas and his coworkers employed melt mixing method to reinforce chemically treated banana fibers and organically modified cloisite 30B. They utilized silane and sodium hydroxide for the chemical modification of banana fibers (Jandas et al., 2013). In this study also, the surface treatment with silane led to the enhancement in thermal properties of the bionanocomposites. They concluded that the silane treatment could result in the increase in molecular weight due to the cross-linking of matrix and fiber and thus enhanced the thermal degradation temperature. The morphological and microstructural analyses through scanning electron microscopy (SEM) confirmed the improved surface topography after the treatment of the fiber with silane (Figure 10.5). They further reported the successful development of biodegradable cutlery based on PLA, banana fiber and nanoclay as shown in Figure 10.6.

10.4.3 COTTON FIBER-BASED BIONANOCOMPOSITES

It is composed of multilayered structures having 88%–97% of cellulose, wax, pectin and proteins (Alomayri et al., 2014; Candido, 2021). The fibrillar structure of cotton fiber is made up of primary cell wall, secondary cell and lumen. Generally, cotton fibers have uneven surfaces due to the convolutions (Dochia et al., 2012). The overall performance of cotton fibers can be improved through functionalization by coating and grafting techniques. The reinforcement of cotton fibers can have a significant impact in altering the mechanical properties of polymer-based composites. Studies reported that the addition of cotton fibers might lead to the decrease in elongation at break and impact strength with an increase in Young's modulus. And these properties will depend on the crystallinity, crystal size and the linkage between crystalline units. The stress–strain curve of banana fibers is exactly similar to glassy solid. Zhang and his coworkers reported the development of copper/cotton fiber and zinc-incorporated bionanocomposites (Zhang et al., 2022). The membrane possesses excellent antimicrobial activity against *Escherichia coli* and *Staphylococcus aureus*. The developed bionanocomposite has shown non-cytotoxic nature, which indicates the good biocompatibility for safer use. Electrical activation of this developed bionanocomposite can also occur through humans breathing water vapor, which confirms its potential use in personal protective devices. The developed bionanocomposites were able to deactivate Enterovirus 71 virions after 15 minutes of contact, which further confirmed its biomedical applications.

10.4.4 SISAL FIBER-BASED BIONANOCOMPOSITES

Natural plant fibers are lignocellulosic, while sisal is one of the most significant lignocellulosic fibers on the planet. The leaves of *Agave sisalana* (sisal) are commonly used for making sisal fibers. It has the potential to be used in a variety of applications due to its higher strength, stretching ability, low cost and easy availability. Due to this, it is always used as reinforcing agents in polymer-based composites (Naveen et al., 2019). Like other natural fibers, chemical modification with acetylation can improve the mechanical properties of the polymer-based composites (Fink, 2013). Sisal fibers have a pivotal role in enhancing the thermal stability of polymer-based composites (Liu et al., 2022).

Kim and his coworker developed alkali-treated sisal fiber-reinforced soy protein resin-based biocomposites. They discovered that the mercerization treatment raised the fracture stress and stiffness, respectively, of the sisal fiber by 12.2% and 36.2%

Wettability studies

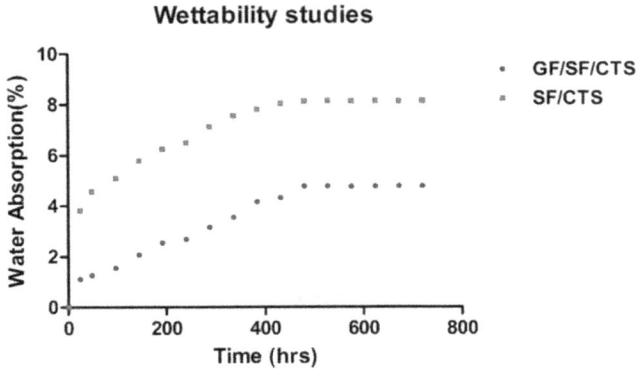

FIGURE 10.7 Water absorption properties of glass fiber (GF)/sisal fiber (SF)/chitosan (CTS) composite. (Arumugam et al., 2020.)

with a decrease in fracture strain and hardness. This could indicate an increase in interfacial adhesion between fibers and PLA. Moreover, the increased surface tension of the fiber allows for better PLA resin spreading on the fibers and thereby improving the fiber/resin interfacial adhesion (Kim & Netravali, 2010). Arumugam and his coworkers studied the effect of hybridization on mechanical properties of glass fiber (GF)/silica fiber (SF)/chitosan (CTS)-reinforced composites (Arumugam et al., 2020). Improved mechanical properties were observed for sandwich-structured composites. Moreover, the GF/SF/CTS composite has shown lower water absorption rate than SF/CTS/epoxy composites (Figure 10.7). The water contact angle of SF/CTS/epoxy composites was 54.28°, whereas GF/SF/CTS had 92.41°. Considering the properties, they suggested its potential application in orthopedic bone fracture plates.

Verma and his coworkers investigated the properties of nanocomposites reinforced with soy protein and sisal fibers by varying the sisal fiber weight percentages from 0% to 3%, 5%, 6%, 7%, and 10% (Verma et al., 2019). They coated the composites with chitosan by immersing the same in solution. They observed that the introduction of sisal fiber at a greater weight percentage and chitosan coating to the thermal tests resulted in an increase in thermal stability. Additionally, the Dynamic Mechanical Analysis (DMA) study revealed that the storage modulus and glass transition temperature for various compositions were larger for chitosan-coated specimens than for uncoated specimens, with the highest values recorded for sisal fiber composites containing 5% by weight.

10.4.5 Date Palm Fiber-Based Bionanocomposites

The date palm (*Phoenix dactylifera* L.) is one of the most ancient fruits in the Arabian Peninsula, North Africa and the Middle East (Chao & Krueger, 2007). In dry and semi-arid parts of the world, the date palm is one of the most widely farmed palms. It is considered as one of the rich sources of cellulosic fibers and is extracted from mesh, leaflets, midribs and spadix stems (Elseify & Midani, 2020). Fibers obtained from date palm are always considered as an alternative for obtaining cellulose nanocrystals. Adel and his coworkers utilized cellulose nanocrystals obtained from date palm sheath fibers for the development of chitosan-based bionanocomposites (Adel et al., 2018). They observed that the surface of chitosan film was changed to a rough surface by the addition of oxidized cellulose nanocrystals (Figure 10.8). The bionanocomposite exhibited a major weight loss at 200°C–400°C, and they suggested this loss is due to the decomposition of chitosan and oxidized cellulose nanocrystals. An improvement in thermal degradation temperature compared to the neat films could be due to the strong hydrogen bonding between the polymeric matrix and the cellulose nanocrystals or the formation of compact structures (Figure 10.9). Bionanocomposites based on chitosan, date palm fiber and montmorillonite were used to develop eco-friendly and non-toxic trays (Semlali Aouragh Hassani et al., 2020). They found that the intercalation of montmorillonite with thiabendazolium could have significant effect in enhancing the biological properties. SEM analysis revealed the better dispersion of montmorillonite, which might have caused the improvement in mechanical properties as well as hydrophobicity of the composites. The coating resulted in the increase of tensile strength from 0.5 MPa to 4 MPa. Moreover, the developed composites have shown excellent antibacterial activity against *E. coli*, *S. aureus* and *Pseudomonas aeruginosa*, which further suggests its potential application as eco-friendly trays.

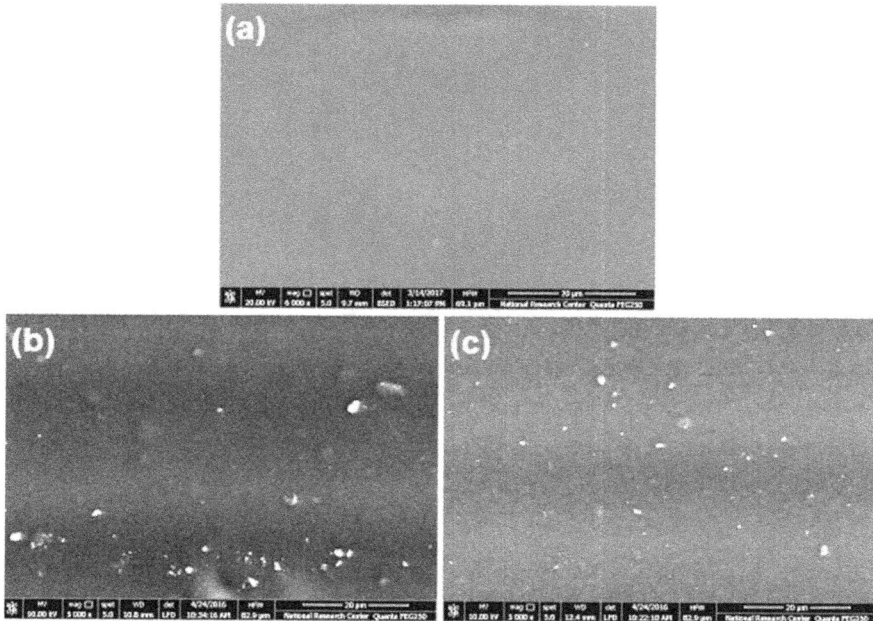

FIGURE 10.8 Scanning electron microscopic images of cellulose nanocrystal-loaded chitosan bionanocomposites: (a) chitosan film; (b) and (c) bionanocomposites containing nanocellulose and chitosan. [Reproduced with permission from Elsevier, License Number: 5277410796427] (Adel et al., 2018.)

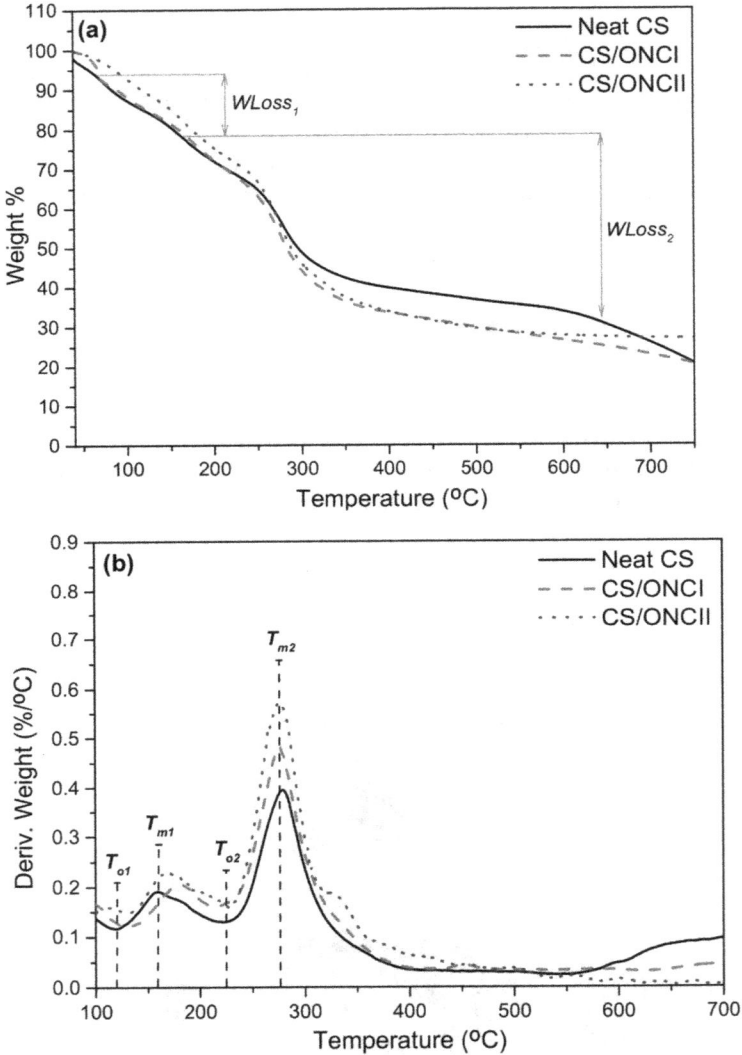

FIGURE 10.9 Thermogravimetric analysis of bionanocomposites based on chitosan and cellulose nanocrystals. (a) Cellulose I (CL) and (b) cellulose II polymorphs of sulfated nanocrystalline cellulose (SNCC) and phosphorylated nanocrystalline cellulose (PNCC). [Reproduced with permission from Elsevier, License Number: 5277410796427] (Adel et al., 2018.)

10.4.6   KAPOK FIBER-BASED BIONANOCOMPOSITES

Kapok fiber is an organic seed fiber obtained from *Ceiba pentandra* that is highly lignified and mostly composed of cellulose, lignin and xylan (McDougall et al., 1993). It is made up of two primary layers, each with a unique microfibrillar orientation. The outer layer is made up of cellulose microfibrils that are transverse to the fiber axis, whereas the inner layer is made up of fibrils that are almost parallel to the fiber axis (Nilsson & Björdal, 2008). In comparison to pure cotton, kapok/cotton-blended fiber exhibits comparable moisture permeability and lower air permeability, as well as superior warmth retention and water vapor transfer capabilities due to its extremely porous structure. Kapok fiber has been utilized as a reinforcement for polymeric resins due to its design flexibility, economic efficiency, absence of health hazards and recycling capabilities. Kapok fiber may be hybridized with glass and sisal fibers in a polyester matrix to significantly increase the impact strength of hybrid composites. The structural and the waxy nature of kapok fibers can be altered to improve its functional properties by several chemical or physical treatments.

10.4.7 BAMBOO FIBER-BASED BIONANOCOMPOSITES

Bamboo is considered as the strongest and fastest-growing plants with good strength and eco-friendly nature. Bamboo fiber has a multilayered structure having thick cell walls, small lumen with fewer pits and microfibrils. Generally bamboo fibers are 1–4 mm in length and 10–30 μm in diameter (Wang & Chen, 2017). They are elastic in nature along with antimicrobial, UV-resistant and biodegradable nature (Imadi et al., 2014). Due to the high mechanical properties, durability, stability and aspect ratio, it is often used as reinforcing agents in polymer-based composites (Yu et al., 2013). The tensile properties of bamboo fibers are highly dependent on the unit size of fibers. The higher the unit size, the lower the mechanical properties which could be due to the weak interfaces present in the middle lamella of bamboo fiber, the damages occurred during the developmental and the modification stages (Wang & Chen, 2017). Strongest bamboo fibers can be produced by treating with 10% NaOH in 48 hours (Chin et al., 2020). Rasheed and his coworkers utilized cellulose nanocrystals (CNC) obtained from bamboo fiber to reinforce PLA/poly(butylene succinate) (PBS) by melt mixing approach (Rasheed et al., 2021). The thermal properties of nanofibers were

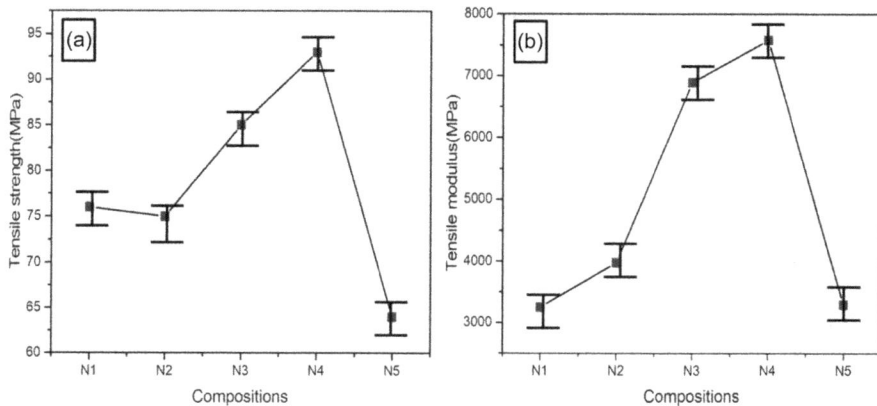

FIGURE 10.10 Graph showing (a) tensile strength and (b) modulus of poly(lactic acid) (PLA)/poly(butylene succinate) (PBS) and cellulose nanocrystals (varying concentrations: N1: 0, N2: 0.5, N3: 0.75, N4: 1, N5: 1.5)-based nanocomposites. (Rasheed et al., 2021.)

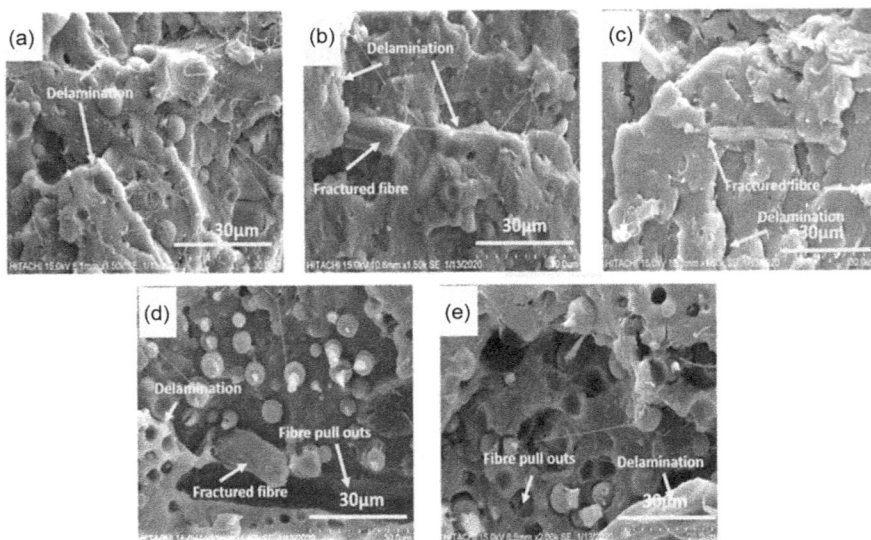

FIGURE 10.11 Scanning electron microscopic images of fractured tensile-tested specimens of poly(lactic acid) (PLA)/poly(butylene succinate) (PBS) and cellulose nanocrystals at varying concentrations. (a) N1: 0, (b) N2: 0.5, (c) N3: 0.75, (d) N4: 1, (e) N5: 1.5 (Rasheed et al., 2021).

reported to get enhanced due to the uniform distribution of cellulose nanocrystals. The initial thermal degradation temperature of the nanocrystal-incorporated composites (PLA-PBS-CNC) was raised to 334°C, compared to that of neat composite (PLA-PBS), which was 331.2°C. The increase in tensile strength of cellulose nanocrystal (different weight percentages)-incorporated composites is represented in Figure 10.10. The SEM images of tensile-tested bionanocomposites are shown in Figure 10.11. The fractured surface has a rough surface with broken fibers. The pulling out of fibers and the delamination were also observed in the SEM images.

10.5   CONCLUSION

The environmental concern related to the synthetic fiber-based composites paved the way for intense research on applications of natural fiber-based bionanocomposites. For improving the properties of natural fibers, several chemical or surface treatments are carried out prior to their use. The major drawbacks associated with natural fiber-based composites can be overcome by the incorporation of nanoparticles. Moreover, the nanomodification strategies can have a good impact in improving the properties of natural fiber-based bionanocomposites. Several studies reported the enhancement of properties like mechanical, thermal, water barrier and other functional properties of natural fiber-based bionanocomposites. Hence, bionanocomposites based on natural fibers can have significant potential in environmental as well as other applications.

ACKNOWLEDGMENT

The author AJ gratefully thanks for the financial support by the King Mongkut's University of Technology North Bangkok (KMUTNB), Thailand, through the Post-Doctoral Program (Grant No. KMUTNB-Post-64-01) and Thailand Science Research and Innovation fund (Grant no: KMUTNB-Basic R-64-16).

REFERENCES

Adel, A., El-Shafei, A., Ibrahim, A., & Al-Shemy, M. (2018). Extraction of oxidized nanocellulose from date palm (*Phoenix dactylifera* L.) sheath fibers: Influence of CI and CII polymorphs on the properties of chitosan/bionanocomposite films. *Industrial Crops and Products, 124*, 155–165. doi:10.1016/j.indcrop.2018.07.073

Al Rashid, A., Khalid, M. Y., Imran, R., Ali, U., & Koc, M. (2020). Utilization of banana fiber-reinforced hybrid composites in the sports industry. *Materials, 13*(14), 3167. doi:10.3390/ma13143167

Alam, M. A., Sapuan, S. M., & Mansor, M. R. (2017). Design characteristics, codes and standards of natural fibre composites. In *Advanced High Strength Natural Fibre Composites in Construction* (pp. 511–528). Woodhead Publishing. doi:10.1016/b978-0-08-100411-1.00020-0

Alomayri, T., Shaikh, F. U. A., & Low, I. M. (2014). Synthesis and mechanical properties of cotton fabric reinforced geopolymer composites. *Composites Part B: Engineering, 60*, 36–42. doi:10.1016/j.compositesb.2013.12.036

Alothman, O. Y., Jawaid, M., Senthilkumar, K., Chandrasekar, M., Alshammari, B. A., Fouad, H., … Siengchin, S. (2020). Thermal characterization of date palm/epoxy composites with fillers from different parts of the tree. *Journal of Materials Research and Technology, 9*(6), 15537–15546. doi:10.1016/j.jmrt.2020.11.020

Ammar, I. M., Huzaifah, M. R. M., Sapuan, S. M., Ishak, M. R., & Leman, Z. B. (2018). Development of sugar palm fiber reinforced vinyl ester composites. In *Natural Fibre Reinforced Vinyl Ester and Vinyl Polymer Composites* (pp. 211–224). Woodhead Publishing. doi:10.1016/b978-0-08-102160-6.00011-1

Arumugam, S., Kandasamy, J., Md Shah, A. U., Hameed Sultan, M. T., Safri, S. N. A., Abdul Majid, M. S., Basri, A. A., & Mustapha, F. (2020). Investigations on the mechanical properties of glass fiber/sisal fiber/chitosan reinforced hybrid polymer sandwich composite scaffolds for bone fracture fixation applications. *Polymers, 12*(7), 1501. doi:10.3390/polym12071501

Asim, M., Paridah, M. T., Chandrasekar, M., Shahroze, R. M., Jawaid, M., Nasir, M., & Siakeng, R. (2020). Thermal stability of natural fibers and their polymer composites. *Iranian Polymer Journal, 29*(7), 625–648. doi:10.1007/s13726-020-00824-6

Candido, R. G. (2021). Recycling of textiles and its economic aspects. In *Fundamentals of Natural Fibres and Textiles* (pp. 599–624). Woodhead Publishing. doi:10.1016/b978-0-12-821483-1.00009-7

Chandrasekar, M., Siva, I., Kumar, T. S. M., Senthilkumar, K., Siengchin, S., & Rajini, N. (2020). Influence of fibre inter-ply orientation on the mechanical and free vibration properties of banana fibre reinforced polyester composite laminates. *Journal of Polymers and the Environment, 28*(11), 2789–2800. doi:10.1007/s10924-020-01814-8

Chao, C. T., & Krueger, R. R. (2007). The date palm (*Phoenix dactylifera* L.): Overview of biology, uses, and cultivation. *HortScience, 42*(5), 1077–1082. doi:10.21273/hortsci.42.5.1077

Chin, S. C., Tee, K. F., Tong, F. S., Ong, H. R., & Gimbun, J. (2020). Thermal and mechanical properties of bamboo fiber reinforced composites. *Materials Today Communications, 23*, 100876. doi:10.1016/j.mtcomm.2019.100876

Djafari Petroudy, S. R. (2017). Physical and mechanical properties of natural fibers. In *Advanced High Strength Natural Fibre Composites in Construction* (pp. 59–83). Woodhead Publishing. doi:10.1016/b978-0-08-100411-1.00003-0

Dochia, M., Sirghie, C., Kozłowski, R. M., & Roskwitalski, Z. (2012). Cotton fibres. In *Handbook of Fiber Science and Technology* (pp. 11–23). doi:10.1533/9780857095503.1.9

Elseify, L. A., & Midani, M. (2020). Characterization of date palm fiber. In *Date Palm Fiber Composites. Composites Science and Technology* (pp. 227–255). doi:10.1007/978-981-15-9339-0_8

Fairuz, A. M., Sapuan, S. M., Zainudin, E. S., & Jaafar, C. N. A. (2016). Effect of filler loading on mechanical properties of pultruded kenaf fibre reinforced vinyl ester composites. *Journal of Mechanical Engineering and Sciences, 10*(1), 1931–1942. doi:10.15282/jmes.10.1.2016.16.0184

Fink, J. K. (2013). Compatibilization. In *Reactive Polymers Fundamentals and Applications* (pp. 373–409). doi:10.1016/b978-1-4557-3149-7.00016-4

Hao, L. C., Sapuan, S. M., Hassan, M. R., & Sheltami, R. M. (2018). Natural fiber reinforced vinyl polymer composites. In *Natural Fibre Reinforced Vinyl Ester and Vinyl Polymer Composites* (pp. 27–70). Woodhead Publishing. doi:10.1016/b978-0-08-102160-6.00002-0

Hasan, K. M. F., Horváth, P. G., & Alpár, T. (2020). Potential natural fiber polymeric nanobiocomposites: A review. *Polymers, 12*(5), 1072. doi:10.3390/polym12051072

Imadi, S. R., Mahmood, I., & Kazi, A. G. (2014). Bamboo fiber processing, properties, and applications. In *Biomass and Bioenergy* (pp. 27–46). doi:10.1007/978-3-319-07641-6_2

Jandas, P. J., Mohanty, S., & Nayak, S. K. (2013). Surface treated banana fiber reinforced poly (lactic acid) nanocomposites for disposable applications. *Journal of Cleaner Production, 52*, 392–401. doi:10.1016/j.jclepro.2013.03.033

Kabir, M. M., Wang, H., Lau, K. T., & Cardona, F. (2012). Chemical treatments on plant-based natural fibre reinforced polymer composites: An overview. *Composites Part B: Engineering, 43*(7), 2883–2892. doi:10.1016/j.compositesb.2012.04.053

Karimah, A., Ridho, M. R., Munawar, S. S., Adi, D. S., Ismadi, Damayanti, R., Subiyanto, B., Fatriasari, W., & Fudholi, A. (2021a). A review on natural fibers for development of eco-friendly bio-composite: characteristics, and utilizations. *Journal of Materials Research and Technology, 13*, 2442–2458. doi:10.1016/j.jmrt.2021.06.014

Karimah, A., Ridho, M. R., Munawar, S. S., Ismadi, Amin, Y., Damayanti, R., Lubis, M. A. R., Wulandari, A. P., Nurindah, Iswanto, A. H., Fudholi, A., Asrofi, M., Saedah, E., Sari, N. H., Pratama, B. R., Fatriasari, W., Nawawi, D. S., Rangappa, S. M., & Siengchin, S. (2021b). A comprehensive review on natural fibers: Technological and socio-economical aspects. *Polymers, 13*(24), 4280. doi:10.3390/polym13244280

Kim, J. T., & Netravali, A. N. (2010). Mercerization of sisal fibers: Effect of tension on mechanical properties of sisal fiber and fiber-reinforced composites. *Composites Part A: Applied Science and Manufacturing, 41*(9), 1245–1252. doi:10.1016/j.compositesa.2010.05.007

Krishnasamy, S., Nagarajan, R., & Thiagamani, S. M. K. (2021). *Mechanical and Dynamic Properties of Biocomposites.* Wiley-VCH, GmbH.

Liu, Y., Liang, Z., Liao, L., & Xiong, J. (2022). Effect of sisal fiber on retrogradation and structural characteristics of thermoplastic cassava starch. *Polymers and Polymer Composites, 30*, 096739112210803. doi:10.1177/09673911221080363

Lu, N., Oza, S., & Tajabadi, M. G. (2015). Surface modification of natural fibers for reinforcement in polymeric composites. In *Surface Modification of Biopolymers* (pp. 224–237). doi:10.1002/9781119044901.ch9

McDougall, G. J., Morrison, I. M., Stewart, D., Weyers, J. D. B., & Hillman, J. R. (1993). Plant fibres: Botany, chemistry and processing for industrial use. *Journal of the Science of Food and Agriculture, 62*(1), 1–20. doi:10.1002/jsfa.2740620102

Nasimudeen, N. A., Karounamourthy, S., Selvarathinam, J., Kumar Thiagamani, S. M., Pulikkalparambil, H., Krishnasamy, S., & Muthukumar, C. (2021). Mechanical, absorption and swelling properties of vinyl ester based natural fibre hybrid composites. *Applied Science and Engineering Progress.* doi:10.14416/j.asep.2021.08.006

Naveen, J., Jawaid, M., Amuthakkannan, P., & Chandrasekar, M. (2019). Mechanical and physical properties of sisal and hybrid sisal fiber-reinforced polymer composites. In *Mechanical and Physical Testing of Biocomposites, Fibre-Reinforced Composites and Hybrid Composites* (pp. 427–440). Woodhead Publishing. doi:10.1016/b978-0-08-102292-4.00021-7

Nilsson, T., & Björdal, C. (2008). The use of kapok fibres for enrichment cultures of lignocellulose-degrading bacteria. *International Biodeterioration & Biodegradation, 61*(1), 11–16. doi:10.1016/j.ibiod.2007.06.009

Rahman, R., & Zhafer Firdaus Syed Putra, S. (2019). 5 - Tensile properties of natural and synthetic fiber-reinforced polymer composites. In *Mechanical and Physical Testing of Biocomposites, Fibre-Reinforced Composites and Hybrid Composites* (pp. 81–102). doi:10.1016/b978-0-08-102292-4.00005-9

Rasheed, M., Jawaid, M., & Parveez, B. (2021). Bamboo fiber based cellulose nanocrystals/poly(lactic acid)/poly(butylene succinate) nanocomposites: morphological, mechanical and thermal properties. *Polymers, 13*(7), 1076. doi:10.3390/polym13071076

Rizal, S., Mistar, E. M., Oyekanmi, A. A., H.P.S, A. K., Alfatah, T., Olaiya, N. G., & Abdullah, C. K. (2021a). Propionic anhydride modification of cellulosic kenaf fibre enhancement with bionanocarbon in nanobiocomposites. *Molecules, 26*(14), 4248. doi:10.3390/molecules26144248

Rizal, S., Mistar, E. M., Rahman, A. A., H.P.S, A. K., Oyekanmi, A. A., Olaiya, N. G., Abdullah, C. K., & Alfatah, T. (2021b). Bionanocarbon functional material characterisation and enhancement properties in nonwoven kenaf fibre nanocomposites. *Polymers, 13*(14), 2303. doi:10.3390/polym13142303

Sabaruddin, F. A., Paridah, M. T., Sapuan, S. M., Ilyas, R. A., Lee, S. H., Abdan, K., Mazlan, N., Roseley, A. S. M., & Abdul Khalil, H. P. S. (2020). The effects of unbleached and bleached nanocellulose on the thermal and flammability of polypropylene-reinforced kenaf core hybrid polymer bionanocomposites. *Polymers, 13*(1), 116. doi:10.3390/polym13010116

Sajna, V. P., Mohanty, S., & Nayak, S. K. (2014). Hybrid green nanocomposites of poly(lactic acid) reinforced with banana fibre and nanoclay. *Journal of Reinforced Plastics and Composites, 33*(18), 1717–1732. doi:10.1177/0731684414542992

Sajna, V. P., Mohanty, S., & Nayak, S. K. (2017). A study on thermal degradation kinetics and flammability properties of poly(lactic acid)/banana fiber/nanoclay hybrid bionanocomposites. *Polymer Composites, 38*(10), 2067–2079. doi:10.1002/pc.23779

Sanadi, A. R., Caulfield, D. F., Jacobson, R. E., & Rowell, R. M. (2002). Renewable agricultural fibers as reinforcing fillers in plastics: mechanical properties of kenaf fiber-polypropylene composites. *Industrial & Engineering Chemistry Research, 34*(5), 1889–1896. doi:10.1021/ie00044a041

Semlali Aouragh Hassani, F.-Z., El Bourakadi, K., Merghoub, N., Qaiss, A. E. K., & Bouhfid, R. (2020). Effect of chitosan/modified montmorillonite coating on the antibacterial and mechanical properties of date palm fiber trays. *International Journal of Biological Macromolecules, 148*, 316–323. doi:10.1016/j.ijbiomac.2020.01.092

Senthil Muthu Kumar, T., Senthilkumar, K., Chandrasekar, M., Karthikeyan, S., Ayrilmis, N., Rajini, N., & Siengchin, S. (2021). Mechanical, thermal, tribological, and dielectric properties of biobased composites. In *Biobased Composites: Processing, Characterization, Properties, and Applications*, John Wiley & Sons, Inc. (pp. 53–73).

Senthilkumar, K., Saba, N., Chandrasekar, M., Jawaid, M., Rajini, N., Siengchin, S., ... Al-Lohedan, H. A. (2021). Compressive, dynamic and thermo-mechanical properties of cellulosic pineapple leaf fibre/polyester composites: Influence of alkali treatment on adhesion. *International Journal of Adhesion and Adhesives, 106*, 102823. doi:10.1016/j.ijadhadh.2021.102823

Senthilkumar, K., Saba, N., Rajini, N., Chandrasekar, M., Jawaid, M., Siengchin, S., & Alotman, O. Y. (2018). Mechanical properties evaluation of sisal fibre reinforced polymer composites: A review. *Construction and Building Materials, 174*, 713–729.

Senthilkumar, K., Siva, I., Rajini, N., & Jeyaraj, P. (2015). Effect of fibre length and weight percentage on mechanical properties of short sisal/polyester composite. *International Journal of Computer Aided Engineering and Technology, 7*(1), 60. doi:10.1504/IJCAET.2015.066168

Senthilkumar, K., Subramaniam, S., Ungtrakul, T., Kumar, T. S. M., Chandrasekar, M., Rajini, N., ... Parameswaranpillai, J. (2022). Dual cantilever creep and recovery behavior of sisal/hemp fibre reinforced hybrid biocomposites: Effects of layering sequence, accelerated weathering and temperature. *Journal of Industrial Textiles, 51*(2_suppl), 2372S–2390S. doi:10.1177/1528083720961416

Shahroze, R. M., Chandrasekar, M., Senthilkumar, K., Senthil Muthu Kumar, T., Ishak, M. R., Rajini, N., ... Ismail, S. O. (2021). Mechanical, interfacial and thermal properties of silica aerogel-infused flax/epoxy composites. *International Polymer Processing, 36*(1), 53 59. doi:10.1515/ipp-2020-3964

Thiruchitrambalam, M., & Shanmugam, D. (2012). Influence of pre-treatments on the mechanical properties of palmyra palm leaf stalk fiber–polyester composites. *Journal of Reinforced Plastics and Composites, 31*(20), 1400–1414. doi:10.1177/0731684412459248

Thomas, S. K., Parameswaranpillai, J., Krishnasamy, S., Begum, P. M. S., Nandi, D., Siengchin, S., ... Sienkiewicz, N. (2021). A comprehensive review on cellulose, chitin, and starch as fillers in natural rubber biocomposites. *Carbohydrate Polymer Technologies and Applications, 2*, 100095. doi:10.1016/j.carpta.2021.100095

Thyavihalli Girijappa, Y. G., Mavinkere Rangappa, S., Parameswaranpillai, J., & Siengchin, S. (2019). Natural fibers as sustainable and renewable resource for development of eco-friendly composites: A comprehensive review. *Frontiers in Materials, 6*. doi:10.3389/fmats.2019.00226

V P, S., Mohanty, S., & Nayak, S. K. (2014). Fabrication and characterization of bionanocomposites based on poly (lactic acid), banana fiber and nanoclay. *International Journal of Plastics Technology, 20*(1), 187–201. doi:10.1007/s12588-014-9088-6

Verma, A., Singh, C., Singh, V. K., & Jain, N. (2019). Fabrication and characterization of chitosan-coated sisal fiber – Phytagel modified soy protein-based green composite. *Journal of Composite Materials, 53*(18), 2481–2504. doi:10.1177/0021998319831748

Vidya Bharathi, S., Vinodhkumar, S., & Saravanan, M. M. (2021). Strength characteristics of banana and sisal fiber reinforced composites. *IOP Conference Series: Materials Science and Engineering, 1055*(1), 012024. doi:10.1088/1757-899x/1055/1/012024

Waghmare, P. M. (2017). Review on mechanical properties of banana fiber biocomposite. *International Journal for Research in Applied Science and Engineering Technology, 5*(10), 847–850. doi:10.22214/ijraset.2017.10120

Wang, G., & Chen, F. (2017). Development of bamboo fiber-based composites. In *Advanced High Strength Natural Fibre Composites in Construction* (pp. 235–255). Woodhead Publishing. doi:10.1016/b978-0-08-100411-1.00010-8

Yu, Y., Wang, H., Lu, F., Tian, G., & Lin, J. (2013). Bamboo fibers for composite applications: A mechanical and morphological investigation. *Journal of Materials Science, 49*(6), 2559–2566. doi:10.1007/s10853-013-7951-z

Zhang, S., Dong, H., He, R., Wang, N., Zhao, Q., Yang, L., Qu, Z., Sun, L., Chen, S., Ma, J., & Li, J. (2022). Hydro electroactive Cu/Zn coated cotton fiber nonwovens for antibacterial and antiviral applications. *International Journal of Biological Macromolecules, 207,* 100–109. doi:10.1016/j.ijbiomac.2022.02.155

Zhou, Y., Fan, M., & Chen, L. (2016). Interface and bonding mechanisms of plant fibre composites: An overview. *Composites Part B: Engineering, 101,* 31–45. doi:10.1016/j.compositesb.2016.06.055

11 Nanocellulose as Reinforcement in Epoxy Composites

Juliana Botelho Moreira, Suelen Goettems
Kuntzler, Larissa Herter Centeno Teixeira,
Priscilla Quenia Muniz Bezerra, Jorge Alberto
Vieira Costa, and Michele Greque de Morais
Federal University of Rio Grande

CONTENTS

11.1 INTRODUCTION

Epoxy resins are an important class of thermosetting polymers that have been widely used in the manufacture of composite materials (Neves et al., 2021; Zhao et al., 2017; Alothman et al., 2020; Chandrasekar et al., 2020). Epoxy has high chemical and mechanical resistance and excellent adhesion (Wongjaiyen et al., 2018; Yue et al., 2019). Furthermore, this resin is easy to process and available at low cost and has good thermal insulation properties (Chen et al., 2019; Neves et al., 2021). Due to these characteristics, in addition to incorporation into composites, epoxy has been used in various applications such as coatings, adhesives, glue materials, electronic packaging, molding materials, foams, and microelectronic encapsulations (Chen et al., 2019; Wongjaiyen et al., 2018; Yue et al., 2019; Zhao et al., 2017).

DOI: 10.1201/9781003271017-11

However, epoxy resins with high crosslink density present brittleness and low fracture toughness. These properties make them vulnerable to the initiation and propagation of cracks, limiting their application in structural materials (Benega et al., 2021; Goyat et al., 2021). In this way, toughening or strengthening agents are applied to improve the physical properties of epoxy composites (Benega et al., 2021). Moreover, the decrease in the availability of petroleum resources and the increase in environmental concerns drive the search for the development of alternative materials from renewable resources such as cellulose (Yue et al., 2019; Krishnasamy et al., 2021; Nasimudeen et al., 2021).

Cellulose is a natural polymer with wide availability and industrial application (Zaman et al. 2020; Zhao et al. 2017; Senthilkumar et al., 2015, 2018, 2021, 2022). This polysaccharide can be isolated from renewable sources, including wood, cotton, and wastes from the food industry (Yue et al., 2019; Senthil Muthu Kumar et al., 2021). One of the main advantages of using cellulose as reinforcement materials is its high crystallinity that improves composite characteristics such as toughness and stiffness. In this perspective, Xiao et al. (2016) developed epoxy composites with sisal cellulose fibers incorporated with hyperbranched polyglycerol. The authors observe improvements in the thermal and mechanical properties of the composites, resulting in a significant increase in impact strength (119.1%), flexural strength (55.2%), tensile strength (45.6%), Young's modulus (43.1%), and toughness (166.1%) compared to pure epoxy. Furthermore, cellulose addition in polymer matrices results in a composite material with renewable, biodegradable, and low-cost characteristics (Neves et al., 2021; Yeo et al., 2017; Shahroze et al., 2021; Thomas et al., 2021).

Nanotechnology is inserted in this context demonstrating that cellulose nanostructures such as nanocrystals and nanofibers incorporated in epoxy composites can improve mechanical, dynamic, static, and thermal properties, in addition to providing multifunctionality (Liu et al. 2018; Rehman et al., 2019; Zhang et al. 2017; Zhao et al. 2017). These characteristics can also be related to the high surface area relative to the volume of nanomaterials (Jahanbaani et al., 2016; Moreira et al., 2020; Saba et al., 2017). The high tensile strength (Wongjaiyen et al., 2018) and the theoretical Young's modulus for nanocellulose suggest the great potential for its application as a reinforcing agent for composite materials (Yue et al., 2019). In addition, Chen et al. (2017) observed that cellulose nanofibers (CNFs) provided an increase in thermal conductivity by 1400% (at a low load of 9.6 vol%) and volumetric electrical resistivity of 10^{15} Ω cm of epoxy composite. The gradual increase in temperature (25°C–125°C) increased the thermal conductivity of the cellulose-reinforced nanocomposites (16 times greater than the material without the polysaccharide) (Chen et al, 2017).

The chapter highlights the potential of nanocellulose for application as a sustainable and renewable reinforcing agent for the development of epoxy composites. Besides, the chapter mainly addresses the effects of the addition of cellulose nanocrystals (CNCs) and CNFs on the physical and thermal properties of epoxy composites.

11.2 CELLULOSE AND DERIVATIVES FOR APPLICATION IN COMPOSITES

11.2.1 OBTAINING CELLULOSE

The cell wall of plants covers the plasma membrane of cells, providing mechanical resistance and protection against pathogens and osmotic stress. This structure is mainly composed of cellulose, hemicellulose, and lignin, forming a complex and compacted structure. Plants also contain pectin and extractive compounds (terpenes, waxes, oils, resins, alkaloids, and tannins) (Poletto et al., 2014; Tarasov et al., 2018).

The main advantage of using cellulose instead of synthetic polymers is its biodegradability. Cellulose can reinforce other polymeric structures such as fibers, hydrogels, aerogels, and films (Zaman et al., 2020; Zhao et al., 2019). Furthermore, cellulose is the most abundant natural polymer (Motaung & Linganiso, 2018) and a renewable resource that can be used sustainably (Hao et al., 2018). In this context, cellulose can be obtained from waste originating in the food industry (Figure 11.1), which represents an environmental problem in many cases because it does not have an adequate destination (Naduparambath et al., 2018; Nunes, 2017; Oliveira et al., 2019; Pereira & Arantes, 2020). However, the most used sources of cellulose in the industry are still wood and cotton (Zhao et al., 2019).

As cellulose is associated with fractions of other compounds (mainly hemicelluloses and lignin), any use of this polymer must be preceded by extraction and purification processes, which occur mainly through mechanical, chemical, and semichemical routes. Biological and enzymatic treatments are alternative methods for obtaining purified cellulose. In general, these processes consist of a lignin removal

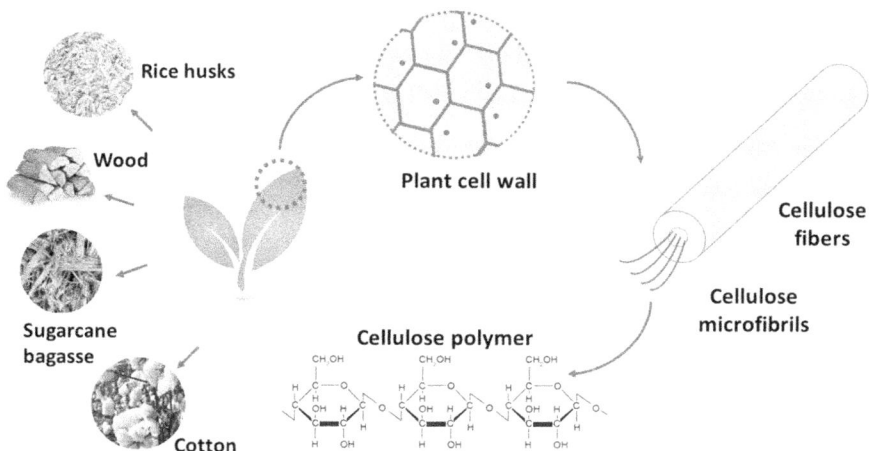

FIGURE 11.1 Illustrative representation of cellulose extraction from plant sources (agroindustrial residues).

step through the use of alkaline inorganic substances (delignification). As in this step, there is no total removal of lignin associated with cellulose. Thus, it is necessary to use oxidizing agents to remove residual lignin, hemicellulose, extractive compounds, impurities, and remaining residues. This second step is called bleaching. Gaseous or elemental chlorine was widely used for this purpose, being gradually replaced by Total Chlorine-Free technologies. These techniques include the use of alkalis, hydrogen peroxide, ozone, and enzymes, in addition, to presenting a lower environmental impact when compared to the use of chlorinated compounds (Bajpai, 2018a, b; Holtzapple, 2003; Sjöström, 1993).

11.2.2 CELLULOSE PROPERTIES

Cellulose is a linear-chain polysaccharide formed by glucose molecules that bond by a condensation reaction, giving rise to D-glucopyranosyl units linked through the β-(1,4) glycosidic linkage with an angle of 180° to each other (French, 2017; Kalász et al., 2020). The degree of polymerization of cellulose varies according to the plant source and the extraction and purification processes, ranging from 300 to 1700 for wood and up to 10,000 for cotton and linen, for example (Shrotri et al., 2017; Zuppolini et al., 2022).

The presence of hydroxyl groups along this polymeric chain is responsible for forming hydrogen bonds that can be of two types: intramolecular, responsible for the rigidity of the polymeric chain, and intermolecular, which organize the cellulose chains in a parallel and orderly manner, forming the cellulose fibrils. This structure contains highly ordered crystalline domains alternating with amorphous regions, characterizing a semicrystalline polymer with a degree of crystallinity between 40% and 60%. These characteristics provide very particular mechanical properties for cellulose, having Young's modulus of 100–140 GPa (for crystalline regions) and tensile strength of 130–200 MPa. In addition, it has a low density, between 1.5 and 1.6 g/cm^3 (Aravamudhan et al., 2014; Van Rie & Thielemans, 2017).

The crystalline domains are formed by a three-dimensional network, whose smallest unit consists of a geometric arrangement of defined dimensions called unit cell. The unit cell is representative of the entire crystal structure in terms of composition and physical, chemical, and optical properties (Moon et al., 2011). There are four possible dimensions for the unit cell, corresponding to four polymorphs named cellulose I, cellulose II, cellulose III, and cellulose IV. In particular, the polymorph Iα is found in algae and bacteria, while Iβ is found in plants. Therefore, the cellulose I is called native cellulose (Nunes, 2017).

Although each glucose unit has three hydroxyl groups in its structure, cellulose is insoluble in water or even in conventional organic and inorganic solvents. The polymer chains are linked through hydrogen bonds. As a result, a highly ordered and compacted structure is formed and preferentially bonded rather than interacting with solvents (O'Brien et al., 2021). Amphoteric solvents, ionic liquids, and combinations such as lithium chloride and dimethylacetamide have been reported as alternative systems capable of dissolving cellulose (Friend et al., 2019; Glasser et al., 2012; Härdelin et al., 2012; Medronho et al., 2012; Zhao et al., 2019).

11.2.3 CELLULOSE DERIVATIVES

Although cellulose is recognized for its mechanical strength, its fibers are not easily processable due to their low solubility. Thus, it cannot interact well with other polymers, limiting its function as a reinforcing agent (Zuppolini et al., 2022). Besides, the high-water holding capacity is estimated to be between 3.5 and 10 times its weight (Ang, 1991), which provides changes in the mechanical properties of the material (Miao & Hamad, 2013). On the other hand, each glucose monomer of cellulose has three hydroxyl groups at positions 2, 3, and 6 (Aravamudhan et al., 2014), allowing obtaining cellulose derivatives, which are generally soluble in water or organic solvents (Oprea & Voicu, 2020). Examples are the esters of inorganic acids (cellulose nitrates), organic acids (cellulose acetates), and ionic and nonionic ethers (carboxymethyl cellulose [CMC], methylcellulose, hydroxyethyl cellulose, and hydroxypropyl cellulose).

Cellulose nitrate, or nitrocellulose, represents esters derived from inorganic acids and are prepared by nitration of cellulose pulp using a mixture of sulfuric acid and nitric acid. The applications of cellulose nitrate vary according to the degree of substitution of the reaction: between 12.6% and 13.3% is classified as explosive; below 12.6% nitrogen, it has good biocompatibility and physicochemical stability (Oprea & Voicu, 2020).

Among the esters of organic acids, cellulose acetate is biodegradable, nontoxic, and non-irritating material. The cellulose acetate is produced by the reaction of cellulose with acetic acid and acetic anhydride in the presence of sulfuric acid. Each glucose molecule has three hydroxyl groups, so the degree of substitution can range from 1 to 3, corresponding to cellulose mono, di, and triacetate, respectively (Aravamudhan et al., 2014; French, 2017; Niaounakis, 2017). Due to its affinity with conventional solvents such as acetic acid, acetone, dimethylformamide, and dimethylacetamide, cellulose acetate is widely used to obtain nanofibers using the electrospinning technique. The purified cellulose is not easily dissolved under these conditions (Zhang et al., 2021). Cellulose acetate is also widely used in the manufacture of membranes applied to matrices for controlled drug delivery (Rodrigues Filho et al., 2016), scaffolds for tissue engineering (Atila et al., 2015), photographic films and coatings (Chen, 2015), membranes to retain contaminants in water (Pandele et al., 2020), and also in the textile industry (Chen, 2015; Wei et al., 2020).

CMC is an anionic and water-soluble cellulose derivative, which is also commonly presented in the form of sodium CMC. CMC is the main ether derived from cellulose, obtained by reaction with sodium monochloroacetate in an alkaline medium (Rachtanapun et al., 2021). Both CMC and its sodium salt are FDA (Food and Drug Administration)-approved for use as a food additive (FDA, 2022; Oprea & Voicu, 2020). They improve palatability, texture, stability, and consistency properties (Ergun et al., 2016). CMC can be used as a disintegrating agent in drugs (Kim et al., 2020), and in the formulation of stimuli-responsive hydrogels used for the controlled release of drugs due to properties such as good gelling capacity and water absorption (Oprea & Voicu, 2020).

11.3 NANOTECHNOLOGY APPLIED TO EPOXY COMPOSITES

Epoxy resins belong to the class of thermosetting polymers and are used as a coating, reinforcement in composites, and encapsulating material (Ahmed et al., 2013; Ayad et al., 2012). However, the application of epoxy in polymer composite is limited because they are considered rigid and brittle materials. Also, these resins exhibit high crosslinking density, making them vulnerable to initiation and propagation of the crack. Thus, the increasing demand for advanced materials with improved properties has led research to develop epoxy composites with nanotechnology (Benega et al., 2021).

The cost-effectiveness, chemical resistance, easy processability, and structural flexibility can be optimized through the appropriate combinations of the nanometric constituents (Hsu et al., 2011; Pascault & Williams, 2010). The main advantage of using nanomaterials to reinforce epoxy composites is to improve the interfacial interaction of the constituents, resulting in enhanced thermal, mechanical, and electrical properties compared to conventional composites (Figure 11.2) (Jahromi et al., 2016; Chammingkwan et al., 2016; Guo et al., 2016; Paran et al., 2019).

Nanostructures can be used in aerospace and marine coating applications due to the improved physicochemical, mechanical, optical, electrical, and corrosive properties of polymer composites with epoxy (Gu et al., 2016). Mohammadkhani et al. (2021) used diamond nanoparticles for the development of polyaniline composite with epoxy resins. The addition of 5% diamond nanoparticles showed better corrosion protection performance in the nanocomposites due to the superior barrier property that prevented the penetration of corrosive agents into the material. Graphene oxide nanosheets and nanofillers also provided the epoxy composites with high corrosion and mechanical properties (Keshmiri et al., 2021; Vinodhini & Xavier, 2022).

The curing behavior of epoxy composites and electrical properties are altered with the addition of nanomaterials (Ahmadi, 2019). Charoeythornkhajhornchai et al. (2021) developed fast curing epoxy composites to improve the electrical and thermal properties of electronic parts processed by 3D printing. The epoxy composites

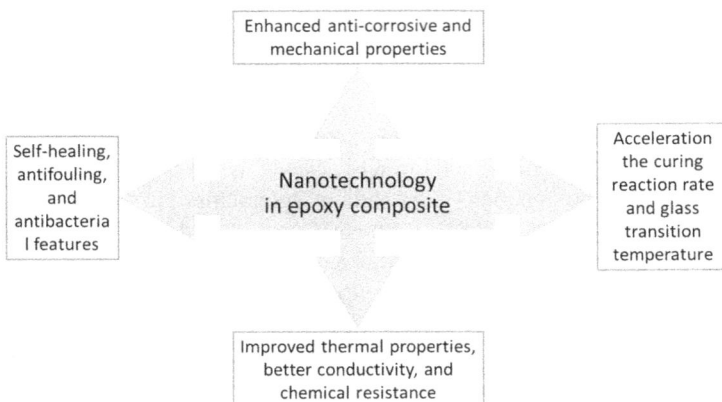

FIGURE 11.2 Schematic illustration of epoxy composites production using nanotechnology.

were produced with multi-walled carbon nanotubes (MWCNTs) and titanium dioxide (TiO_2) nanoparticles. Complete curing of the composite containing 75% of TiO_2 nanoparticles and 3–5 part per hundreds of resin loading of MWCNTs occurred after 3–4 minutes of UV exposure on the material. In addition, the MWCNTs imparted high thermal conductivity and increased the dielectric constant and electrical conductivity of the epoxy composites. Thus, nanomaterials are strategies to produce epoxy composites with superior properties, expanding their applications and multifunctionality (Goyat et al., 2021).

11.4 NANOCELLULOSE AS A REINFORCING AGENT IN EPOXY COMPOSITES

The advancement of nanotechnology has boosted research on nanocellulosic materials, which can be obtained from natural, sustainable, and renewable sources, as well as have the features of low cost, non-toxicity, and biocompatibility (Mokhena & John, 2020). These nanomaterials exhibit high surface area to volume ratio, tensile strength, high Young's modulus, low thermal expansion coefficient, and hydrogen bonding ability (Figure 11.3) (Salimi et al., 2019). Nanocellulose used as reinforcement in epoxy composites can prevent the displacement of constituents. Modified cellulose provides better adhesion between matrix and reinforcement agents with superior properties. Most studies with nanocellulose applied in epoxy composites are related to CNCs and CNFs (Dai et al., 2013; Rehman et al., 2019).

11.4.1 NANOCRYSTALS

CNCs are materials like nanoparticle-elongated, rigid, and highly crystalline rod-shaped with a width between 4 and 70 nm and 54% to 88% crystallinity index (Lavoine & Bergström, 2017; Liu et al., 2016; Naz et al., 2019). This nanomaterial is usually produced from cellulosic fibers undergoing acid hydrolysis. After this step, the

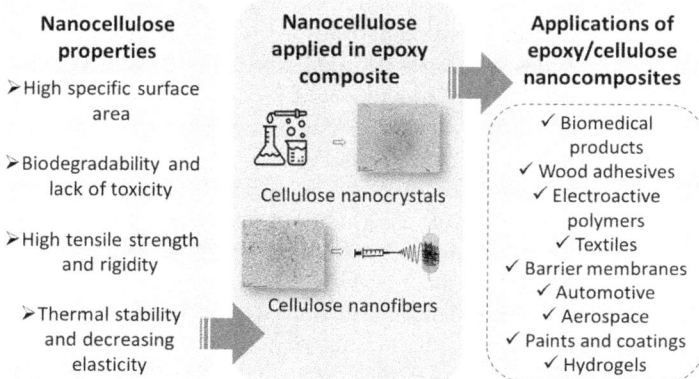

FIGURE 11.3 The main characteristics of nanocrystals and nanofibers cellulose and their applications in epoxy composites.

nanostructure undergoes transverse cleavage of amorphous regions resulting in small rod-shaped particles with enhanced crystallinity (Moberg et al., 2017). CNCs originate only from the amorphous phase of cellulose since hydrolysis of the crystalline part does not occur due to the strong hydrogen bonding of the OH groups (Habibi et al., 2010; Siró & Plackett, 2010). In the literature, CNCs can also be found as cellulose whiskers, cellulose nanowhiskers (CNWs), and nanocrystalline cellulose (Charreau et al., 2020).

CNCs are interesting as reinforcement in epoxy composites due to characteristics such as high specific surface area at around $150\,m^2/g$, modulus of elasticity close to 150 GPa attributing to high crystallinity around 50%–90%, and tensile strength of 10 GPa (Kaboorani & Riedl 2015; Lin & Dufresne 2014; Voronova et al., 2015; Wei et al., 2017). Nanocrystals exhibit hydrophilic surfaces without water swelling ability, surface charge, chiral nematic, and amphiphilic properties, and serve to modify the rheological, optical, and electrical properties of polymers (De France et al., 2017).

Thermoplastic and thermosetting polymers can be reinforced with CNCs to produce high-quality and cost-effective nanomaterials (Lu et al., 2014). The mechanical characteristics of such nanocomposites are affected by the interfacial adhesion between the nanocrystals and the polymer matrix (Gopi et al., 2019). In addition, the combination of CNCs and epoxy resins in nanocomposites provides better corrosion resistance and stiffness and reduces internal stresses during the curing process (Miao & Hamad, 2019). Ayrilmis et al. (2019) developed an epoxy composite with CNWs obtained from acid hydrolysis with fibers from *Moringa oleifera* fruit. The influence of different concentrations (0.06%, 0.12%, or 0.18% by weight) of CNW on the mechanical and thermal characteristics of the epoxy composites was investigated. The thermal stability of the epoxy nanocomposites increased by 0.18% CNW compared to the pure epoxy composite. In addition, tensile (1682.5 MPa) and (3145.9 MPa) flexural modulus of the epoxy composite containing 0.18% CNW increased significantly compared to pure epoxy composite (1280.3 and 2275.4 MPa, respectively). Yue et al. (2018) used CNCs from ramie fibers to produce epoxy nanocomposites. The thermomechanical properties such as storage modulus and glass transition temperature increased to 151.5 MPa and 136.3°C, respectively, for the epoxy nanocomposites with 10% amine-functionalized CNCs compared to pure resin.

Cai et al. (2019) produced epoxy composites with CNCs and nanofiber membranes. The mechanical properties such as tensile modulus (3.53 ± 0.08 GPa), tensile strength (76.42 ± 1.16 MPa), and flexural modulus (2.87 ± 0.18 GPa) and flexural strength (138.91 ± 1.47 MPa) of epoxy composites containing 0.5% CNCs increased compared to the pure material (67.56 ± 2.13 MPa; 3.00 ± 0.03 GPa; 128.24 ± 0.97 MPa; 2.62 ± 0.0 GPa, respectively). The study by Wang et al. (2019) shows the effect of CNCs on the interlaminar fracture toughness of epoxy composites with nanofiber membranes. The material containing 3% CNCs showed a 28% and 20% increase in mode I and II interlaminar fracture, respectively, compared to the pure composite.

Incorporating CNCs in composites can improve mechanical, optical, electrical, and thermal properties, photodynamic activity, nanoporous character, and high adsorption capacity. These characteristics promote the application of cellulose nanocomposites in diverse applications, such as the biomedical and food packaging sector, environment including water purification and collection, photoelectric materials, and electronic devices (Bacakova et al., 2020; Trache et al., 2020).

11.4.2 NANOFIBERS

Nanofibers are materials that can be produced by different techniques, including drawing, template synthesis, phase separation, self-assembly, and electrospinning (Moreira et al., 2018). The electrospinning process has gained prominence in different areas (Moreira et al., 2020; Nioradze et al., 2021; Wang et al., 2022; Zhang et al., 2022) by offering flexibility in handling process conditions and the possibility of industrial-scale application (Moreira et al., 2018). Furthermore, electrospinning can be applied to synthetic polymers and biopolymers such as cellulose (Jahanbaani et al., 2016; Polez et al., 2021; Silva et al., 2021).

In this context, the electrospinning technique has become interesting to produce CNFs as reinforcing agents for composite materials. CNF sheets can be used interchangeably with epoxy layers to form a mechanically robust laminated architecture (Nissilä et al., 2018). Thus, to improve the mechanical properties of the epoxy resin, Jahanbaani et al. (2016) used a mat of CNFs to prepare a laminated epoxy composite. The study found that the results for tensile strength, modulus, and elongation at break of nanofiber-laminated composite were higher (an increase of 28.7 MPa, 1737.5 MPa, and 0.14%, respectively) than those found for epoxy resin. The nanofiber sheet improved impact resistance (202.30 KJ/m^2). Crystallinity and chemical characteristics of cellulose also contributed to the increase in Young's modulus of the composites. Furthermore, the nanofiber layer controlled and limited the propagation of cracks in the fracture zone of the epoxy matrix (Jahanbaani et al., 2016).

Gabr et al. (2014) observed that the epoxy composite prepared with 0.3% of CNFs showed an increase of approximately 15% in the critical stress intensity factor and 24% in the critical strain energy release rate compared to the pure resin. On the other hand, the addition of 0.1% of CNFs provided a 20% increase in the storage modulus at 30°C. The gradual increase in stiffness and flexural strength observed for the 0.1% CNF nanocomposites indicated that the stresses were efficiently transferred across the interface. The authors also found that CNFs contributed to hampering crack propagation.

In another study, nets of CNFs were prepared via ice-templating and used as preforms for impregnation with epoxy resin. The composite produced with 13% of CNFs showed improvement in mechanical properties, which showed an increase in flexural modulus (1.9 GPa), strength (18 MPa), and storage modulus (1.4 GPa). The flexural and storage moduli were approximately 25% higher in the longitudinal direction. This statement indicates that the structure of the network of CNFs is anisotropic, and the CNFs showed an oriented nature within the epoxy matrix (Nissilä et al., 2019). Nissilä et al. (2021) also prepared CNFs by ice-templating. Thin filaments were formed from CNFs and impregnated with an epoxy resin via vacuum infusion. The process resulted in cellulose nanocomposites with a strong fiber–matrix interface and oriented structure. The orientation indices (0.6 and 0.53 for the CNFs in the filament mats and the composites, respectively) indicated a significant alignment.

Saba et al. (2017) evaluated the effect of CNF loading (0.5%, 0.75%, and 1%) on the mechanical thermal and dynamic analysis of epoxy composites as a function of temperature. The thermal stability, char content, storage modulus, loss modulus, and glass transition temperature increase for all nanocomposites compared to pure epoxy. Moreover, based on viscoelastic properties, thermal stability, decomposition temperature, and residual content, 0.75% CNFs were considered the best condition for the production of epoxy nanocomposite.

In general, the studies associate the improvement of thermomechanical properties of epoxy/cellulose composites to the reinforcement with CNFs, cellulose's crystallinity, as well as the formation of interfacial interaction by hydrogen bonds between the CNFs and the epoxy matrix that results in intima adhesion of the epoxy resin to the cellulose fibers.

11.5   FUTURE PERSPECTIVES

The natural origin of cellulose is one of the main advantages of using this biopolymer to develop composites. After its degradation, the principal substance released is glucose, which does not compromise the environment. Furthermore, cellulose is the most abundant polymer on Earth, with practically unlimited resources (Oprea & Voicu, 2020). These aspects have aroused the interest of researchers in the development of different composite materials.

One of the major challenges in nanostructure-reinforced composites is to achieve a homogeneous dispersion of particles during processing (Chanda & Bajwa, 2021). Although cellulose has several advantages related to biodegradability and sustainability, its application for the development of new composite materials depends on the chemical modification of the biomolecule. The use of chemically modified cellulose for the production of epoxy composites can contribute to the adhesion of cellulose to the resin, in addition to improving cellulose dispersion and the final properties of the composite (Neves et al., 2021). However, depending on the chemical treatment used, the process can become expensive and not ecologically correct. On the other hand, physical techniques are not fully explored to make them industrially scalable (Chanda & Bajwa, 2021).

Besides, few reports are found in the literature using cellulose as a reinforcing material for epoxy composites. Therefore, further studies are needed to investigate the reaction mechanisms at the fiber/epoxy interface. Modification methods can also be further explored to achieve more economical and ecological approaches with less impact on the environment (Neves et al., 2021).

11.6   CONCLUSIONS

The use of cellulose as a reinforcement in epoxy composites improves the mechanical and thermodynamic properties of the material. The main advantages were to give the composite multifunctionality and sustainable characteristics. Furthermore, studies with cellulose nanocrystals and nanofibers show that these nanomaterials improve corrosion resistance and stiffness, accelerate the curing process, and increase Young's modulus of the epoxy composite. These new features in the functional properties were attributed to the high surface area the thermomechanical, optical, and electrical resistance of nanocellulose. Thus, the chapter contributes to stimulating research in the use of polymeric matrices of natural, renewable, and low-cost origin, under the scope of producing materials that will cause less impact to the environment.

ACKNOWLEDGMENTS

This study was financed in part by the Coordenação de Aperfeiçoamento de Pessoal de Nível Superior – Brasil (CAPES) – Finance Code 001. This research was developed within the scope of the Capes-PrInt Program (Process # 88887.310848/2018-00).

REFERENCES

Ahmadi, Z. 2019. Epoxy in nanotechnology: a short review. *Prog. Org. Coat.* 132:445–448.

Ahmed, K., S. S. Nizami, and N. Z. Raza. 2013. Characteristics of natural rubber hybrid composites based on marble sludge/carbon black and marble sludge/rice husk derived silica. *J. Ind. Eng. Chem.* 19:1169–1176.

Alothman, O.Y., Jawaid, M., Senthilkumar, K., Chandrasekar, M., Alshammari, B.A., … S. Siengchin. 2020. Thermal characterization of date palm/epoxy composites with fillers from different parts of the tree. *J. Mater. Res. Technol.* 9(6):15537–15546.

Ang, J. F. 1991. Water retention capacity and viscosity effect of powdered cellulose. *J. Food Sci.* 56:1682–1684.

Aravamudhan, A., D. M. Ramos, A. A. Nada, and S. G. Kumbar. 2014. Natural polymers: polysaccharides and their derivatives for biomedical applications. In *Natural and Synthetic Biomedical Polymers*, eds. S. G. Kumbar, C. T. Laurencin and M. Deng, pp. 67–89. Amsterdam: Elsevier, Inc.

Atila, D., D. Keskin, and A. Tezcaner. 2015. Cellulose acetate based 3-dimensional electrospun scaffolds for skin tissue engineering applications. *Carbohydr. Polym.* 133:251–261.

Ayad, M. M., A. Abu El-Nasr, and J. Stejskal. 2012. Kinetics and isotherm studies of methylene blue adsorption onto polyaniline nanotubes base/silica composite. *J. Ind. Eng. Chem.* 18:1964–1969.

Ayrilmis, N., F. Ozdemir, O. B. Nazarenko, and P. M. Visakh. 2019. Mechanical and thermal properties of Moringa oleifera cellulose-based epoxy nanocomposites. *J. Compos. Mater.* 53:669–675.

Bacakova, L., J. Pajorova, M. Tomkova, R. Matejka, A. Broz, J. Stepanovska, S. Prazak, A. Skogberg, S. Siljander, and P. Kallio. 2020. Applications of nanocellulose/nanocarbon composites: focus on biotechnology and medicine. *Nanomaterials* 10:196.

Bajpai, P. 2018a. Pulp bioprocessing. In *Biermann's Handbook of Pulp and Paper*, ed P. Bajpai, vol. 1, pp. 583–602. Amsterdam: Elsevier, Inc.

Bajpai, P. 2018b. Pulp bleaching. In *Biermann's Handbook of Pulp and Paper*, ed P. Bajpai, vol. 1, pp. 465–491. Amsterdam: Elsevier, Inc.

Benega, M. A. G., W. M., Silva, M. C., Schnitzler, R. J. E., Andrade, and H. Ribeiro. 2021. Improvements in thermal and mechanical properties of composites based on epoxy-carbon nanomaterials-A brief landscape. *Polym. Test.* 98:107180.

Cai, S., Y. Li, H. Y. Liu, and Y. W. Mai. 2019. Effect of electrospun polysulfone/cellulose nanocrystals interleaves on the interlaminar fracture toughness of carbon fiber/epoxy composites. *Compos. Sci. Technol.* 181:107673.

Chammingkwan, P., K. Matsushita, T. Taniike, and M. Terano. 2016. Enhancement in mechanical and electrical properties of polypropylene using graphene oxide grafted with end-functionalized polypropylene. *Materials* 9:240.

Chanda, S., and D. S. Bajwa. 2021. A review of current physical techniques for dispersion of cellulose nanomaterials in polymer matrices. *Rev. Adv. Mater. Sci.* 60:325–341.

Chandrasekar, M., I. Siva, T.S.M. Kumar, K. Senthilkumar, S. Siengchin, and N. Rajini. 2020. Influence of fibre inter-ply orientation on the mechanical and free vibration properties of banana fibre reinforced polyester composite laminates. *J. Polym. Environ.* 28(11):2789–2800.

Charoeythornkhajhornchai, P., K. Tedsree, and R. Chancharoen. 2021. Effect of carbazole coating on TiO_2 nanoparticles as a photosensitizer and MWCNTs on the performance of epoxy composites. *J. Sci.: Adv. Mater. Devices.* 6:425–434.

Charreau, H., E. Cavallo, and M. L. Foresti. 2020. Patents involving nanocellulose: analysis of their evolution since 2010. *Carbohyd. Polym.* 237:116039.

Chen, H. 2015. Lignocellulose biorefinery process engineering. In *Lignocellulose Biorefinery Engineering*, ed H. Chen, pp. 167–217. Amsterdam: Elsevier, Inc.

Chen, J., X. Huang, Y. Zhu, and P. Jiang. 2017. Cellulose nanofiber supported 3D interconnected BN nanosheets for epoxy nanocomposites with ultrahigh thermal management capability. *Adv. Funct. Mater.* 27:1604754.

Chen, R., K. Hu, H. Tang, J. Wang, F. Zhu, and H. Zhou. 2019. A novel flame retardant derived from DOPO and piperazine and its application in epoxy resin: flame retardance, thermal stability and pyrolysis behavior. *Polym. Degrad. Stab.* 166:334–343.

Dai, D., M. Fan, and P. Collins. 2013. Fabrication of nanocelluloses from hemp fibers and their application for the reinforcement of hemp fibers. *Ind. Crops. Prod.* 44:192–199.

De France, K. J., T. Hoare, and E. D. Cranston. 2017. Review of hydrogels and aerogels containing nanocellulose. *Chem. Mater.* 29:4609–4631.

Ergun, R., J. Guo, and B. Huebner-Keese. 2016. Cellulose. In *Encyclopedia of Food and Health*, eds B. Caballero, P. M. Finglas and F. Toldrá, pp. 694–702. Amsterdam: Elsevier, Inc.

FDA. 2022. Food Additive Status List. https://www.fda.gov/food/food-additives-petitions/food-additive-status-list#ftnC (accessed March 04, 2022).

French, A. D. 2017. Glucose, not cellobiose, is the repeating unit of cellulose and why that is important. *Cellulose* 24:4605–4609.

Friend, D. F. L., M. E. L. González, M. M. Caraballo, and A. A. A. de Queiroz. 2019. Biological properties of electrospun cellulose scaffolds from biomass. *J. Biomater. Sci. Polym.* 30:1399–1414.

Gabr, M., N. T. Phong, K. Okubo, K. Uzawa, I. Kimpara, and T. Fujii. 2014. Thermal and mechanical properties of electrospun nano-cellulose reinforced epoxy nanocomposites. *Polym. Test.* 37:51–58.

Glasser, W. G., R. H. Atalla, J. Blackwell, M. M. Brown, W. Burchard, A. D. French, D. O. Klemm, and Y. Nishiyama. 2012. about the structure of cellulose: debating the lindman hypothesis. *Cellulose* 19:589–598.

Gopi, S., P. Balakrishnan, D. Chandradhara, D. Poovathankandy, and S. Thomas. 2019. General scenarios of cellulose and its use in the biomedical field. *Mater. Today Chem.* 13:59–78.

Goyat, M. S., A. Hooda, T. K. Gupta, K. Kumar, S. Halder, P. K. Ghosh, and B. S. Dehiya. 2021. Role of non-functionalized oxide nanoparticles on mechanical properties and toughening mechanisms of epoxy nanocomposites. *Ceram. Int.* 47:22316–22344.

Gu, H., C. Ma, J. Gu, J. Guo, X. Yan, J. Huang, Q. Zhang, and Z. Guo. 2016. An overview of multifunctional epoxy nanocomposites. *J. Mater. Chem.* C 4:5890–5906.

Guo, J., J. Long, D. Ding, Q. Wang, Y. Shan, A. Umar, X. Zhang, B. L. Weeks, S. Wei, and Z. Guo. 2016. Significantly enhanced mechanical and electrical properties of epoxy nanocomposites reinforced with low loading of polyaniline nanoparticles. *RSC Adv.* 6:21187–21192.

Habibi, Y., L. A. Lucia, and O. J. Rojas. 2010. Cellulose nanocrystals: chemistry, self-assembly, and applications. *Chem. Rev.* 110:3479–3500.

Hao, L. C., S. M. Sapuan, M.R. Hassan, R.M. Sheltami. 2018. Natural fiber reinforced vinyl polymer composites. In *Natural Fibre Reinforced Vinyl Ester and Vinyl Polymer Composites*, eds. S.M. Sapuan, H. Ismail and E.S. Zainudin, pp. 27–70. Amsterdam: Elsevier, Inc.

Härdelin, L., J. Thunberg, E. Perzon, G. Westman, P. Walkenström, and P. Gatenholm. 2012. Electrospinning of cellulose nanofibers from ionic liquids: the effect of different cosolvents. *J. Appl. Polym. Sci.* 125:1901–1909.

Holtzapple, M.T. 2003. Cellulose. In *Encyclopedia of Food Sciences and Nutrition*, ed. B. Caballero, pp. 998–1007. Amsterdam: Elsevier, Inc.

Hsu, L., C. Weder, and S. J. Rowan. 2011. Stimuli-responsive, mechanically-adaptive polymer nanocomposites. *J. Mater. Chem.* 21:2812–2822.

Jahanbaani, A. R., Behzad, T., Borhani, S., and M. H. K. Darvanjooghi. 2016. Electrospinning of cellulose nanofibers mat for laminated epoxy composite production. *Fibers Polym.* 17:1438–1448.

Jahromi, A. E., H. R. E. Jahromi, F. Hemmati, M. R. Saeb, V. Goodarzi, and K. Formela. 2016. Morphology and mechanical properties of polyamide/clay nanocomposites toughened with NBR/NBR-g-GMA: a comparative study. *Compos. Part B: Eng.* 90:478–484.

Kaboorani, A., and B. Riedl. 2015. Surface modification of cellulose nanocrystals (CNC) by a cationic surfactant. *Ind. Crop Prod.* 65:45–55.

Kalász, H., M. Báthori, and K. L. Valkó. 2020. Basis and pharmaceutical applications of thin-layer chromatography. In *Handbook of Analytical Separations*, ed. K. l. Valkó, vol. 8, pp. 523–585. Amsterdam: Elsevier B.V.

Keshmiri, N., P. Najmi, M. Ramezanzadeh, and B. Ramezanzadeh. 2021. Designing an eco-friendly lanthanide-based metal organic framework (MOF) assembled graphene-oxide with superior active anti-corrosion performance in epoxy composite. *J. Clean. Prod.* 319:128732.

Kim, S., D.-H. Cho, D.-K. Kweon, E.-H. Jang, J.-Y. Hong, S.-T. Lim. 2020. Improvement of mechanical properties of orodispersible hyaluronic acid film by carboxymethyl cellulose addition. *Food Sci Biotechnol.* 29:1233–1239.

Krishnasamy, S., R. Nagarajan, S.M.K. Thiagamani, S. Siengchin. 2021. *Mechanical and Dynamic Properties of Biocomposites*. Hoboken: Wiley-VCH.

Lavoine, N., and L. Bergström. 2017. Nanocellulose-based foams and aerogels: processing, properties, and applications. *J. Mater. Chem. A* 5:16105–16117.

Lin, N., and A. Dufresne. 2014. Nanocellulose in biomedicine: current status and future prospect. *Eur. Polym. J.* 59:302–325.

Liu, C., B. Li, H. Du, D. Lv, Y. Zhang, G. Yu, X. Mu, and H. Peng. (2016). Properties of nanocellulose isolated from corncob residue using sulfuric acid, formic acid, oxidative and mechanical methods. *Carbohyd. Polym.* 151:716–724.

Liu, S., V. S. Chevali, Z. G. Xu, D. Hui, and H. Wang. 2018. A review of extending performance of epoxy resins using carbon nanomaterials. *Compos. B Eng.* 136:197–214.

Lu, Y., H. L. Tekinalp, C. C. Eberle, W. Peter, A. K. Naskar, and S. Ozcan. 2014. Nanocellulose in polymer composites and biomedical applications. *Tappi J.* 13:47–54.

Medronho, B., A. Romano, M. G. Miguel, L. Stigsson, and B. Lindman. 2012. Rationalizing cellulose (in)solubility: reviewing basic physicochemical aspects and role of hydrophobic interactions. *Cellulose* 19:581–587.

Miao, C., and W. Y. Hamad. 2013. Cellulose reinforced polymer composites and nanocomposites: a critical review. *Cellulose* 20:2221–2262.

Miao, C., and W. Y. Hamad. 2019. Critical insights into the reinforcement potential of cellulose nanocrystals in polymer nanocomposites. *Curr. Opin. Solid State Mater. Sci.* 23:100761.

Moberg, T., K. Sahlin, K. Yao, S. Geng, G. Westman, Q. Zhou, K. Oksman, and M. Rigdahl. 2017. Rheological properties of nanocellulose suspensions: effects of fibril/particle dimensions and surface characteristics. *Cellulose* 24:2499–2510.

Mohammadkhani, R., A. Shojaei, P. Rahmani, N. P. Tavandashti, and M. Amouzegar. 2021. Synthesis and characterization of polyaniline/nanodiamond hybrid nanostructures with various morphologies to enhance the corrosion protection performance of epoxy coating. *Diam. Relat. Mater.* 120:108672.

Mokhena, T. C., and M. J. John. 2020. Cellulose nanomaterials: new generation materials for solving global issues. *Cellulose* 27:1149–1194.

Moon, R. J., A. Martini, J. Nairn, J. Simonsen, and J. Youngblood. 2011. Cellulose nanomaterials review: structure, properties and nanocomposites. *Chem. Soc. Rev.* 40:3941–3994.

Moreira, J. B., S. G. Kuntzler, A. L. M. Terra, J. A. V. Costa, and M. G. Morais. 2020. Electrospun nanofibers: fundamentals, food packaging technology, and safety. In *Food Packaging Advanced Materials, Technologies, and Innovations*, ed. S. M. Rangappa, J. Parameswaranpillai, S. M. K. Thiagamani, S. Krishnasamy, and S. Siengchin, pp. 223–254. Boca Raton: CRC Press.

Moreira, J. B., M.G. Morais, E.G. Morais, B.S. Vaz, and J.A.V. Costa. 2018. Electrospun polymeric nanofibers in food packaging. In *Impact of Nanoscience in the Food Industry*, ed. A. M. Grumezescu and A. M. Holban, pp. 387–417. Amsterdam: Elsevier Inc.

Motaung, T. E., and L. Z. Linganiso. 2018. Critical review on agrowaste cellulose applications for biopolymers. *Int. J. Plast. Technol.* 22:185–216.

Naduparambath, S., T.V. Jinitha, V. Shaniba, M.P. Sreejith, A. K. Balan, and E. Purushothaman. 2018. Isolation and characterisation of cellulose nanocrystals from sago seed shells. *Carbohydr. Polym.* 180:13–20.

Nasimudeen, N.A., Karounamourthy, S., Selvarathinam, J., Kumar Thiagamani, S.M., Pulikkalparambil, H., ... C. Muthukumar. 2021. Mechanical, absorption and swelling properties of vinyl ester based natural fibre hybrid composites. *Appl. Sci. Eng. Prog.* 14:680–688.

Naz, S., J. S. Ali, and M. Zia. 2019. Nanocellulose isolation characterization and applications: a journey from non-remedial to biomedical claims. *Biodesign Manuf.* 2:187–212.

Neves, R. M., H. L. Ornagh Jr, A. J. Zattera, and S. C. Amico. 2021. Recent studies on modified cellulose/nanocellulose epoxy composites: a systematic review. *Carbohydr. Polym.* 255:117366.

Niaounakis, M. 2017. The problem of marine plastic debris. In *Management of Marine Plastic Debris*, ed M. Niaounakis, pp. 1–55. Norwich: William Andrew Publishing.

Nioradze, N., Ciornii, D., Kölsch, A., Göbel, G., Khoshtariya, D. E., Zouni, A., and F. Lisdat. 2021. Electrospinning for building 3D structured photoactive biohybrid electrodes. *Bioelectrochemistry* 142:107945.

Nissilä, T., Hietala, M., and K. Oksman. 2019. A method for preparing epoxy-cellulose nanofiber composites with an oriented structure. *Compos. Part A* 125:105515.

Nissilä, T., Karhula, S. S., Saarakkala, S., and K. Oksman. 2018. Cellulose nanofiber aerogels impregnated with bio-based epoxy using vacuum infusion: structure, orientation and mechanical properties. *Compos. Sci. Technol.* 155:64–71.

Nissilä, T., Wei, J., Geng, S., Teleman, A., and K. Oksman. 2021. Ice-templated cellulose nanofiber filaments as a reinforcement material in epoxy composites. *Nanomaterials* 11:490.

Nunes, R.C.R. 2017. Rubber nanocomposites with nanocellulose. In *Progress in Rubber Nanocomposites*, eds. S. Thomas and H. J. Maria, pp. 463–494. Cambridge: Woodhead Publishing.

O'Brien, C. T., T. Virtanen, S. Donets, J. Jennings, O. Guskova, A. H. Morrell, M. Rymaruk, L. Ruusuvirta, J. Salmela, H. Setala, J.U. Sommer, A.J. Ryan, and O.O. Mykhaylyk. 2021. Control of the aqueous solubility of cellulose by hydroxyl group substitution and its effect on processing. *Polymer* 223:123681.

Oliveira, J. P., G. P. Bruni, S. L. M. el Halal, F. C. Bertoldi, A. R. G. Dias, and E. R. Zavareze. 2019. Cellulose nanocrystals from rice and oat husks and their application in Aerogels for Food Packaging. *Int. J. Biol. Macromol.* 124:175–184.

Oprea, M., and S. I. Voicu. 2020. Recent advances in composites based on cellulose derivatives for biomedical applications. *Carbohydr. Polym.* 247:116683.

Pandele, A. M., H. Iovu, C. Orbeci, C. Tuncel, F. Miculescu, A. Nicolescu, C. Deleanu, and S. I. Voicu. 2020. Surface modified cellulose acetate membranes for the reactive retention of tetracycline. *Sep. Purif. Technol.* 249:1–9.

Paran, S. M. R., H. Vahabi, M. Jouyandeh, F. Ducos, K. Formela, and M. R. Saeb. 2019. Thermal decomposition kinetics of dynamically vulcanized polyamide 6-acrylonitrile butadiene rubber-halloysite nanotube nanocomposites. *J. Appl. Polym. Sci.* 136:47483.

Pascault, J. P., and R. J. J. Williams. 2010. *Epoxy Polymers: New Materials and Innovations*. Hoboken: John Wiley & Sons.

Pereira, B., and V. Arantes. 2020. Production of cellulose nanocrystals integrated into a biochemical sugar platform process via enzymatic hydrolysis at high solid loading. *Ind. Crops Prod.* 152:112377.

Poletto, M., H. Ornaghi, and A. Zattera. 2014. Native cellulose: structure, characterization and thermal properties. *Materials* 7:6105–6119.

Polez, R. T., B. V. M. Rodrigues, O. A. El Seoud, and E. Frollini. 2021. Electrospinning of cellulose carboxylic esters synthesized under homogeneous conditions: effects of the ester degree of substitution and acyl group chain length on the morphology of the fabricated mats. *J. Mol. Liq.* 339:116745.

Rachtanapun, P., W. Klunklin, P. Jantrawut, N. Leksawasdi, K. Jantanasakulwong, Y. Phimolsiripol, P. Seesuriyachan, T. Chaiyaso, W. Ruksiriwanich, S. Phongthai, S.R. Sommano, W. Punyodom, A. Reungsang, and T.M.P Ngo. 2021. Effect of monochloroacetic acid on properties of carboxymethyl bacterial cellulose powder and film from nata de coco. *Polymers* 13:488.

Rehman, M. M., M. Zeeshan, K. Shaker, and Y. Nawab. 2019. Effect of micro-crystalline cellulose particles on mechanical properties of alkaline treated jute fabric reinforced green epoxy composite. *Cellulose* 26:9057–9069.

Rodrigues Filho, G., F. Almeida, S. D. Ribeiro, T. F. Tormin, R. A.A. Muñoz, R. M.N. Assunção, and H. Barud. 2016. Controlled release of drugs from cellulose acetate matrices produced from sugarcane bagasse: monitoring by square-wave voltammetry. *Drug Dev. Ind. Pharm.* 42:1066–1072.

Saba, N., A. Safwan, M. L. Sanyang, F. Mohammad, M. Pervaiz, M. Jawaid, O. Y. Alothman, and M. Sain. 2017. Thermal and dynamic mechanical properties of cellulose nanofibers reinforced epoxy composites. *Int. J. Biol. Macromol.* 102:822–828.

Salimi, S., R. Sotudeh-Gharebagh, R. Zarghami, S. Y. Chan, and K. H. Yuen. 2019. Production of nanocellulose and its applications in drug delivery: a critical review. *ACS Sustain. Chem. Eng.* 7:15800–15827.

Senthil Muthu Kumar, T., K. Senthilkumar, M. Chandrasekar, S. Karthikeyan, N. Ayrilmis, … S. Siengchin. 2021. Mechanical, thermal, tribological, and dielectric properties of biobased composites. In *Biobased Composites: Processing, Characterization, Properties, and Applications*, ed. A. Khan, S. M. Rangappa, S. Siengchin, A. M. Asiri, pp. 53–73. Hoboken: John Wiley & Sons.

Senthilkumar, K., N. Saba, M. Chandrasekar, M. Jawaid, N. Rajini, … H.A. Al-Lohedan. 2021. Compressive, dynamic and thermo-mechanical properties of cellulosic pineapple leaf fibre/polyester composites: influence of alkali treatment on adhesion. *Int. J. Adhes. Adhes.* 106:102823.

Senthilkumar, K., N. Saba, N. Rajini, M. Chandrasekar, M. Jawaid, … O.Y. Alotman. 2018. Mechanical properties evaluation of sisal fibre reinforced polymer composites: a review. *Constr. Build. Mater.* 174:713–729.

Senthilkumar, K., I. Siva, N. Rajini, and P. Jeyaraj. 2015. Effect of fibre length and weight percentage on mechanical properties of short sisal/polyester composite. *Int. J. Comput. Aided Eng. Technol.* 7(1):60.

Senthilkumar, K., S. Subramaniam, T. Ungtrakul, T.S.M. Kumar, M. Chandrasekar, … J. Parameswaranpillai. 2022. Dual cantilever creep and recovery behavior of sisal/hemp fibre reinforced hybrid biocomposites: effects of layering sequence, accelerated weathering and temperature. *J. Ind. Text.* 51(2_suppl):2372S–2390S.

Shahroze, R.M., M. Chandrasekar, K. Senthilkumar, T. Senthil Muthu Kumar, M.R. Ishak, … S.O. Ismail. 2021. Mechanical, Interfacial and thermal properties of silica aerogel-infused flax/epoxy composites. *Int. Polym. Process.* 36(1):53–59.

Shrotri, A., H. Kobayashi, and A. Fukuoka. 2017. Catalytic conversion of structural carbohydrates and lignin to chemicals. In *Advances in Catalysis*, ed. C. Song, pp. 59–123. Amsterdam: Elservier Inc.

Silva, B. A., Cunha, R. S., Valério, A., Junior, A. D. N., Hotza, D., and S. Y. G. González. 2021. Electrospinning of cellulose using ionic liquids: an overview on processing and applications. *Eur. Polym. J.* 147:110283.

Siró, I., and D. Plackett. 2010. Microfibrillated cellulose and new nanocomposite materials: a review. *Cellulose* 17:459–494.

Sjöström, E. 1993. Pulp Bleaching. In *Wood Chemistry* (Second Edition), ed. E. Sjöström, pp. 165–203. Amsterdam: Elservier Inc.

Tarasov, D., M. Leitch, and P. Fatehi. 2018. Lignin–carbohydrate complexes: properties, applications, analyses, and methods of extraction: a review. *Biotechnol. Biofuels* 11:269.

Thomas, S.K., J. Parameswaranpillai, S. Krishnasamy, P.M.S. Begum, D. Nandi, … N. Sienkiewicz. 2021. A comprehensive review on cellulose, chitin, and starch as fillers in natural rubber biocomposites. *Carbohydr. Polym. Technol. Appl.* 2:100095.

Trache, D., A. F. Tarchoun, M. Derradji, T. S. Hamidon, N. Masruchin, N. Brosse, and M. H. Hussin. 2020. Nanocellulose: from fundamentals to advanced applications. *Front. Chem.* 8:392.

Van Rie, J., and W. Thielemans. 2017. Cellulose–gold nanoparticle hybrid materials. *Nanoscale* 9:8525–8554.

Vinodhini, S. P., and J. R. Xavier. 2022. Effect of graphene oxide wrapped functional silicon carbide on structural, surface protection, water repellent, and mechanical properties of epoxy matrix for automotive structural components. *Colloids Surf. A: Physicochem. Eng. Asp.* 639:128300.

Voronova, M. I., O. V. Surov, S. S. Guseinov, V. P. Barannikov, and A. G. Zakharov. 2015. Thermal stability of polyvinyl alcohol/nanocrystalline cellulose composites. *Carbohyd. Polym.* 130:440–447.

Wang, J., T. R. Pozegic, Z. Xu, R. Nigmatullin, R. L. Harniman, and S. J. Eichhorn. 2019. Cellulose nanocrystal-polyetherimide hybrid nanofibrous interleaves for enhanced interlaminar fracture toughness of carbon fibre/epoxy composites. *Compos. Sci. Technol.* 182:107744.

Wang, J., Z. Wang, J. Ni, and L. Li. 2022. Electrospinning for flexible sodium-ion batteries. *Energy Stor. Mater.* 45:704–719.

Wei, L., U. P. Agarwal, K. C. Hirth, L. M. Matuana, R. C. Sabo, and N. M. Stark. 2017. Chemical modification of nanocellulose with canola oil fatty acid methyl ester. *Carbohyd. Polym.* 169:108–116.

Wei, W., Y. Zhu, Q. Li, Z. Cheng, Y. Yao, Q. Zhao, and P. Zhang. 2020. An Al_2O_3-cellulose acetate-coated textile for human body cooling. *Sol. Energy Mater. Sol. Cells* 211:110525.

Wongjaiyen, T., W. Brostow, and W. Chonkaew. 2018. Tensile properties and wear resistance of epoxy nanocomposites reinforced with cellulose nanofibers. *Polym. Bull.* 75: 2039–2051.

Xiao, X., S. Lu, L. Pan, C. Zeng, Z. He, J. Gao, and J. Yu. 2016. Enhanced thermal and mechanical properties of epoxy composites by addition of hyperbranched polyglycerol grown on cellulose fibers. *J. Polym. Res.* 23:72.

Yeo, J. S., O. Y. Kim, and S. H. Hwang. 2017. The effect of chemical surface treatment on the fracture toughness of microfibrillated cellulose reinforced epoxy composites. *J. Ind. Eng. Chem.* 45:301–306.

Yue, L., A. Maiorana, F. Khelifa, A. Patel, J. M. Raquez, L. Bonnaud, R. Gross, P. Dubois, and I. Manas-Zloczower. 2018. Surface-modified cellulose nanocrystals for biobased epoxy nanocomposites. *Polymer* 134:155–162.

Yue, L., F. Liu, S. Mekala, A. Patel, R. A. Gross, and I. Manas-Zloczower. 2019. High performance biobased epoxy nanocomposite reinforced with a bacterial cellulose nanofiber network. *ACS Sustainable Chem. Eng.* 7:5986–5992.

Zaman, A., F. Huang, M. Jiang, W. Wei, and Z. Zhou. 2020. Preparation, properties, and applications of natural cellulosic aerogels: a review. *Energy Built Environ.* 1:60–76.

Zhang, X., L. Xie, X. Wang, Z. Shao, and B. Kong. 2022. Electrospinning super-assembly of ultrathin fibers from single- to multi-Taylor cone sites. *Appl. Mater. Today* 26:101272.

Zhang, Y., C. Zhang, and Y. Wang. 2021. Recent progress in cellulose-based electrospun nanofibers as multifunctional materials. *Nanoscale Adv.* 3:6040–6047.

Zhang, Z., Y. Li, and C. Chen. 2017. Synergic effects of cellulose nanocrystals and alkali on the mechanical properties of sisal fibers and their bonding properties with epoxy. *Compos. Part A Appl. Sci. Manuf.* 101:480–489.

Zhao, G., X. Lyu, J. Lee, X. Cui, and W.-N. Chen. 2019. Biodegradable and transparent cellulose film prepared eco-friendly from durian rind for packaging application. *Food Packag. Shelf Life* 21:100345.

Zhao, J., Q. Li, X. Zhang, M. Xiao, W. Zhang, and C. Lu. 2017. Grafting of polyethylenimine onto cellulose nanofibers for interfacial enhancement in their epoxy nanocomposites. *Carbohydr. Polym.* 157:1419–1425.

Zuppolini, S., A. Salama, I. Cruz-Maya, V. Guarino, and A. Borriello. 2022. Cellulose amphiphilic materials: chemistry, process and applications. *Pharmaceutics* 14:386.

12 Fatigue Behavior of Natural Fiber-Based Epoxy Composites

P. Sathish Kumar, J. Arulmozhivarman,
L. Rajeshkumar, M. R. Sanjay,
and Suchart Siengchin
King Mongkut's University of Technology
North Bangkok (KMUTNB)

M. Ramesh
KIT-Kalaignarkarunanidhi Institute of Technology

CONTENTS

12.1 INTRODUCTION

A polymer serves as the matrix phase in a polymer matrix composite (PMC), whereas fibers, microparticles or nanoparticles might indeed serve as reinforcements. The three most often used polymer matrices in engineering applications are polyesters, epoxies, and vinyl esters. Due of their availability, polyesters and vinyl esters are primarily employed with glass fiber-reinforced composites. Typically, in structural, automobile, and aerospace applications, epoxy resins, particularly those of the diglycidyl ether of bisphenol A (DGEBA) type, have been cured using a range

DOI: 10.1201/9781003271017-12

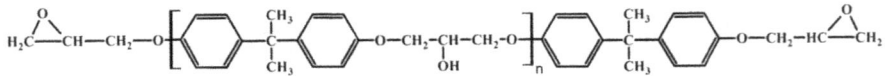

FIGURE 12.1 Structure of epoxy (DGEBA) resin [2].

of curing agents, including aliphatic amines, anhydrides, polyamides, and aromatic amines. The maximum application temperature for epoxy resins may vary significantly depending on the monomers and curing agents. Petroleum products usually come in liquid or solid state, with different degrees of viscosity, as epoxies. These resins establish three-dimensional (3D) lattice structures with excellent mechanical characteristics and exceptional resistance to solvents, bases, and acids upon reacting with several curing agents and various hardeners [1]. DGEBA, which is the reaction product of epichlorohydrin and bisphenol A, is used to make the majority of commercially available epoxies (Figure 12.1) [2]. The opening of the epoxy ring produces hydroxyls (OH), which enable additional cross-linking among chemical reagents. Furthermore, the presence of OH groups speeds up the initial reaction rate [3,4] while acting as a catalyst.

Epoxies exhibit exceptional resistance to reactor neutrons with degassing characteristic as well as stability. Epoxy is primarily used as a matrix in the atomic energy sector, which includes floor coating of nuclear fuel components and radiation equipment [5]. Epoxy also has good resilience against gamma radiations. Composites having a polymer matrix are among the many low Z materials that are excellent at reducing high-energy neutrons. Epoxy resin is more valuable than metal matrix composites (MMCs) or thermoplastic polymers due to its low weight and simplicity in the production process. Due to relatively minimum contraction, epoxy resin exhibits great qualities such as dimensional stability [6], excellent adherence to a variety of reinforcements with specific improvement in mechanical strength, heat resistance, and resistance to chemical solvents [7]. Unless the polymer contains an aromatic ring, epoxy possesses extraordinary resistance to neutron and gamma radiation [8]. As a result, epoxy resin is often utilized as a reinforcement component in radiation equipment, and various studies have investigated composite materials with epoxy polymers as a radiation-shielding material [9–12]. The erosion rates of several polymeric materials have been examined by certain authors [13,14]; however, there is very little research on polymer coatings for the potential usage of PMCs in tribological applications. With steels, epoxy polymers demonstrate their suitability as a coating material [15].

Epoxy has to be cross-linked via a curing process in order to create polymers with 3D networks. The ultimate qualities of the cured epoxies depend on the kind of curing agent and epoxies used, the curing cycle, and the amount of cross-linkage. Numerous scientific techniques, including differential scanning calorimetry and dynamic mechanical analysis, have been utilized to characterize the curing response and continuously monitor the thermoset curing processes [16–18]. Software may be used to link the glass transition temperature, T_g, to the extent of epoxy polymerization. As with zeolite-based epoxy composites [19] and graphite-epoxy preimpregnated fibers [20,21], the synthesis of epoxy-based composites has utilized Kamal's approach. The technique is also applicable to the curing process, in which the original monomer quickly disappears and lowers the free volume [22]. A nominal proportion of functional groups get "nested" in the networks if the conversion hits the threshold value, which either completely or partly restricts their mobility. Diffusion now begins to influence the response, strengthening the continuous 3D network. Other researchers who studied the resin made from epoxidized soybean oil also noticed these same results [23]. This technique was determined by researchers to be a suitable way for predicting kinetic properties of glass fiber-reinforced epoxy matrix composite along with other approaches [24]. The material properties in a modeling framework for the assessment of stresses created as a result of curing thermoset composites may also be used as input from this data [25]. Viscosity modeling may also be done using these kinetic parameters [26]. The cure kinetics of amine-epoxy resins is also susceptible to diffusion-control modeling [27]. Epoxy resins display a remarkable quality combined with good endurance in severe conditions, such as when exposed to neutron and gamma rays [28,29]. In composites, the epoxy binds the particles together and transmits load. As a result, the durability and macroscopic characteristics of the developed composites are determined by a complex interaction between chemical and physical variables that occur at the interface areas of matrices and reinforcements [30]. It might be beneficial to use composites in structural applications to comprehend their fatigue characteristics.

S–N graphs are often employed to compare and estimate the fatigue life of different materials [31]. As shown in equation (12.1), Mandell [32] suggested a model wherein the fatigue life cycle is proportional to the fatigue strength. As long as the total number of cycles is about 10^2 and 10^6, such a perfect correlation ensures a satisfactory fit.

$$\sigma = \sigma_{ult} - BlogN_f \qquad (12.1)$$

In this equation, N_f is the number of cycles until failure, B is the coefficient of fatigue strength, σ is the load range, and σ_{ult} is the ultimate tensile strength (UTS). Caprino and Giorleo [33] presented a normalized approach, as indicated in equation (12.2), to more accurately evaluate the fatigue characteristics of various materials.

$$\frac{\sigma}{\sigma_{ult}} = 1 - BlogN_f \qquad (12.2)$$

Although fatigue has been the most frequent failure in almost all of the structural parts, the fatigue characteristics of natural fiber-based composites have recently attracted a lot of attention from scientific researchers. Fifty percent of these failures are found to be due to fatigue [34]. Experimental research into the fatigue characteristics of non-woven, hemp fiber mat-based polyester matrix composites with random orientation was conducted by Yuanjian and Isaac [35]. The fatigue behavior of composites reinforced with natural fibers has been the subject of a number of investigations. In lightweight structure applications like aircraft industry and automotives, PMCs reinforced with carbon, kevlar, and glass fibers were frequently utilized. Epoxies in specific are used because of their high mechanical properties, less viscosity, low volatile nature, and low shrinkage, in comparison to many thermoset polymers [36–39]. The reinforcing is something that gives the composite its outstanding strength and rigidity [40,41]. Epoxy is often used in the production of composites incorporating reinforcement from natural fibers. Epoxies may be used in a variety of applications; however, they do not always perform as expected [42]. Epoxy's toughness and strength may be improved by combining it with reinforcements such as carbon fillers, silica, eggshell powder, zinc oxide, and zinc powder, as well as by incorporating other synthetic and natural fillers. It is a matrix that strengthens the composites' mechanical and physical characteristics, making it useful in a wide variety of disciplines, including high-performance transport facilities, aeronautical, and automotive, as well as in biological engineering and artificial organ transplanting [43].

12.2   FATIGUE TEST METHODS

A stress–strain versus time plot of a fatigue experiment is shown in Figure 12.2. Between two fixed stress or strain extremes (σ_{max} and σ_{min}), a cyclical load is imposed. R ratio refers to the relationship between the lowest and highest stresses. Other crucial measurements are the mean stress, stress amplitude, and cycle frequency. The sinusoidal or triangular cyclic stress mode is both feasible and desirable depending on future practice. C–C fatigue, T–T fatigue, flexural fatigue, interlaminar shear fatigue, torsion fatigue, and torsional fatigue are the six most common forms of fatigue testing methods. Axial tension–tension cycling has been used for a wide range of fatigue studies on fiber-reinforced polymer composites. However, failure by compressive buckling may occur in thin laminates, and T–C and C–C cycles are not often employed. During fatigue testing, the tension and compression are cycled in the opposite direction. Interlaminar shear fatigue and in-plane shear fatigue experiments were conducted, although only in small quantities. S–N curves are frequently used to represent the fatigue strength (S) and fatigue life (N) with constant amplitude loading. The peak stress or the stress amplitude throughout the course of a cycle is depicted. For a given stress cycle, the number of cycles to failure is shown on a logarithmic axis. The most common way of displaying data clearly shows how steadily the amplitude fatigue degrades the material parameters. All materials, including polymers, metals, and composites, have negative S–N curve slopes. Different materials have unique curve patterns.

12.3   PREDICTIONS FOR FATIGUE LIFE

Kang et al. [45] carried out research work to enhance noise, vibration, and crashworthiness. Structural adhesives are extensively used in the automotive sector to enhance mechanical durability. Tension samples have been fabricated and bonded by employing any adhesive, either by itself or in combination with a spot-welding contact. The samples were loaded to the adhesive junction plane at an angle of 45°. Sheet steels such as ultra-high strength steel and mild steel were used to fabricate the samples. Fatigue tests were performed on the specimens, and a stress-life curve was generated using a structural stress technique. The Monte Carlo simulation is frequently

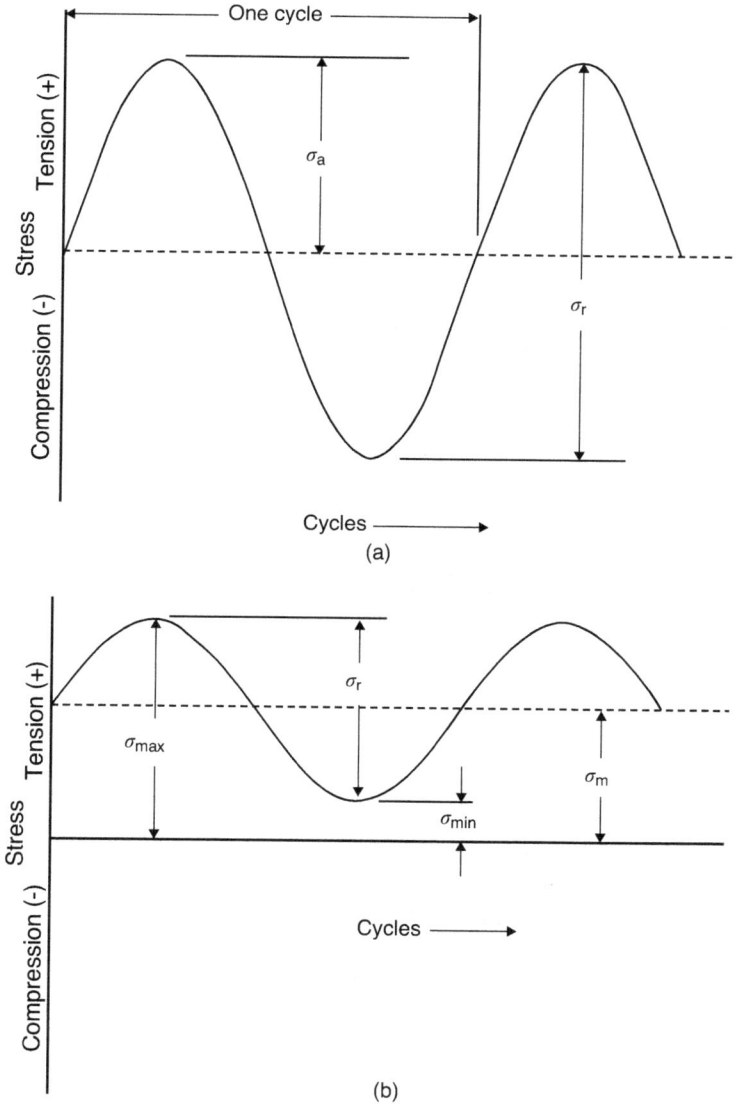

FIGURE 12.2   (a, b) Stress–strain versus time plot in fatigue tests [44].

employed in the research done by Pashah et al. [46] to ascertain how random vari-ability in the key parameters affects the anticipated fatigue behavior of a heat sink. Using the measured parameter uncertainties, a suitable comparison is then made between the predicted fatigue life of laminated boosters constructed with similar adhesive and placed in bending joint configurations.

Nonlinear computational modeling of a single-lap bonded joint was investigated by Kumar and Pandey [47]. The modified Coffin–Manson equations for bonded joints were used to forecast fatigue performance with respect to geometrical and

TABLE 12.1

Fatigue Characteristics Upon Various Hybridizations for Natural Fiber-Reinforced Composites [49,50]

Materials	Operating Temperature	Amplitude of Applied Stress	Loading Type	No. of Cycles to Failure (N_f)
Flax-reinforced epoxy	23°C	40%–80% of UTS	T–T	2058×10^6 cycles
Glass-reinforced epoxy	23°C	40%–80% of UTS	T–T	700×10^6 cycles

material nonlinearities. It has been proven through an explicit approach for calculating the typical fatigue durability of welded joints subjected to random loading that causes weld toe stresses in the critical area with a stress record distributed evenly in accordance with the Weibull law. However, Pang et al. [48] demonstrate the inelastic strain range or inelastic strain energy density fatigue-producing stress variables generated through finite element analysis (FEA) results. Natural fiber composites' fatigue characteristics upon various hybridizations were depicted in Table 12.1.

12.4 FATIGUE LIFE PREDICTION USING ARTIFICIAL NEURAL NETWORKS (ANNs)

With neural networks, Lee et al. were able to accurately depict the fatigue performance of epoxy composites reinforced with carbon and glass fibers at ranges of stress ratio from 0.1 to 10. They discovered that the precision of the output variable was significantly affected by stress ratios and the number of hidden layers during development [51]. Scientists used many distinct neural network architectures in an effort to fine-tune the predictive ability of their models [52]. According to their findings, for unidirectionally oriented glass/epoxy composites, modular neural networks offer the best accurate relationship between design variables with a reduced amount of data sets. Prediction of the fatigue behavior of glass fibre reinforced epoxy (GFRE) composites has been performed using the Levenberg–Marquardt (LM) method. The loads applied to the composites are a combination of rotational bending moments. In order to simulate real-world conditions, fatigue testing in a rotating bending configuration at 10 Hz, with completely reversible loading cycles, was conducted. The fatigue lifetime of the GFRE matrix composite reduced as the stress levels increased, according to the findings of the fatigue tests. Additionally, this research used ANN technique to demonstrate how stress levels affect the fatigue strength of GFRE composites, and their corresponding failure modes were determined. With a mean-squared error of 1.38E−21 obtained during development; the results of the ANN predictions show good agreement in relation to experimental results (Figure 12.3a). An absolute R^2 of 0.99857 was computed, proving that ANN can predict and forecast the fatigue lifetime of glass fiber-based composite materials with the use of the LM method. Even though just a minimal sample of experimental data was used to train the neural network, it outperforms more conventional fatigue life predictions. A conjugate-gradient backpropagation method using a pattern recognition module

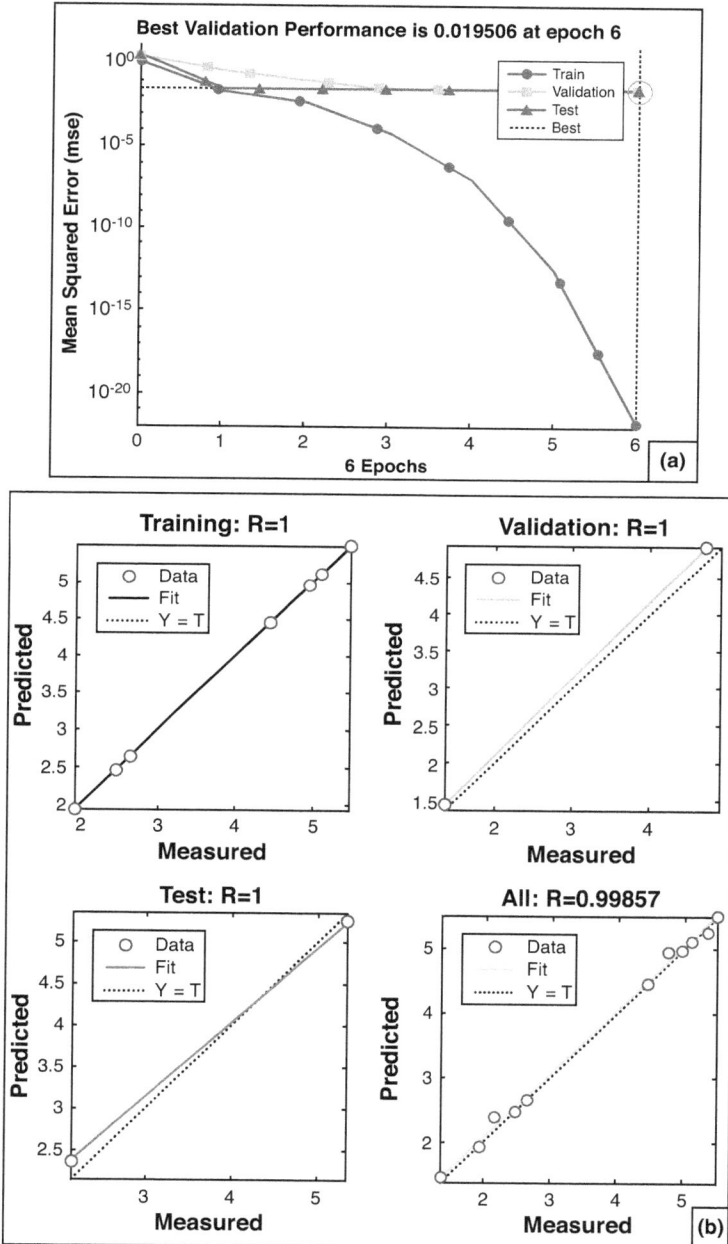

FIGURE 12.3 (a) The mean square error graph for the output parameter according to the iteration of ANN; (b) relationship between GFRE composites' experimental and anticipated failure cycles [53].

in MATLAB was successfully used to detect various types of faults under diverse beginning circumstances with 100% accuracy. At low stresses, matrix cracking and interface debonding have been the dominant failure mechanism, while with high stresses, fiber interfacial splitting contributed to a catastrophic event of glass fiber-reinforced composite materials under low and high cycle fatigue testing.

12.5 FATIGUE DAMAGE MODELING OF FIBER-REINFORCED EPOXY COMPOSITES

Glass fabric-reinforced epoxy composites were fatigue-loaded as a cantilever beam under displacement control, and their behavior during fatigue was modeled with the help of a finite element method. Dissimilar damage distribution over the thickness and throughout the length of the specimen caused stresses to be continually redistributed throughout the fatigue life [54]. The implementation using finite elements successfully replicated the same. Caprino and Giorleo [33] recently utilized their model to analyze the fatigue of glass fabric-reinforced epoxy composites under four-point bending, while Caprino [55] employed the residual strength model to analyze the fatigue of carbon fiber-reinforced composites under tension. Caprino found that the model could estimate the fatigue performance of carbon fiber-reinforced composites, but the strength properties as determined empirically did not follow the trend as predicted by the residual strength law. In such a scenario, the model needs to be seen as a fatigue life model rather than a residual strength model.

Short fatigue fracture propagation at notches in woven glass-reinforced epoxy composites was studied by Biner and Yuhas [56]. It was shown that the effective stress intensity factor range (ΔK_{eff}) effectively describes the initiation and propagation rate of small fractures propagating from blunt notches. Using conformal mapping methods, researchers determined the value of ΔK_{eff} for a notch plus crack structure, applicable to short cracks. With long sufficient fracture lengths, ΔK_{eff} converged into ΔK. The mechanical characteristics in the warp direction were much greater than those in the weft direction, which was a subject of research for Xiao and Bathias [57,58]. While no fatigue evolution rule was proposed, fatigue ratios were suggested for evaluating the consistency of the available field data. The findings indicated that the fatigue strength ratios of the unnotched and notched laminates for the three stacking sequences investigation were equivalent to their respective static strength ratio analysis.

According to their findings, the fatigue life is affected by the fiber orientation. When 90° layers are confined by 0° layers, damage in the 90° layers is less likely to spread over the interface to the other plies. Due to this, the damage trace is very curvy as it progresses through the material. Coats and Harris [59] utilized the internal stress factors theory to investigate tension–tension fatigue of unnotched and notched graphite-reinforced bismaleimide laminates, whereas Lee et al. [60] used the same theory to predict stiffness losses in glass-reinforced and graphite-reinforced epoxy laminates with matrix deformation. The tensile fatigue behavior of notched carbon-reinforced epoxy crossply laminates has been extensively researched by Beaumont and Spearing [61], Spearing and Beaumont [62,63], and Spearing et al [64,65]. Splitting in the 0° plies, delamination regions at the 90°/0° interfaces whose size is proportional to the split length, and transverse ply cracking in the 90° plies, were the most often observed damage types. The split length and delamination angle define the idealized damaged patterns. Few experiments have been conducted to verify the validity of the suggested residual strength equation; however, the microcrack density has been experimentally validated for tension–tension fatigue testing on $[0_2, 90_2]_s$ carbon-reinforced epoxy composites.

12.6   FATIGUE PROPERTIES OF NATURAL FIBER-REINFORCED COMPOSITES

Like all other mechanical characteristics, the fatigue behavior of composites made from natural fibers has not been extensively investigated. Gassan et al. [66] investigated the behavior of thermoplastics and thermosets. These flax and jute composites were evaluated for tension–tension fatigue using epoxy, polyester, and polypropylene (PP). Fiber mechanical characteristics, quantity of fiber, textile architecture, fiber–matrix adhesion, and tension–tension fatigue behavior were all studied. Composite damping was shown to be significantly different between unidirectionally oriented flax and jute-epoxy composites, with the flax/epoxy composites showing an increase in damping of around twofold. This difference may be attributed to the fibers' fine structure and surface morphology. The development of damage in both composite combinations was quite comparable. Using jute-epoxy composites with both unidirectional and weaving reinforcement, researchers studied the effect of textile architecture in material properties. A lower critical load for damage propagation and failure, as well as quicker damage propagation, was found for woven reinforcing composites. It was shown that the fatigue behavior of reinforced brittle polyesters and ductile PP matrices was significantly impacted by the quality of fiber–matrix interaction.

The critical load for crack development and the rate of damage propagation were lower for composites reinforced with untreated jute woven mats. It was discovered that the fiber strength and modulus had an impact on the critical load for crack initiation, the degree of damage generation, and the load at fracture in unidirectionally oriented flax/epoxy composites. The critical load and the failure load both increase with the mechanical fiber characteristics, and their corresponding damage propagation is minimized. Below and beyond the damage initiation critical load, damage propagated at similar rates and composite damping decreased with increasing fiber percentage. These modified jute-PP composites have significantly greater dynamic strength, as measured by stress at fracture in the load increasing test, than the composites with jute fibers that are untreated, even at comparable fiber contents. This is because the coupling agent, in this case, maleic anhydride (MAH)-grafted PP, improves fiber–matrix bonding. Damage progression for non-modified jute-PP composites is almost irrespective of the fiber volume fraction, leading to independent peak stresses owing to the increased fiber–matrix bonding, produced by the MAH-grafted PP coupling agent, and the resulting better force transmission. When compared to untreated jute-PP composites, using the coupling agent increases dynamic strength by 40% at similar fiber quantities (40 vol.%). The deterioration of jute-PP composites (both modified and unmodified) does not happen randomly, but rather in a linear way with increasing load.

Exceptional fatigue behaviors of bamboo fiber-reinforced PP composites were reported by Thwe and Liao [67]. They found that all composite materials evaluated at loads lower than 80% of UTS survived 1000000 cycles of tension–tension fatigue (R = 0.1, f = 5 Hz). Composites evaluated at 80% UTS showed excellent durability, passing the 50000 to 500000 cycle limits. Additionally, they found that the residual stiffness for bamboo fiber incorporated PP composites evaluated in tension–tension fatigue decreased gradually between 80% and 65% of UTS, 50% and 35% of UTS, and 35% and 20% of UTS. Such composites were reported to display the typical three-stage stiffness reduction with synthetic fiber composites. With just 5% of bamboo fibers replaced by glass fibers, hybrid composites showed improved fatigue resistance. Yuanjian [68] looked at the fatigue characteristics of polyester composites reinforced with hemp fiber at a weight fraction of 44%. At constant relative stresses, the fatigue lifetime of glass fiber composites was shown to be exceeding those of hemp fiber-based composites.

12.7 FATIGUE PROPERTIES OF NATURAL FIBER-REINFORCED EPOXY COMPOSITES

Carbon fiber-reinforced polymer composites (CFRPCs) are generally employed in aeronautical, military, construction, and automobile sectors because of fatigue-prone structures. This chapter describes a novel approach that, by enhancing the interface, extends the fatigue life of CFRPC by an average of 96.56%. The common bending cycle fatigue test was used to assess the fatigue performance of CFRPC composites. The CFRPCs were made using a specifically developed compression molding machine and the hand lay-up procedure. Statistical analysis was used to assess the influence of the volume concentration changes of the treated/untreated composites on the cyclic fatigue strength (CFS). The numerical simulation model was developed using the dynamic simulation of COMSOL multi-physics to evaluate the deformation and forces for treated/untreated composites using a single strand of composite materials. To verify and complete the finite element (FE) model, experiments were conducted on treated and untreated composites. The relationships between stress, deformation, and fatigue life for carbon fiber-based composites were determined. The structure–property interaction was established and the causes of the increase in fatigue strength of CFRPC were explored using the scanning electron microscopy images. The validation of the experimental findings was confirmed by using ANOVA [69].

In their study [70] on short boron fiber-reinforced epoxy composites, Lavengood and Gulbransen found that fatigue strength tends to increase with the length of the fiber and achieves a maximum with a particular aspect ratio. This research examined the fatigue behavior of compression-molded epoxy-based composite materials reinforced using short carbon fibers. The major findings are that, in accordance with the Mixtures Law, mechanical characteristics, toughness, and tensile properties are closely related to fiber content, but that, for high-volume fractions, they are less efficient in terms of strengths. Composites with a 17.5 vol.% displayed the greatest result, with an increase in tensile strength of around 52% and stiffness of about 400% when compared to neat resin. The strain at the breaking point reduces as the fiber content increases. The fiber dispersion and porosity have the most effects on fatigue life.

The difference in fatigue life and the degree of porosity flaws were found to be clearly correlated. Surface roughness has little impact on fatigue dispersion. The experimentally measured fatigue life and the projected life based on Mortazavian and Fatemi's model [71] were validated with experimental conformity.

Towo and Ansell [72,73] investigated the fatigue characteristics of epoxy and polyester composites reinforced with unidirectionally oriented sisal fiber. In T–T fatigue, it was discovered that epoxy-based composites exhibited higher fatigue lifetime compared to polyester-based composite materials. Chemical pre-treatments of fibers significantly enhanced the fatigue lifetime of polyester-based composite materials, which had little to no impact on fatigue behavior of epoxy-based composite materials. Constant-life diagrams for the epoxy matrix composites demonstrated that the alkali treatment significantly enhanced the fatigue life under cyclic loading. Additionally, the authors discovered variations on the failure mode during T–T and T–C fatigue loading with a different failure pattern. T–T fatigue behavior of $[0/90]_{3S}$ and $[\pm45]_{3S}$ flax and glass fiber-reinforced epoxy matrix composites were discussed by Liang et al. in their study [74]. The higher static strength of the glass fiber-based composites made them more resistant to fatigue loading, but the S–N plots showed a predominant reduction in fatigue properties as the number of cycles gets increased. Also, in the case of $[\pm45]_{3S}$ lay-ups, the stiffness of glass/epoxy composites decreased under fatigue loading substantially higher when compared to flax-reinforced epoxy composites. Glass/epoxy samples for $[0/90]_{3S}$ lay-ups dropped up to 25% of their original modulus, which is remarkable, because flax/epoxy composites showed only moderate increase in modulus (up to roughly 2%) as fatigue loading increases.

Tension–tension fatigue results with a decrease in stiffness, and Figure 12.4 displays the results for hemp fiber composites under a range of stresses. At any stress

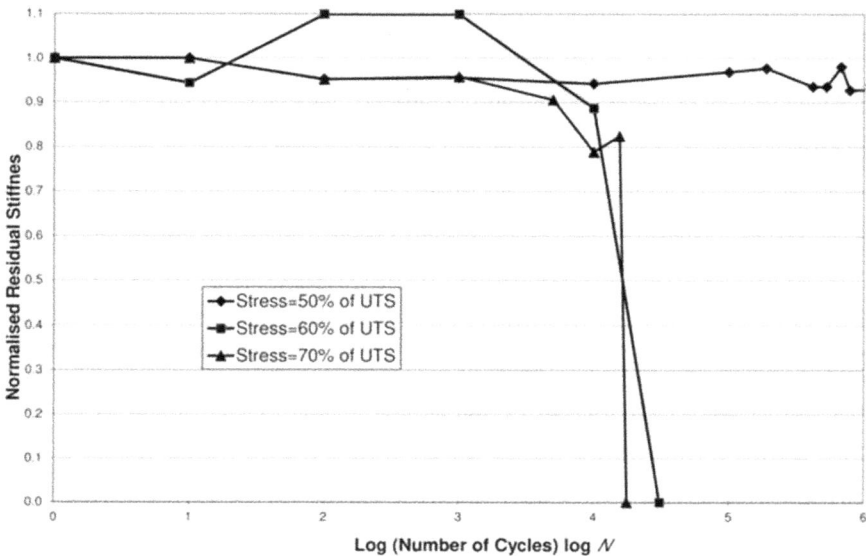

FIGURE 12.4 Difference in stiffness under cyclic loading for hemp fiber-based composites [76].

level, these composites demonstrated almost no stiffness reductions (and in some instances, slight increment) with increasing fatigue cycles. For $[0/90]_{3S}$ stitched non-crimp flax-reinforced epoxy matrix composites in T–T fatigue, Liang et al. [74] showed a similar trend of low stiffness reductions (in fact, slight increases), demonstrating the resilience of natural fiber-based composites during fatigue cycling loading. Fatigue testing revealed a slight increment in residual stiffness up to roughly 10000 cycles for composites made with hemp fiber. Liang et al. [74] found that $[0/90]_{3S}$ flax fiber-based epoxy had a minor but statistically significant improvement in stiffness, whereas $[\pm 45]_{3S}$ did not. These progressive enhancements are caused by small structural transformation, like the restructuring of wavy natural fabrics or the flattening of helically bundled microfibrils even when the flax fibers were stretched in tensile stresses. According to previous research, the modulus of $[0/90]_{3S}$ glass fiber-based epoxy matrices significantly degrade with decreasing stress levels, but the modulus of $[\pm 45]_{3S}$ lay-ups hardly varies at all (see, for example, Liang et al. [74]). However, the reduction in stiffness for glass fiber-reinforced composite resides between [0/90] and $[\pm 45]_{3S}$ lay-ups. This result is different from that of hemp fiber composites, which demonstrated negligible degradation in stiffness at normalized stress levels. Over the course of the 10^6 cycles, the mechanical stiffness of the glass fiber-reinforced composites was reduced by around 20% when subjected to a maximum fatigue stress. The surface of the sample was examined after it had been subjected to 10^6 cycles without cracking. There were no visible cracks; however, there was significant whitening from fiber/matrix debonding. These are in line with the theory that debonding is the primary damage process of fatigue cyclic loading in glass fiber-based composite materials [75]. The maximal stress intensity of 30% ultimate tensile strength (UTS) inside the experimental range of 10^6 cycles is insufficient to initiate matrix deformation.

Hyakutake et al. [77] explained that the notch-root radius affects the fatigue resistance of a glass fiber-reinforced epoxy laminates. According to Boller [78], fiber-reinforced epoxy composites exhibit superior fatigue properties than epoxy, polyester, silicone, e-glass and phenolic laminates. Few authors [79] examined the fatigue properties of unidirectionally oriented epoxy and vinyl ester-based composites reinforced with glass fiber. Unidirectionally oriented e-glass-reinforced vinyl ester and e-glass-reinforced epoxy specimens revealed comparable fatigue damages under low and high levels of flexural stresses. Due to much more repeated cracks, the e-glass-reinforced vinyl ester samples reached the same failure mechanism as the e-glass/epoxy material. The fatigue behavior of many fiber reinforced polymer (FRP) composites and hybrid composites was explored by many researchers [80]. Carbon, glass, and basalt fiber in epoxies were employed to fabricate the laminates for this investigation. Samples of hybrid fiber composites were made from layers of two distinct materials. Single carbon material was joined with single layer of basalt fiber or glass fiber with a 50% normalized fiber content to create carbon-reinforced basalt fiber and carbon fiber-reinforced glass FRP layers, respectively. According to the research by Wu et al., CFRP and FRP composites outperformed glass fibre reinforced polymer (GFRP) and banana fibre reinforced polymer (BFRP) composite laminates in terms of fatigue behavior. When compared to the BFRP composites, the CFRP/BFRP hybrid material showed significantly enhanced fatigue resistance.

TABLE 12.2

Summary of Materials [81]

Matrix	Fiber	Fiber wt./vol.%	Ref.
Epoxy	Carbon and glass	58, 64	[82]
Epoxy, vinyl ester	Glass	48, 46	[79]
Epoxy	Carbon	60	[83]
Epoxy	Glass	-	[84]
Epoxy	Glass		[85]
Epoxy	Carbon and glass	60	[86]
Epoxy	Carbon	30	[87]
Epoxy	Carbon	-	[88]
Epoxy	Carbon	-	[89]
Epoxy	Glass	-	[90]
Epoxy	Carbon	47	[91]
Epoxy	Glass	-	[92]
Epoxy	Glass	57	[93]
Epoxy	Glass	50	[94]
Epoxy	Glass	-	[95]
Epoxy	Glass	34.4, 38.6, 48.9, 57	[96]
Epoxy	Carbon	30, 55	[97,98]

During the testing, a run-out cycle length of 2106 cycles were carried out. CFRP and FRP composites revealed maximum fatigue range of 83.7% and 76.7% of the UTS, accordingly. The comparable limit for glass fiber- and basalt fiber-reinforced composites was 61.3% and 55%, respectively. As for fatigue life, carbon banana fiber reinforced polymer (CBFRP) composite outperformed carbon-glass reinforced polymer (C-GFRP) hybrid composite (with a limit of 73.6%) while the latter (with a limit of 58%) was only good for the tensile load. Table 12.2 depicts different matrix and fibers with varying weight/volume ratios.

Wang et al. [99] evaluated the influence of sample size on randomly oriented short glass fiber epoxy composites for load-controlled fatigue behavior. Shorter, broader samples with a length-to-width ratio of three or less had greater fatigue resistance. Lavengood and Gulbransen [70] studied the influence of fiber aspect ratio on carrying out three-point flexural fatigue tests of a short boron fiber-reinforced epoxy composites. Fatigue life is highly reliant on fibers aspect ratio, growing fast with aspect ratio up to 200 before plateauing. Matrix breaking starts at fiber ends and spreads across the fiber interfaces. In contrast to microscopic cracks, fiber brittle fracture at 45° direction with respect to fiber length was detected. Underhill and DuQuesnay [100] evaluated the influence of silane pre-treatment on fatigue behavior of epoxy composites. Epoxy joints of silane pre-treated aluminum alloy had a one-order-of-magnitude longer fatigue performance than those without surface modification [101].

Goumghar et al. [102] emphasizes on T–T fatigue performance analysis of flax-glass-reinforced hybrid epoxy composites. Stiffness loss, hysteresis slopes, and energy dissipation capacity were measured. Acoustic emission analysis has been used to detect fatigue damage mechanisms and their progression over time. In 2011, few experimenters produced unidirectionally oriented kenaf fiber-reinforced composites. By adding fibers, they were able to improve the epoxy resin's mechanical properties and stiffness. There was a significant enhancement in tensile strength from 32.19 MPa to 57.95 MPa and 100.53 MPa when 15% and 45% reinforcement was added, accordingly. The same group of researchers followed up on their first investigation to look into how the percentage of kenaf fiber in the composite affected the fatigue behavior of the material. However, 45% kenaf fiber-loaded epoxy composites outperformed a 15% addition of kenaf fiber-based epoxy composite materials in terms of load-bearing capability and degradation rates. Increased degradation rates were attributed to factors including poor fiber distributions within the matrices and the presence of void spaces in higher content of kenaf fiber loading [103,104].

According to Liang et al. [74], both flax fiber-based epoxy composites and glass fiber-based epoxy composites exhibit similar fatigue behavior. They found that perhaps the fatigue stress of the $[0/90]_{3S}$ stacking patterns in flax-reinforced epoxy composite was lesser compared to glass fiber-reinforced epoxy composites across the examined life spectrum (2×10^6), whereas the fatigue stress of the $[0/45]_{3S}$ in flax-reinforced epoxy composite was higher. The fatigue behavior of a woven hemp fiber-based epoxy composite was explored by Vasconcellos et al. [105]. They evaluated $[0°/45°]_7$ with $[0°/90°]_7$, two possible stacking sequences. It was shown that fatigue resistance was greater in $[0°/45°]_7$ lay-ups compared to $[0°/90°]_7$. Impact breakage and low-velocity impacts were investigated, as they pertained to the quasi-static tensile and the fatigue parameters of woven hemp fiber-reinforced epoxy matrix composites by Vasconcellos et al. [106]. Fracture at the fiber interface and microscopic cracks in the matrices were determined to have been the direct result of impact tests. From a limited data set consisting of fatigue behavior of non-impacted samples and low-energy impact testing, they were able to quantify the corresponding composites residual fatigue performance using an analytical method. The fatigue behavior of flax fiber-based epoxy composites was investigated by Liang et al. [107]. Their findings indicate that composites provide superior fatigue resistance and static toughness. The ratio of fibers stacked vertically to the load direction determines the increase in longitudinal Young's modulus of composites [108].

Biaxially oriented flax-reinforced epoxy matrix composites have a decreased fatigue behavior compared to glass fiber-based epoxy composites, as explained by Liang et al. [74]. Although the tensile characteristics of NFCs are greatly influenced by factors like type of fiber, quality of fiber, proportion of fiber and matrix, and textile pattern, fatigue behavior is slightly impacted by these factors. The fatigue behavior of woven hemp fiber-based epoxy composites was found to be higher for a 45° weave, as stated by de Vasconcellos et al. [105]. Local shear forces occurring at 45° composite laminates are responsible for this variation in damaging mechanics. The developed damage mechanisms for the woven hemp-reinforced epoxy matrix composites followed the same three-stage bending fatigue impact pattern as occurred in carbon-reinforced epoxy and glass-reinforced epoxy composites. Currently utilizing thermoplastic and thermoset materials, Gassan et al. [66,109] conducted in-depth analyses of the effect of the matrix and interface between the fiber and matrix on the fatigue characteristics of NFCs. T–T fatigue testing was carried out on composites including flax fibers and jute fibers-reinforced polyester, vinyl ester, epoxy and PP systems. It was discovered that the fatigue behavior of brittle and ductile PP-reinforced polyesters is significantly influenced by the quality of fiber–matrix interaction. Untreated woven jute fiber-based composites had a lower critical load for crack development and faster damage dissemination. A coupling agent, like MAH-grafted PP, strengthened the fiber–matrix compaction, which resulted in significantly increased dynamic endurance. Utilizing the coupling agent resulted in a 40% improvement in dynamic modulus compared to untreated jute-PP composites while maintaining similar fiber proportions. Stress-induced failure of the jute-PP composites (both modified and unmodified) was a continuous process rather than a discrete occurrence.

Unidirectionally oriented sisal fiber-based epoxy and polyester-based composite materials were investigated for their fatigue characteristics by Towo and Ansell [72,73]. Results from a tension–tension fatigue test showed that epoxy-based composites materials had much greater fatigue lifespan than polyester-based matrix composites. The fatigue behavior of polyester-based matrix composites was greatly improved by appropriate chemical pre-treatment of fibers, whereas the epoxy-based materials were hardly affected. When compared to $[0/45]_{3S}$ flax fiber-based epoxy matrix composite materials, the stiffness of glass fiber-based epoxy composite materials was shown to drastically decrease under cyclic loadings, as reported by Liang et al. [74]. As loading increased, the modulus of glass fiber-reinforced epoxy samples for $[0/90]_{3S}$ lay-ups decreased by as much as 25%; meanwhile, the modulus of flax fiber-reinforced epoxy matrix composites increased by up to 2%. Research suggests that NFCs' fatigue behavior might be enhanced by soaking up in water. Unidirectionally oriented flax fiber-based epoxy matrix laminates were subjected to hygrothermal conditioning at a temperature of 70°C and 85% RH, and the impact of moisture absorption with respect to mechanical behavior was investigated by Berges et al. [110]. In a surprising finding, conditioned specimens outperformed dry ones in terms of fatigue strength and stiffness across a wider range of cycles. The authors claim that water sorption triggered a laminate hardening process. Table 12.3 depicts the results of fatigue strength of some of the natural fiber-reinforced composites with 10^6 cycles to understand the degradation of fatigue strength with respect to increase in loading cycles.

Experimental research into the fatigue characteristics of hemp fiber mat-reinforced PE composites (HFRC) that are non-woven and randomly oriented was conducted by Yuanjian and Isaac [35]. As a means of differentiating between hemp fiber reinforced composites (HFRC) and $\pm45°$ glass fiber reinforced composites (GFRC) with a comparable fiber wt.%, a correlation was done among the two. Hemp fiber-based composites were found to have similar fatigue life cycles comparable to those of GFRC. Figure 12.5 demonstrates the S–N graphs of both hemp fiber-based composites and glass fiber-based composites. Even though the fatigue properties of hemp fiber composites degraded at a faster pace than that of glass fiber composites, their overall strength was greater. Impact loading was also observed to decrease the fatigue behavior of the hemp-based composite materials without seemingly altering the damage process. Fotouh et al. [114] carried out an experimental investigation to determine the fatigue behavior of hemp fiber-based high density polyethylene (HDPE) composites with various fiber weight fractions (20% and 40%).

TABLE 12.3

Tensile and Tensile Fatigue Properties of Various Natural Fiber-Based Composites [111]

Fiber	wt./vol.% of Fiber	Matrix	Fiber Form	Fabrication Method	Fatigue Property at 10^6 Cycles (MPa)	Ref.
Flax	-	Epoxy	Long, biaxial (BA) [±45]	Prepreg	41	[112]
Flax	-	Epoxy	Long, unidirectional	Prepreg	152	[112]
Hemp	36	Epoxy	Long, BA woven [0,90]	Resin transfer molding	45	[105]
Hemp	36	Epoxy	Long, BA woven [±45]	Resin transfer molding	38	[109]
Flax	44	Epoxy	Long, BA stitched [0,90]	Compression molding	70	[74]
Hemp	44	Epoxy	Long, BA stitched [±45]	Compression molding	46	[74]
Flax	28	Epoxy	Long, unidirectional	Vacuum infusion	59	[113]
Flax	27	Epoxy	Long, unidirectional	Vacuum infusion	115	[113]
Jute	32	Epoxy	Long, unidirectional	Vacuum infusion	85	[113]
Hemp	36	Epoxy	Long, unidirectional	Vacuum infusion	83	[113]
Sisal	72	Epoxy	Long, unidirectional	Compression molding	125	[72]

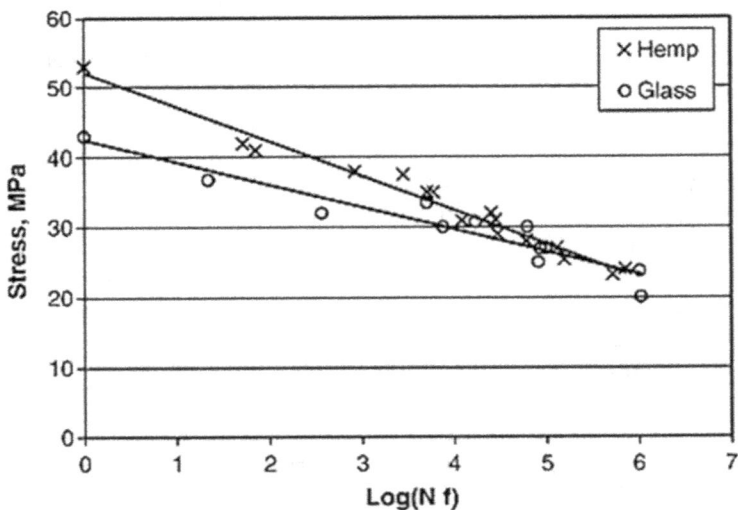

FIGURE 12.5　S–N curves of HFRC and GFRC at fiber orientation of ±45° [35].

Hemp fiber addition boosts fatigue strength but has no impact on fatigue sensitivity in composite materials. As the fiber content was raised, brittle fatigue failure developed as opposed to the ductile behavior of unreinforced HDPE composites. After being water-immersed for 35 days, the fatigue strength of the composites got decreased. The reduction in adhesion between fiber and matrix was considered to be the cause of the above reason. Taking into account stress ratios, fiber composition, and moisture uptake, a novel model has been established to predict the composites fatigue life. The fatigue behavior of polyester composites reinforced with sisal fiber was experimentally studied by Towo and Ansell [72] at stress ratios of 0.1 and −1. Since the adhesion between fiber and matrix was enhanced after treating it with an alkali solution, fatigue performance has improved. The fatigue behavior of flax fiber-based epoxy matrix composites with different fiber structures was the primary focus of Bensadoun et al. [115].

Fiber-based composites with improved static strength and modulus were shown to have superior fatigue properties. At the initial stage of fatigue cyclic loading at lower stress level, composites were observed to become stiff. The fatigue behavior of oil palm fruit bunch fiber reinforced with epoxy composites and carbon fiber-reinforced epoxy composites was compared by Kalam et al. [116]. Fixed at 35% and 55%, respectively, the volume fractions of oil palm fiber composites were lower than those of carbon fiber composites, which were 65%. However, during fatigue test, two distinct stress ratios (0.1 and 0.5) were held constant in order to analyze their impacts. When the stress ratio was improved, it was shown that the fatigue resistance of both materials got enhanced. Because the oil palm fibers diminished the tensile behavior of the developed composites when compared to the plain epoxy, the higher vol.% of fiber in the oil palm-based composites significantly decreased the fatigue life.

Through S–N plots, Shah et al. [112] calculated the fatigue lifetime of cellulosic fibers. It can be seen in the provided evidence that the S–N graphs of flax fiber-based PE composite materials are located above the jute fiber-reinforced and hemp fiber-reinforced PE composites, which is indicative of their superior fatigue behavior. However, the rate of deterioration of fatigue properties was greater in the flax fiber composites, as shown by the steep gradient relative to that of the hemp and jute-based composites. The S–N slopes of the plant fiber composites were, however, generally flatter than those of the GFRC. This result suggested that the rate of damage growth and fatigue strength degradation was lesser than that of glass fiber-based composites. Further, it was shown that the fatigue strength may be increased while maintaining a constant fatigue strength coefficient if the fiber volume percentage was increased. It was found that the fatigue strength was at its highest when the fiber orientation was perpendicular to the plane of loading, but that reducing the off-axis angle slowed the pace at which the fatigue strength degraded. Stress ratio increases were shown to have a similar effect on S–N curve flattening as those seen in the earlier research. This is because the stress amplitude of the fatigue loading is slowly decreasing, which decreases the rate at which cracks develop and degrades the fatigue strength.

In addition, Liang et al. [74] did an investigation to compare how flax fiber and GF-based epoxy matrix composites with two fiber orientations held up over time. The 0°/90° oriented flax fiber-based epoxy matrix composites were found to have less fatigue resistance than the glass fiber-reinforced epoxy. At high cyclic loading, the $\pm45°$ oriented flax fiber-based epoxy composites had the same fatigue resistance as the glass fiber-reinforced composites. Once density has been considered, the flax fiber-based composites with a fiber orientation of $\pm45°$ had a higher specific strength than glass fiber-based epoxy composites. This was because the matrix controlled the behavior of the flax fibers at this fiber orientation, and the flax fibers were less dense. When the flax fibers were oriented with $0°/90°$, the composites got stiffer, and the modulus climbed up by about 2%. Abdullah et al. [104] said that the amount of kenaf fiber in epoxy matrix composites significantly influences the fatigue life over time. However, epoxy composites with more fiber have an increased load-carrying capacity, but they also degrade more quickly.

Shahzad [117] researched the impact of alkali pre-treatment on mechanical characteristics of hemp-based polyester matrix composites. In addition, hemp fibers were pre-treated for 24 hours at 23°C with 1%, 5%, and 10% sodium hydroxide. The 10% pre-treated hemp fiber composite materials exhibited low static strength and equivalent fatigue behavior compared to untreated samples. The endurance strength of hemp fiber composites pre-treated with 1% and 5% NaOH solution improved by around 50% over untreated composites. Alkali-treated fiber composites showed improved fatigue sensitivity over untreated composite samples. Thus, 5% pre-treated hemp fiber composite improved the most, followed by the treated composite samples. Fatigue life investigations of hybrid fiber-reinforced composite materials are very less. Few studies have also focused on the fatigue life of plant fiber-based hybrid composites. Different kenaf fiber structures were studied by Sharba et al. [118] to determine fatigue life of a hybrid composite made of polyester and glass fibers. Fatigue responses of hybrid fiber-reinforced composites were shown to be substantially architecture-dependent. Composites made from unidirectionally oriented kenaf fiber were found to have a greater static and fatigue strength, followed by hybrid composites made from woven and non-woven kenaf fiber. However, the fatigue strength coefficient also increased for unidirectionally oriented woven and non-woven kenaf fiber composites, when glass fibers were hybridized with kenaf fibers. Fatigue testing on bamboo/glass-based PP hybrid composite materials was performed by Thwe and Liao [67].

Researchers found that the hybrid composites performed better than pure bamboo composites in terms of fatigue life and damage resistance. T–T fatigue behavior of woven flax and glass fabric-based hybrid epoxy composites was experimentally investigated by Asgarinia et al. [119]. Parallel to this, the fatigue behavior of flax-reinforced epoxy composites with varying fiber areal densities was studied and compared. As can be seen in Figure 12.6, fatigue strength was greatest for flax-reinforced epoxy composites with 224 g/m^2 of areal weight. Stress concentration under fatigue loading may have been caused by the flax fibers' high crimp level and poor yarn areal weight. The fatigue strength of the hybrid fiber composites was also increased because of the incorporation of glass fibers into the laminate. Several S–N curves for hybrid composites including flax and glass fibers are shown in Figure 12.7.

The hybrid laminates kept their modulus quite well owing to the flax fibers' stiffening effect. Using a thermographic study and morphological evaluation, Bagheri et al. [120] projected a high cyclic loading to determine the fatigue behavior of carbon and flax fiber-based hybrid epoxy composites for medical sector. The thermographic

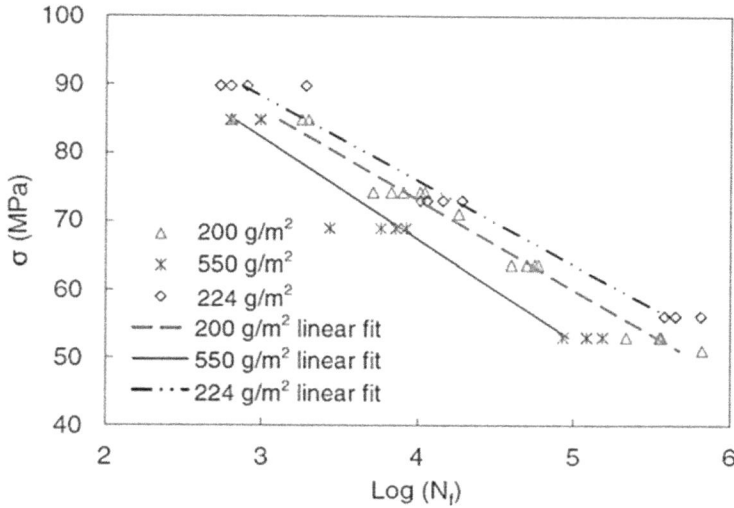

FIGURE 12.6   S–N graphs of varying areal densities for flax-reinforced epoxy matrix composites [119].

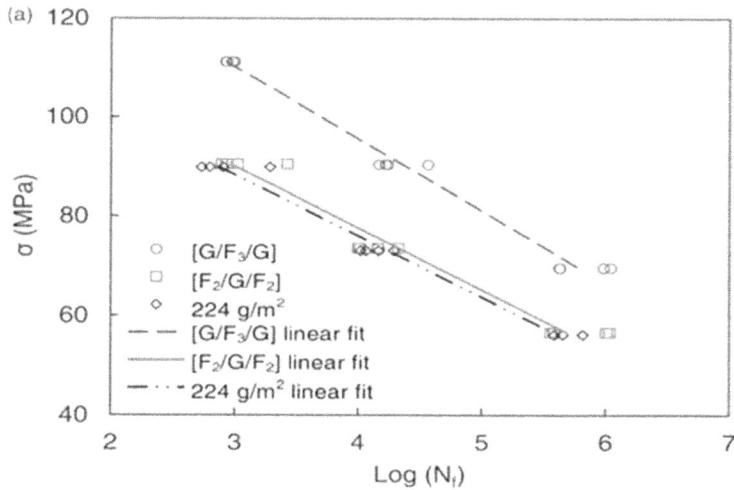

FIGURE 12.7   S–N plots of flax fiber-based epoxy matrix composites with varying fiber configurations [119].

analysis was shown to correspond well with the standard fatigue test, allowing for the accurate determination of fatigue strength. After studying the fatigue life cycles of hybrid laminates, it was found that the modulus degradation remained fairly stable. Avoiding debris generation is very important in medical applications, and the hybrid composites' steady modulus of evolution is a key factor to determine. Fatigue characteristics of hemp and glass fiber-based hybrid polyester composites were discovered by Shahzad [121]. Based on the findings, it was clear that the fatigue characteristics of the developed hybrid laminates were mostly determined by the hemp fibers. The fatigue strength of the hybrid composites was enhanced by the addition of glass fiber to the laminate, while the fatigue sensitivity coefficient remained same. The fatigue behavior of two distinct fiber configurations of thermoset-reinforced and thermoplastic-based kenaf and glass woven ply composites was compared by Sivakumar et al. [122].

Fatigue strength was shown to be greater for hybrid composites with glass fibers on the outside compared to those with glass fibers in the inner layer. After evaluating the fatigue behavior of hybrid composites made from two distinct matrices, it was discovered that the thermoset-based hybrid composites performed better than their thermoplastic counterparts. S–N curves of hybrid reinforced composites with kenaf and glass fibers are shown in Figure 12.8. The thermoplastic-based hybrid composites that had a higher kenaf domination performed better than the glass-dominant ones when evaluated using the fatigue sensitivity coefficient. This, however, was not the case with hybrid composites based on thermosets. The report on fatigue testing for natural fiber-based composite materials is summarized in Table 12.4.

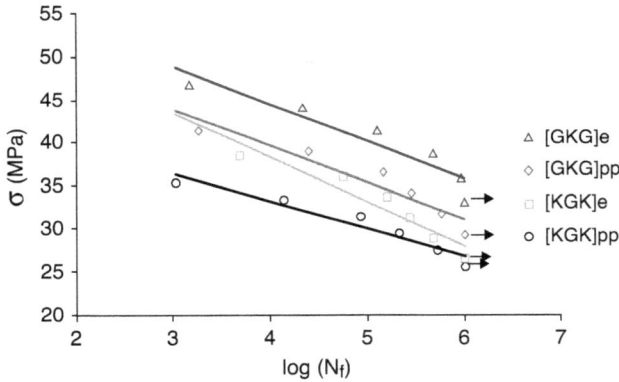

FIGURE 12.8 S–N plots of kenaf/glass-based epoxy matrix composites and polypropylene derived reinforced composite materials [122].

TABLE 12.4
Fatigue Test Report on Natural Fiber-Reinforced Composites

Reinforcements	Matrix	Category	Factors	Ref.
Flax	Epoxy	Non-hybrid	Comparison of fiber architectures	[115]
Oil palm		Non-hybrid	Impact of volume fraction and stress ratio	[116]
Flax and glass		Non-hybrid	Comparison to glass fibers	[74]
Kenaf		Non-hybrid	Fiber compositions	[104]
Flax-reinforced glass		Hybrid	Impact of fiber areal densities and hybridization	[119]
Flax-reinforced carbon		Hybrid	Thermography analysis	[120]
Kenaf-reinforced glass	PP/Epoxy	Hybrid	Impact of fiber configuration and type of matrix	[122]

12.8 NANOPARTICLES-REINFORCED EPOXY COMPOSITES

Knoll et al. [123] examined the influence of carbon-based nanoparticles on fatigue properties of carbon-reinforced epoxy composites. Jen and Wang [124] compared the fatigue performance of multi-walled carbon nanotubes (MWCNT) incorporated on epoxy nanocomposites. Silica particles' morphology significantly affects mechanical characteristics such as fatigue resistance, tensile and fracture characteristics, as found by Yamamoto et al. [125]. Epoxy resin mixed with non-spherical crystalline silica particles was found to provide the ideal balance for engineering properties. Under such a typical helicopter rotor spectrum load, Jagannathan et al. [126] investigated the fatigue behavior of glass fiber-based epoxy matrix composites with incorporation of rubber microparticles and silica nanoparticles. Kinloch et al. [127] produced GFRP composites of two types, one with neat epoxy and the other with hybrid combination of rubber microparticles and silica nanoparticles, and found that adding microparticles and nanoparticles in both the epoxy matrix composites with spectrum load tripled their fatigue lifetime. Compared to GFRP-neat hybrid composites, the GFRP-hybrid composites showed fatigue lifetime that was four to five times longer. Polymethyl methacrylate (PMMA) resin filler incorporated with nano-hydroxyapatite and ZrO_2 particles was studied by Salih et al. [128] for its compressive and fatigue resistance in prosthetic denture components. A hybrid composite material made of PMMA fiber and 5% kevlar had a fatigue strength of 90 MPa, which was much higher than that of laminated specimens made of PMMA fiber and 3% nano-hydroxyapatite (80 MPa) and pure PMMA resin (55 MPa) [129].

Another substance with immense potential for enhancing the mechanical performance of epoxies is SiO_2 nanoparticles. It has been shown that modifying epoxy resin with SiO_2 nanoparticles not only improves the epoxies' toughness, but also does so without significantly increasing the resins' viscosity. Epoxy resin concentrations containing 20 nm SiO_2 nanoparticles with a restricted particle size distribution have indeed been commercially marketed [130]. Many qualities, including strength, modulus, toughness, and fatigue behavior, may be enhanced by incorporating these SiO_2 nanoparticles inside an epoxy matrix. A variety of mechanical characteristics are enhanced by the incorporation of SiO_2 nanoparticles, and this effect and the processes are responsible for the enhancements [131]. Effects of silica nanoparticles on certain shear behavior and fatigue endurance of epoxy-bonded joints were studied by Sarac et al. [132]. Mechanical stirring was used to distribute SiO_2 nanoparticles throughout the system. Adherends made of stainless-steel plates were utilized to fix the joints. Adherends were polished, rinsed them with soapy water, and then cleaned them with acetone. Incorporating SiO_2 particles up to 6 wt.%, as shown by the shear study results, elevated the failure stress from 5 kN to 10 kN. Incorporating untreated nano-Al_2O_3 into the epoxy resin with filler proportions below 1 vol.% massively improved the shear behavior and fatigue endurance of joint surfaces. However, the enhanced mechanical characteristics of the bulk epoxy resin and the higher chemical compatibility among the nano-Al_2O_3/epoxy resin with the aluminum substrate were acclaimed for the increased shear behavior and fatigue life of the joints after the incorporation of these nanomaterials.

The impact of CNT distribution on shear behavior and fatigue properties of bonded joints in aluminum was studied by Mactabi et al. [133]. Calendaring was used to incorporate carbon nanotubes (CNTs) into the epoxy resin at a proportion of 0.5, 1.0, and 2.0 wt.%,. After being washed with acetone, the aluminum substrates were etched in a chromic acidic medium. When comparison to joints made with neat epoxy, the 0.5 wt.% CNT-added composites showed a 12% improvement in shear strength. There was more variation in the findings for the fatigue properties of CNT-reinforced epoxy joints. However, in situ electrical signals of shear joints were used to assess the sensing capacity of the CNT networks, allowing for continuous monitoring of joint integrity even when subjected to fatigue loading. Since CNT networks formed in the epoxy matrix, it was possible to detect the crack formation, growth, and failure of cracking. Manjunatha et al. [92] examined the T–F mechanism of neat epoxy and hybrid composites having 9 wt.% of rubber microparticles and silica nanoparticles. The testing frequencies were kept low to minimize heat effects for fatigue life. Hybrid epoxy composites have ten times the cyclic fatigue life of pristine-epoxy matrix composites. Reduced fracture formation in hybrid epoxy polymers extended their fatigue life. This crack minimization in the hybrid epoxy matrix results in less GFRP degradation under fatigue stress than that in the neat composite.

The effectiveness of nanofiller in increasing the tensile strength of epoxy matrix composites was the subject of extensive investigation by a number of researchers [134]. For instance, Withers et al. [135] used organo-modified nanoclays to strengthen epoxy-based composites. Researchers demonstrated that adding 2 wt.% of organo-modified nanoclays to glass fiber-reinforced epoxy increased its tensile stress by 11.7% and its tensile rigidity by 10.5%, using monotonic tensile testing. Further, when compared to its clean counterpart throughout a simulated 10-year cycle life, the nanofiller-reinforced composite exhibits 7.9% better fatigue strength and a fatigue life of roughly a decade longer. Many research evidence suggests that TiO_2 nanoparticles may improve the resilience to fatigue crack propagation (FCP) when subjected to dynamic stress [136]. FCP models show the relationship between crack growth rate and stress applied to push the crack further [137,138], which also provided visual evidence of this phenomenon. The speed at which a fracture expands as a function of stress intensity is shown by the steep slopes of the line in such representations. When crack propagation is started once, a zone of steady crack propagation is achieved [139]. The authors of the cited researchers found that comparing fracture propagation rates at a constant amount of stress intensity is a good medium of expression for understanding the reinforcing impact, as per the provided approach for studying the fatigue properties.

12.9   NANOFIBERS-REINFORCED EPOXY COMPOSITES

In fatigue loading, crack development occurs under cyclic stress along weaker planes. Strength and toughness are the two primary factors that determine a materials' rate of growth. Toughening is necessary to reduce the crack growth and the threshold damage onset load in laminated composite areas, which are areas of vulnerability [140]. In this part, the authors highlight the limited research on nanofiber reinforcements that have focused on this type of nanoscale damage. Nylon nanofibers-reinforced carbon-based epoxy hybrid composites in a double cantilever beam (DCB) configuration were the subject of fatigue-based delamination study by Brugo et al. [141]. Mid-plane reinforcement was added to a 14-ply laminate, and the material was subjected to a constant loading ratio. Figure 12.9a depicts the fatigue test results for reinforced and unreinforced composites. The fatigue graph of nanofiber composites showed a value of 2.5 times greater than that of pristine composite materials, indicating an immediate effect of delamination fatigue.

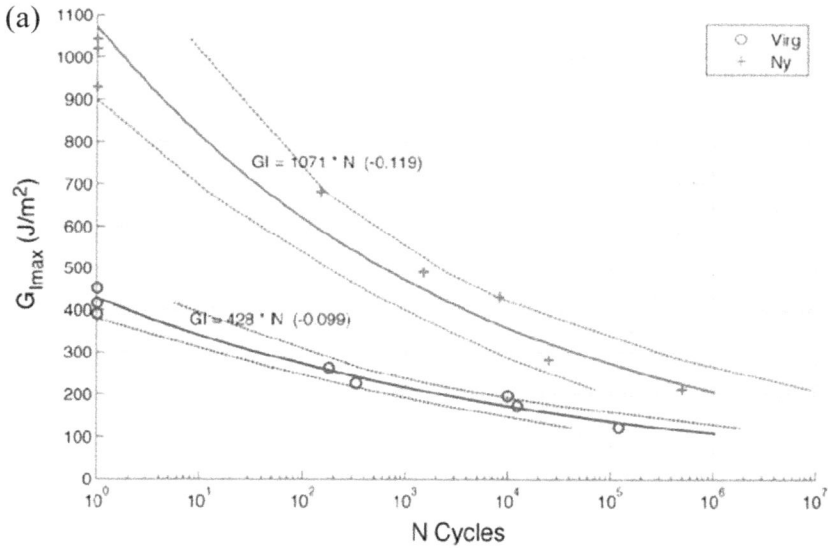

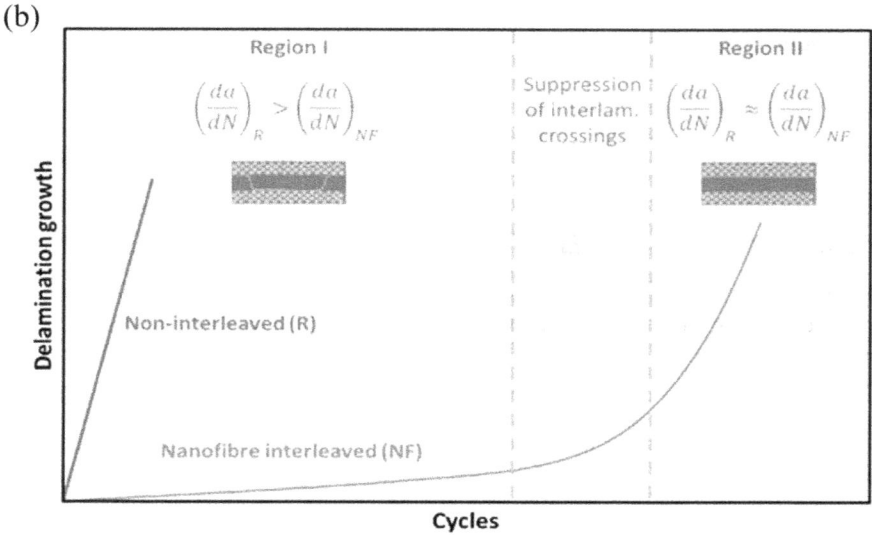

FIGURE 12.9   (a) Fatigue graph and (b) growth rate of delamination in nanofiber-reinforced composites with respect to fatigue load [141, 142].

Another interesting finding was that the nanofiber composite had a threshold energy release rate that was 90% greater than that of virgin laminate (109 J/m^2). The crack propagation between the nanofiber and carbon fiber was found to be retarded by the use of fracture tomography, which was not seen in virgin laminate. Daelemans et al. used a central cut-ply approach to examine the fatigue analysis of glass fiber-based epoxy matrix composites with nanofiber mats interleaved between the plies [142]. Eight-ply laminates were constructed for testing with respect to tension-fatigue loading at constant stress ratios. After a few millimeters of fracture propagation, the delamination growth rate increased as that of virgin specimens (Figure 12.9b). The reason was the fiber–matrix debonding at the interface and the numerous crack crossovers seen in region I but not in region II. Nanofibers slowed down the rate of delamination development by as much as 15% when compared to pristine laminates. Brugo found that carbon-based epoxy composites reinforced with nylon nanofibers have a specific behavior in terms of the relationship between crack propagation rate and energy release rate [143].

A graph of crack development rate versus maximum stress (Gmax) reveals that nanofiber laminate develops cracks more slowly under stress. The maximum allowable energy absorption capacity, or Gmax, for neat is 287 J/m^2 and for reinforced laminates is 565 J/m^2 at a threshold of 5×10^5. Also, for pristine and interleaved laminate, the fracture propagation rates are 2.4×10^4 and 7.9×10^6 mm/cycle at Gmax = 400 J/m^2, respectively. Further, mode of energy absorption was found to be the initiation of a new fracture in the interface of fiber and matrix. Phong et al. impregnated carbon fiber laminate with epoxy containing certain weight percentage of polyvinyl alcohol (PVA) nanofiber and then investigated its tension–tension fatigue characteristics [144]. Stress levels resulted in a 10–30-fold increase in fatigue life compared to unmodified composites. Microscopical inspection showed many distinct forms of damage, including as delamination at the point of intersection, weft fractures at longitudinal and transverse directions, and finally, fiber breaking, as explained in Figure 12.10. There have been several fatigue tests conducted on nanofiber-reinforced epoxy composites, and Table 12.5 provides a concise overview of these results. To enhance the fatigue performance of composites, the primary mechanism is slowing down the growth rate of delamination.

12.10   3D WOVEN FIBER-BASED EPOXY MATRIX COMPOSITES

Due to their strong resistance to debonding or delamination, 3D woven fiber-based epoxy composites have gathered a lot of interest in the construction and manufacturing sectors. Cyclic loading situations are common in the construction industry, and they have a negative impact on the durability and resilience of engineering structures. Weaved textiles were implanted in an epoxy matrix to improve their mechanical characteristics, and the fatigue behaviors of these resulting 3D woven fiber-based epoxy composite materials were examined. Vertical and oblique repeat units were used to categorize 3D woven materials. However, the fatigue characteristics of the appropriate composite materials have been examined based on the testing methodology. Different observation techniques were used to examine the effect of stress factors for the fatigue characteristics, and their failure mechanisms were studied.

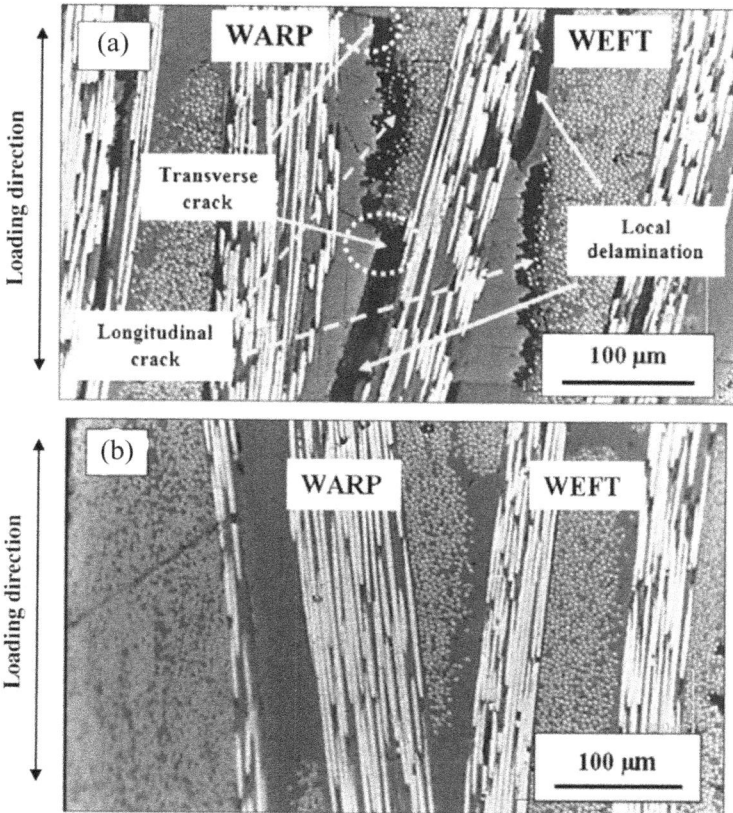

FIGURE 12.10 Scanning electron micrograph of (a) unmodified carbon-based epoxy composites and (b) modified carbon-based epoxy composites [144].

Future trends and difficulties were explored after reviewing the theoretical prediction models based on testing methodology. Hence, the authors have reviewed the test methodology and predicted the fatigue characteristics of 3D woven fiber-based epoxy matrix composites. When subjected to cyclic loading, the fatigue characteristics of 3D woven-based composites will be analyzed and reported. Experimental studies were carried out using different observation techniques based on cyclic loading, which includes T–T, T–C, C–C, and T–B. The failure mechanisms are examined, and research is conducted into how internal and exterior factors affect fatigue characteristics. Finally, cyclic loading is used to assess the theoretical prediction models and explore future trends and problems.

Damage progression from T–T fatigue experiments performed on e-glass fiber-reinforced epoxy 3DOWC at both directions (weft and warp) was described by Carvelli et al. [145]. The rectangular samples were measured as 250 mm × 252.58 mm, and the grip portions had bonded metal tabs. If indeed the fatigue sample broke more than 2 cm away from the tabs, then the test was declared "valid," removing the potential

TABLE 12.5
Fatigue Behavior of Epoxy Composites Reinforced with Nanofibers

Matrix	Methods	Type of Nanofiber	Nanofiber Content (%)	Nanofiber Configuration	Improvement (%)	Ref.
Glass-based epoxy	T–T fatigue	polycarbolactone (PCL), polyamide (PA)	Not provided	Interleaved	15 times reduction in delamination growth	[142]
Carbon-based epoxy	T–T fatigue	PVA	0.03, 0.05, and 0.1 wt.%	Short fiber	$N_f = 3 \times 10^6$ (30 times)	[144]
Carbon-based epoxy	T–T fatigue	Nylon	40 μm	Interleaved	36–27 times reduction in delamination growth	[143]
Carbon-based epoxy	T–T fatigue	Nylon	40 μm	Interleaved	90% increment in threshold energy release rate	[141]

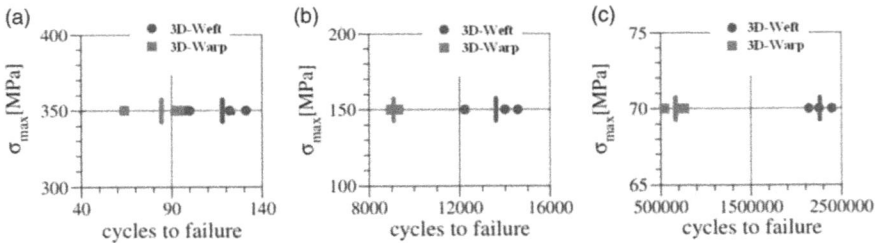

FIGURE 12.11   Fatigue life comparison at different stress levels: (a) 350 MPa, (b) 150 MPa, and (c) 70 MPa in both directions [145].

for bias from tab zones and grip pressure. Figure 12.11 demonstrates that regardless of the loading condition, the weft-oriented composites had a higher cycle to failure ratio than the warp directions. The vertical lines displayed the averaged fatigue life during T–T cyclic stress, and the R1:40.1 ratio was represented by the horizontal lines. Backlit observations show that in the event of weft-directional loading, the lack of crimp and uneven ply nesting has a significant beneficial influence on fatigue life. Delamination of 3DOWC did not occur during fatigue loading.

The T–T loading characteristics of carbon-reinforced epoxy matrix-based 3DOWC under varying testing frequencies were also experimentally investigated by Karahan et al. [146]. The average usable length, width, and thickness were 170 mm, 24.3 mm, and 2.75 mm, respectively, across the aluminum end tabs. The fatigue life appears to be almost three times shorter under warp-directed loading compared

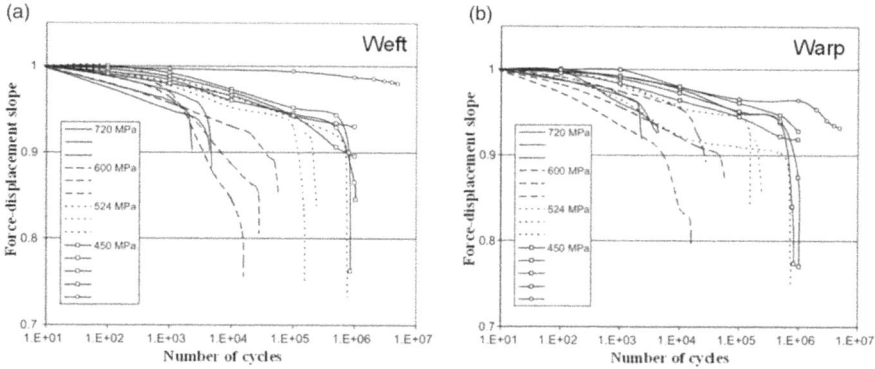

FIGURE 12.12 Force–displacement versus number of cycles in (a) weft direction and (b) warp direction [146].

to that under weft-directional loading for the same maximum cycle stress level. Maximum cycle stress was found to be in the same general range (between 412 and 450 MPa) for both loading orientations, corresponding to a life of 30,000,000 cycles. The range of stresses examined was found to be in excess of the initial static damage threshold and the threshold at which static damage begins to occur. This range included the second static damage threshold. Figure 12.12 displays the variation in specimen stiffness as demonstrated by shifts in the slope of the load–displacement curve that occurred during cyclic loading. It was determined that the modulus did not fluctuate by more than 5% from its "virgin" value before the fatigue breakdown occurred.

Figure 12.13 displays four primary curves to characterize the usual 3D of e-glass fiber-based epoxy matrix composites [147]. An S–N curve (shown in Figure 12.13a) was created by using the equation to match the experimental data. The bending fatigue limit was found to be about 35%. Under a stress of 35%, fatigue damage seems unable to occur. It was seen from the fitted curve that the degree of dispersion increased as the stress levels dropped. The samples at 60% stress level and 22,821 cycles were used as an example in the characterization of stiffness degradation. Figure 12.13b depicts the hysteresis loops for the bending moment relationship. In Figure 12.13c, the elastic modulus of specimens may be computed after subjecting them to a predetermined cycle loading. There were three distinct phases toward the improvement in residual stiffness. Epoxy cracks experienced rapid growth in stage I, drastically reducing the materials rigidity. In stage II, fractures propagate throughout the composite, most noticeably at the fiber/resin interface. Stage III noticed a dramatic decrease in stiffness that persisted until the last impact which is due to fiber failure at fatigue life of 97%. There was an increase in the amplitude of fluctuation and a decrease in the load-bearing capability of the specimens. Figure 12.13d displays the three-stage changes in maximum deflection and minimum deflection.

To estimate fatigue lifetime of 3D fiber-reinforced epoxy matrix composites, a simple model was suggested. Fatigue resistance of 3D composites was shown to be

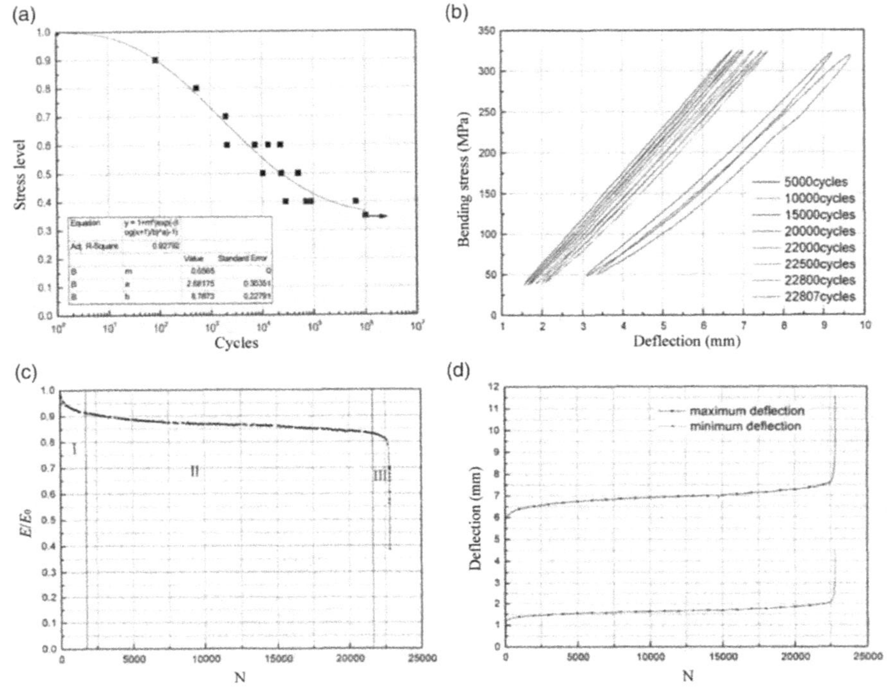

FIGURE 12.13   (a–d) Graphs of 3D for e-glass-reinforced epoxy matrix composites [147].

directly proportional to that of 2D composites of the same type. However, the apparent shortcoming of the model prevented it from accurately predicting their fatigue behaviors at all instances. Fatigue life prediction models for 3D woven fiber-based epoxy matrix composites were primarily presented with respect to experimental data. Different modeling approaches for 3D woven fiber-reinforced epoxy composites subjected to cyclic loading were discussed. The most popular fatigue test for PMC materials is the uniaxial T–T fatigue test, which follows ASTM and ISO standards for testing [148–150]. Fatigue stress values of natural fiber-reinforced epoxy matrix composites were typically under consistent cyclic loading. The fatigue life estimation theory of composite materials is divided into two parts: S–N graphs principle and fatigue damage accumulation theory. According to the principle, two different types of fatigue life prediction models were suggested. Residual stress, fatigue modulus, toughness, dissipation energy, fatigue progressive damage, and other nonlinear fatigue cumulative damage models were the key components of the fatigue cumulative damage theories. T–T fatigue parameters of 3D woven fiber-based epoxy matrix composite materials had been predicted by means of residual strength and stiffness models. The tension–tension fatigue features and damage mechanism were studied through a multi-scale geometry and material theory [151]. The epoxy matrix utilized nonlinear viscoelastic model for determining the decrease in stiffness values over time in micro- and meso- units, which also helped to minimize the computation time [152].

12.11 CONCLUSION

The overwhelming amount of the available research demonstrated that reinforcement of natural fibers increased the epoxy composites stiffness during fatigue loading. This longer fatigue life is a direct outcome of the stiffening effect, which revealed a rise in the modulus of the natural fibers. Substantial reductions in modulus up to the critical threshold contributed to fatigue damage rather than a loss of strength. In reality, after undergoing cyclic stress, the microfibrils in natural fibers straighten and reposition themselves at a certain angle, causing the material to stiffen. Peeling off of the fiber structure may account for natural fibers apparent insensitivity to fatigue stress. Therefore, the damaged layer need not prevent the load from being transmitted to the layers above and below it. The stiffness of the composite materials decreases below a certain threshold which leads to fatigue failure. To be sure, stiffness decay rather than strength deterioration governs the fatigue characteristics of epoxy-based composite materials.

REFERENCES

1. Chinnasamy, V., Subramani, S.P., Palaniappan, S.K., Mylsamy, B. and Aruchamy, K., 2020. Characterization on thermal properties of glass fiber and kevlar fiber with modified epoxy hybrid composites. *Journal of Materials Research and Technology*, *9*(3), pp. 3158–3167.
2. Abenojar, J., Encinas, N., Del Real, J.C. and Martínez, M.A., 2014. Polymerization kinetics of boron carbide/epoxy composites. *Thermochimica Acta*, *575*, pp. 144–150.
3. Aruchamy, K., Subramani, S.P., Palaniappan, S.K., Sethuraman, B. and Kaliyannan, G.V., 2020. Study on mechanical characteristics of woven cotton/bamboo hybrid reinforced composite laminates. *Journal of Materials Research and Technology*, *9*(1), pp. 718–726.
4. Aruchamy, K., Subramani, S.P., Palaniappan, S.K., Pal, S.K., Mylsamy, B. and Chinnasamy, V., 2022. Effect of blend ratio on the thermal comfort characteristics of cotton/bamboo blended fabrics. *Journal of Natural Fibers*, *19*(1), pp. 105–114.
5. Sathishkumar, T.P., de-Prado-Gil, J., Martínez-García, R., Rajeshkumar, L., Rajeshkumar, G., Mavinkere Rangappa, S., Siengchin, S., Alosaimi, A.M. and Hussein, M.A., 2022. Redeemable environmental damage by recycling of industrial discarded and virgin glass fiber mats in hybrid composites—An exploratory investigation. *Polymer Composites*. https://doi.org/10.1002/pc.27047
6. Ramesh, M., Rajeshkumar, L., Balaji, D. and Sivalingam, S., 2022 Mechanical and viscoelastic properties of wool composites. In: S. Thomas and S. Jose (eds.) *Wool Fiber Reinforced Polymer Composites*. Elsevier, pp. 299–318.
7. Paluvai, N.R., Mohanty, S. and Nayak, S.K., 2015. Studies on thermal degradation and flame-retardant behavior of the sisal fiber reinforced unsaturated polyester toughened epoxy nanocomposites. *Journal of Applied Polymer Science*, *132*(24), 42068.
8. Chilton, A.B., Shultis, J.K. and Faw, R.E., 1984. *Principles of Radiation Shielding*, Prentice Hall Inc., United States.
9. Aygün, B., Korkut, T., Karabulut, A., Gencel, O. and Karabulut, A., 2015. Production and neutron irradiation tests on a new epoxy/molybdenum composite. *International Journal of Polymer Analysis and Characterization*, *20*(4), pp. 323–329.
10. Ramesh, M., Bhuvaneswari, V., Balaji, D. and Rajeshkumar, L. (2022) Self-healable conductive and polymeric composite materials. In: T.A. Inamuddin and S.M. Adnan (eds.) *Aerospace Polymeric Materials*. Wiley-Scrivener LLC, pp. 231–258. http://dx.doi.org/10.1002/9781119905264.ch10

11. Ramesh, M., Tamil Selvan, M., Rajeshkumar, L., Deepa, C. and Ahmad, A., 2022. Influence of *Vachellia nilotica* Subsp. *indica* tree trunk bark nano-powder on properties of milkweed plant fiber reinforced epoxy composites, *Journal of Natural Fibers*, https://doi.org/10.1080/15440478.2022.2106341

12. Korkut, T., Gencel, O., Kam, E. and Brostow, W., 2013. X-ray, gamma, and neutron radiation tests on epoxy-ferrochromium slag composites by experiments and Monte Carlo simulations. *International Journal of Polymer Analysis and Characterization*, *18*(3), pp. 224–231.

13. Iwai, Y., Okada, T. and Tanaka, S., 1989. A study of cavitation bubble collapse pressures and erosion part 2: Estimation of erosion from the distribution of bubble collapse pressures. *Wear*, *133*(2), pp. 233–243.

14. Hattori, S. and Itoh, T., 2011. Cavitation erosion resistance of plastics. *Wear*, *271*(7–8), pp. 1103–1108.

15. Nagappan, S., Subramani, S.P., Palaniappan, S.K. and Mylsamy, B., 2021. Impact of alkali treatment and fiber length on mechanical properties of new agro waste Lagenaria Siceraria fiber reinforced epoxy composites. *Journal of Natural Fibers*, *19*(13), pp. 6853–6864.

16. Palaniappan, S.K., Pal, S.K., Rathanasamy, R., Kaliyannan, G.V. and Chinnasamy, M., 2020. Experimental investigations in the drilling of hybrid fiber composites. In: (Anish Khan et al., eds.) *Hybrid Fiber Composites: Materials, Manufacturing, Process Engineering*, Wiley-VCH Verlag GmbH, pp. 69–85.

17. Palaniappan, S.K., Chinnasamy, M., Rathanasamy, R. and Pal, S.K., 2020. Synthetic binders for polymer division. In: (Inamuddin et al., eds.) *Green Adhesives: Preparation, Properties and Applications*, Wiley-VCH Verlag GmbH, pp. 227–272.

18. Bhaskaran, P.E., Subramaniam, T., Kaliyannan, G.V., Palaniappan, S.K. and Rathanasamy, R., 2020. Green adhesive for industrial applications. In: (Inamuddin et al., eds.)*Green Adhesives: Preparation, Properties and Applications*, Wiley-VCH Verlag GmbH, pp. 57–84.

19. Erdoğan, B., Seyhan, A., Ocak, Y., Tanoğlu, M., Balköse, D. and Ülkü, S., 2008. Cure kinetics of epoxy resin-natural zeolite composites. *Journal of Thermal Analysis and Calorimetry*, *94*(3), pp. 743–747.

20. Kim, J., Moon, T.J. and Howell, J.R., 2002. Cure kinetic model, heat of reaction, and glass transition temperature of AS4/3501-6 graphite–epoxy prepregs. *Journal of Composite Materials*, *36*(21), pp. 2479–2498.

21. Priyadharshini, M., Balaji, D., Bhuvaneswari, V., Rajeshkumar, L., Sanjay, M. R., Siengchin, S., 2022, Fiber reinforced composite manufacturing with the aid of artificial intelligence – A state-of-the-art review. *Archives of Computational Methods in Engineering*, https://doi.org/10.1007/s11831-022-09775-y.

22. Macan, J., Brnardić, I., Ivanković, M. and Mencer, H.J., 2005. DSC study of cure kinetics of DGEBA-based epoxy resin with poly (oxypropylene) diamine. *Journal of Thermal Analysis and Calorimetry*, *81*(2), pp. 369–373.

23. Mylsamy, B., Palaniappan, S.K., Subramani, S.P., Pal, S.K. and Aruchamy, K., 2019. Impact of nanoclay on mechanical and structural properties of treated *Coccinia indica* fiber reinforced epoxy composites. *Journal of Materials Research and Technology*, *8*(6), pp. 6021–6028.

24. Bhuvaneshwaran, M., Subramani, S.P., Palaniappan, S.K., Pal, S.K. and Balu, S., 2021. Natural cellulosic fiber from *Coccinia indica* stem for polymer composites: Extraction and characterization. *Journal of Natural Fibers*, *18*(5), pp. 644–652.

25. Jayanth, D., Kumar, P.S., Nayak, G.C., Kumar, J.S., Pal, S.K. and Rajasekar, R., 2018. A review on biodegradable polymeric materials striving towards the attainment of green environment. *Journal of Polymers and the Environment*, *26*(2), pp. 838–865.

26. Shankar, R.S., Srinivasan, S.A., Shankar, S., Rajasekar, R., Kumar, R.N. and Kumar, P.S., 2014. Review article on wheat flour/wheat bran/wheat husk-based bio composites. *International Journal of Scientific and Research Publication*, 4(1), 1-9.

27. Corezzi, S., Fioretto, D., Santucci, G. and Kenny, J.M., 2010. Modeling diffusion-control in the cure kinetics of epoxy-amine thermoset resins: An approach based on configurational entropy. *Polymer*, 51(24), pp. 5833–5845.

28. Ramakrishnan, S., Krishnamurthy, K., Prasath, M.M., Kumar, R.S., Dharmaraj, M., Gowthaman, K., Kumar, P.S. and Rajasekar, R., 2015. Theoretical prediction on the mechanical behavior of natural fiber reinforced vinyl ester composites. *Applied Science and Advanced Materials International*, 1(3), pp. 85–92.

29. Kumar, A.M., Parameshwaran, R., Kumar, P.S., Pal, S.K., Prasath, M.M., Krishnaraj, V. and Rajasekar, R., 2017. Effects of abaca fiber reinforcement on the dynamic mechanical behavior of vinyl ester composites. *Materials Testing*, 59(6), pp. 555–562.

30. Mohankumar, A., Parameshwaran, R., Prasath Mm, S.S., Kumar Ps, M.C. and Rajasekar, R., 2018. A theoretical study on the physico-mechanical behavior of polyester composites using different classes of natural fiber reinforcements. In (S. Thomas et al., eds.) *Functionalized Engineering Materials and Their Applications*, Apple Academic Press, p. 20.

31. Feng, N.L., Malingam, S.D., Jenal, R., Mustafa, Z. and Subramonian, S., 2020. A review of the tensile and fatigue responses of cellulosic fiber-reinforced polymer composites. *Mechanics of Advanced Materials and Structures*, 27(8), pp. 645–660.

32. Mandell, J.F., Huang, D.D. and McGarry, F.J., 1981. Fatigue of glass and carbon fiber reinforced engineering thermoplastics. *Polymer Composites*, 2(3), pp. 137–144.

33. Caprino, G. and Giorleo, G., 1999. Fatigue lifetime of glass fabric/epoxy composites. *Composites Part A: Applied Science and Manufacturing*, 30(3), pp. 299–304.

34. Anandraj, M.K., Rathanasamy, R., Rathinasamy, P. and Palaniappan, S.K., 2022. Effect of Cloisite 15A on the mechanical properties of an abaca-based composite. *Materials Testing*, 64(1), pp. 125–131.

35. Yuanjian, T. and Isaac, D.H., 2007. Impact and fatigue behaviour of hemp fiber composites. *Composites Science and Technology*, 67(15–16), pp. 3300–3307.

36. Capela, C., Oliveira, S.E. and Ferreira, J.A.M., 2019. Fatigue behavior of short carbon fiber reinforced epoxy composites. *Composites Part B: Engineering*, 164, pp. 191–197.

37. Devarajan, B., LakshmiNarasimhan, R., Venkateswaran, B., Mavinkere Rangappa, S. and Siengchin, S., 2022. Additive manufacturing of jute fiber reinforced polymer composites – A concise review of material forms and methods. *Polymer Composites*, 43(10), pp. 6735–6748. https://doi.org/10.1002/pc.26789.

38. Deepa, C., Rajeshkumar, L. and Ramesh, M., 2022. Preparation, synthesis, properties and characterization of graphene-based 2D nano-materials for biosensors and bio-electronics. *Journal of Materials Research & Technology*, 19, 2657–2694. https://doi.org/10.1016/j.jmrt.2022.06.023.

39. Agarwal, G., Patnaik, A. and Sharma, R.K., 2014. Mechanical and thermo–mechanical properties of bi-directional and short carbon fiber reinforced epoxy composites. *Journal of Engineering Science and Technology*, 9(5), pp. 590–604.

40. Sethuraman, B., Subramani, S.P., Palaniappan, S.K., Mylsamy, B. and Aruchamy, K., 2020. Experimental investigation on dynamic mechanical and thermal characteristics of *Coccinia indica* fiber reinforced polyester composites. *Journal of Engineered Fibers and Fabrics*, 15, p. 1558925020905831.

41. Mylsamy, B., Chinnasamy, V., Palaniappan, S.K., Subramani, S.P. and Gopalsamy, C., 2020. Effect of surface treatment on the tribological properties of *Coccinia indica* cellulosic fiber reinforced polymer composites. *Journal of Materials Research and Technology*, 9(6), pp. 16423–16434.

42. Mylsamy, B., Palaniappan, S.K., Subramani, S.P., Pal, S.K. and Sethuraman, B., 2020. Innovative characterization and mechanical properties of natural cellulosic *Coccinia indica* fiber and its composites. *Materials Testing, 62*(1), pp. 61–67.

43. Girisha, C., Sanjeevamurthy, S., Gunti, R. and Manu, S., 2012. Mechanical performance of natural fiber reinforced epoxy hybrid composites. *International Journal of Engineering Research and Applications, 2*(5), pp. 615–619.

44. Shahzad, A., 2009. *Impact and Fatigue Properties of Natural Fiber Composites.* Swansea University, Swansea, UK.

45. Kang, H., Li, Z., Khosrovaneh, A.K., Kang, B. and Li, Z., 2015. Fatigue life predictions of adhesive joint of sheet steels. *Procedia Engineering, 133*, pp. 518–527.

46. Pashah, S. and Arif, A.F.M., 2014. Fatigue life prediction of adhesive joint in heat sink using Monte Carlo method. *International Journal of Adhesion and Adhesives, 50*, pp. 164–175.

47. Kumar, S. and Pandey, P.C., 2011. Fatigue life prediction of adhesively bonded single lap joints. *International Journal of Adhesion and Adhesives, 31*(1), pp. 43–47.

48. Pang, J.H.L., 2012. Fatigue life prediction models. In: *Lead Free Solder.* Springer, New York, pp. 49–63.

49. Teja, J.K., Biju, P.C.D., Varma, C.G.S.A., Kumar, R.R., Singh, D. and Gaur, P., 2022. Fatigue behavior of hybrid eco-composites: A review. *Materials Today: Proceedings, 72*, pp. 2245–2250.

50. Arumugam, S., Kandasamy, J., Sultan, M.T.H., Shah, A.U.M. and Safri, S.N.A., 2021. Investigations on fatigue analysis and biomimetic mineralization of glass fiber/sisal fiber/chitosan reinforced hybrid polymer sandwich composites. *Journal of Materials Research and Technology, 10*, pp. 512–525.

51. Lee, J.A., Almond, D.P. and Harris, A.B., 1999. The use of neural networks for the prediction of fatigue lives of composite materials. *Composites Part A: Applied Science and Manufacturing, 30*(10), pp. 1159–1169.

52. El Kadi, H. and Al-Assaf, Y., 2002. Prediction of the fatigue life of unidirectional glass fiber/epoxy composite laminae using different neural network paradigms. *Composite Structures, 55*(2), pp. 239–246.

53. Kumar, H. and Swamy, R.P., 2021. Fatigue life prediction of glass fiber reinforced epoxy composites using artificial neural networks. *Composites Communications, 26*, p. 100812.

54. Degrieck, J. and Van Paepegem, W., 2001. Fatigue damage modeling of fiber-reinforced composite materials. *Applied Mechanics Reviews, 54*(4), pp. 279–300.

55. Caprino, G., 2000. Predicting fatigue life of composite laminates subjected to tension-tension fatigue. *Journal of Composite Materials, 34*(16), pp. 1334–1355.

56. Biner, S.B. and Yuhas, V.C., 1989. Growth of short fatigue cracks at notches in woven fiber glass reinforced polymeric composites. *Journal of Engineering Materials and Technology, 111*(4), pp. 363–367.

57. Xiao, J. and Bathias, C., 1994. Fatigue behaviour of unnotched and notched woven glass/epoxy laminates. *Composites Science and Technology, 50*(2), pp. 141–148.

58. Mohankumar, D., Rajeshkumar, L., Muthukumaran, N., Ramesh, M., Aravinth, P., Anith, R., Balaji, S.V., 2022. Effect of fiber orientation on tribological behaviour of Typha angustifolia natural fiber reinforced composites. *Materials Today: Proceedings, 62*(4), pp. 1958–1964. https://doi.org/10.1016/j.matpr.2022.02.

59. Coats, T.W. and Harris, C.E., 1995. Experimental verification of a progressive damage model for IM7/5260 laminates subjected to tension-tension fatigue. *Journal of Composite Materials, 29*(3), pp. 280–305.

60. Lee, J.W., Allen, D.H. and Harris, C.E., 1989. Internal state variable approach for predicting stiffness reductions in fibrous laminated composites with matrix cracks. *Journal of Composite Materials, 23*(12), pp. 1273–1291.

61. Beaumont, P.W.R. and Spearing, S.M., 1991. Development of fatigue damage mechanics for application to the design of structural composite components. In: (A. Vautrin and H. Sol, eds.) *Mechanical Identification of Composites*. Springer, Dordrecht, pp. 311–326.
62. Spearing, S.M. and Beaumont, P.W., 1992. Fatigue damage mechanics of composite materials. I: Experimental measurement of damage and post-fatigue properties. *Composites Science and Technology*, 44(2), pp. 159–168.
63. Sathish Kumar, T. P., Satheesh Kumar, S. and Rajesh Kumar, L., 2022. Jute fibers, their composites, and applications. In: S.M. Rangappa, J. Parameswaranpillai, S. Siengchin, T. Ozbakkaloglu and H. Wang (eds.) *Plant Fibers, their Composites, and Applications*. Elsevier. pp. 253–282. http://dx.doi.org/10.1016/B978-0-12-824528-6.00020-5.
64. Spearing, S.M., Beaumont, P.W. and Ashby, M.F., 1992. Fatigue damage mechanics of composite materials. II: A damage growth model. *Composites Science and Technology*, 44(2), pp. 169–177.
65. Sathishkumar, T.P., Rajeshkumar, L., Rajeshkumar, G., Sanjay, M.R., Siengchin, S. and Thakrishnan, N., 2022. Improving the mechanical properties of jute fiber woven mat reinforced epoxy composites with addition of zinc oxide filler. In: *E3S Web of Conferences*. EDP Sciences, vol. 355, p. 02006.
66. Gassan, J., 2002. A study of fiber and interface parameters affecting the fatigue behaviour of natural fiber composites. *Composites Part A: Applied Science and Manufacturing*, 33(3), pp. 369–374.
67. Thwe, M.M. and Liao, K., 2003. Durability of bamboo-glass fiber reinforced polymer matrix hybrid composites. *Composites Science and Technology*, 63(3–4), pp. 375–387.
68. Tong, Y., 2004. *The Low Velocity Impact Fatigue and Stress Relaxation Behaviour of Composite Materials* (Doctoral dissertation, University of Wales Swansea).
69. Vedrtnam, A., 2019. Novel method for improving fatigue behavior of carbon fiber reinforced epoxy composite. *Composites Part B: Engineering*, 157, pp. 305–321.
70. Lavengood, R.E. and Gulbransen, L.B., 1969. The effect of aspect ratio on the fatigue life of short boron fiber reinforced composites. *Polymer Engineering & Science*, 9(5), pp. 365–369.
71. Mortazavian, S. and Fatemi, A., 2015. Fatigue behavior and modeling of short fiber reinforced polymer composites: A literature review. *International Journal of Fatigue*, 70, pp. 297–321.
72. Towo, A.N. and Ansell, M.P., 2008. Fatigue of sisal fiber reinforced composites: Constant-life diagrams and hysteresis loop capture. *Composites Science and Technology*, 68(3–4), pp. 915–924.
73. Ramesh, M., Balaji, D., Rajeshkumar, L. and Bhuvaneswari, V., 2022 Manufacturing methods of elastomer blends and composites. In: MR Sanjay, P. Jyotishkumar, S. Siengchin and T. Ozbakkaloglu (eds.) *Elastomer Blends and Composites - Principles, Characterization, Advances, and Applications*. Elsevier, pp. 11–25. https://doi.org/10.1016/B978-0-323-85832-8.00011-0
74. Liang, S., Gning, P.B. and Guillaumat, L., 2012. A comparative study of fatigue behaviour of flax/epoxy and glass/epoxy composites. *Composites Science and Technology*, 72(5), pp. 535–543.
75. Mandell, J.F., 1991. Fatigue behavior of short fiber composite materials. In: (K. L. Reifsnider, ed.) *Composite Materials Series*. Elsevier, vol. 4, pp. 231–337.
76. Shahzad, A. and Isaac, D.H., 2014. Fatigue properties of hemp and glass fiber composites. *Polymer Composites*, 35(10), pp. 1926–1934.
77. Hyakutake, H., Hagio, T. and Yamamoto, T., 1993. Fatigue failure criterion for notched FRP plates. *JSME International Journal. Ser. A, Mechanics and Material Engineering*, 36(2), pp. 215–219.

78. Boller, K., 1969. Fatigue fundamentals for composite materials. In: (K. H. Boller, ed.) *Composite Materials: Testing and Design.* ASTM International, USA.

79. Ramesh, M., Rajeshkumar, L., Balaji, D. and Bhuvaneswari, V. (2022) Influence of moisture absorption on mechanical properties of biocomposites reinforced surface modified natural fibers. In: C. Muthukumar, S. Krishnasamy, S.M.K. Thiagamani and S. Siengchin (eds.) *Aging Effects on Natural Fiber-Reinforced Polymer Composites. Composites Science and Technology.* Springer, Singapore, pp. 17–34. https://doi.org/10.1007/978-981-16-8360-2_2.

80. Wu, Z., Wang, X., Iwashita, K., Sasaki, T. and Hamaguchi, Y., 2010. Tensile fatigue behaviour of FRP and hybrid FRP sheets. *Composites Part B: Engineering, 41*(5), pp. 396–402.

81. Ansari, M.T.A., Singh, K.K. and Azam, M.S., 2018. Fatigue damage analysis of fiber-reinforced polymer composites—A review. *Journal of Reinforced Plastics and Composites, 37*(9), pp. 636–654.

82. Jones, C.J., Dickson, R.F., Adam, T., Reiter, H. and Harris, B., 1984. The environmental fatigue behaviour of reinforced plastics. *Proceedings of the Royal Society of London. A. Mathematical and Physical Sciences, 396*(1811), pp. 315–338.

83. Ahlborn, K., 1988. Fatigue behaviour of carbon fiber reinforced plastic at cryogenic temperatures. *Cryogenics, 28*(4), pp. 267–272.

84. El Kadi, H. and Ellyin, F., 1994. Effect of stress ratio on the fatigue of unidirectional glass fiber/epoxy composite laminae. *Composites, 25*(10), pp. 917–924.

85. Soden, P.D., Hinton, M.J. and Kaddour, A.S., 2004. Lamina properties, lay-up configurations and loading conditions for a range of fiber reinforced composite laminates. In: (M. J. Hinton et al., eds.) *Failure Criteria in Fiber-Reinforced-Polymer Composites.* Elsevier, pp. 30–51.

86. Shan, Y. and Liao, K., 2001. Environmental fatigue of unidirectional glass–carbon fiber reinforced hybrid composite. *Composites Part B: Engineering, 32*(4), pp. 355–363.

87. Barron, V., Buggy, M. and McKenna, N.H., 2001. Frequency effects on the fatigue behaviour on carbon fiber reinforced polymer laminates. *Journal of Materials Science, 36*(7), pp. 1755–1761.

88. Kawai, M., Yajima, S., Hachinohe, A. and Takano, Y., 2001. Off-axis fatigue behavior of unidirectional carbon fiber-reinforced composites at room and high temperatures. *Journal of Composite Materials, 35*(7), pp. 545–576.

89. Hojo, M., Matsuda, S., Tanaka, M., Ochiai, S. and Murakami, A., 2006. Mode I delamination fatigue properties of interlayer-toughened CF/epoxy laminates. *Composites Science and Technology, 66*(5), pp. 665–675.

90. Kumagai, S., Shindo, Y. and Inamoto, A., 2005. Tension–tension fatigue behavior of GFRP woven laminates at low temperatures. *Cryogenics, 45*(2), pp. 123–128.

91. Zhuang, H. and Wightman, J.P., 1997. The influence of surface properties on carbon fiber/epoxy matrix interfacial adhesion. *The Journal of Adhesion, 62*(1–4), pp. 213–245.

92. Manjunatha, C.M., Sprenger, S., Taylor, A.C. and Kinloch, A.J., 2010. The tensile fatigue behavior of a glass-fiber reinforced plastic composite using a hybrid-toughened epoxy matrix. *Journal of Composite Materials, 44*(17), pp. 2095–2109.

93. Bizeul, M., Bouvet, C., Barrau, J.J. and Cuenca, R., 2011. Fatigue crack growth in thin notched woven glass composites under tensile loading. Part I: Experimental. *Composites Science and Technology, 71*(3), pp. 289–296.

94. Huh, Y.H., Lee, J.H., Kim, D.J. and Lee, Y.S., 2012. Effect of stress ratio on fatigue life of GFRP composites for WT blade. *Journal of Mechanical Science and Technology, 26*(7), pp. 2117–2120.

95. Mini, K.M., Lakshmanan, M., Mathew, L. and Mukundan, M., 2012. Effect of fiber volume fraction on fatigue behaviour of glass fiber reinforced composite. *Fatigue & Fracture of Engineering Materials & Structures, 35*(12), pp. 1160–1166.

96. Nixon-Pearson, O.J., Hallett, S.R., Withers, P.J. and Rouse, J., 2013. Damage development in open-hole composite specimens in fatigue. Part 1: Experimental investigation. *Composite Structures, 106*, pp. 882–889.

97. Brunbauer, J., Stadler, H. and Pinter, G., 2015. Mechanical properties, fatigue damage and microstructure of carbon/epoxy laminates depending on fiber volume content. *International Journal of Fatigue, 70*, pp. 85–92.

98. Brunbauer, J. and Pinter, G., 2015. Effects of mean stress and fiber volume content on the fatigue-induced damage mechanisms in CFRP. *international Journal of Fatigue, 75*, pp. 28–38.

99. Wang, A.S.D., Tung, R.W. and Sanders, B.A., 1980. Size effect on strength and fatigue of a short fiber composite material. In: (J. R. Vinson, ed.) *Emerging Technologies in Aerospace Structures, Design, Structural Dynamics and Materials*, California, pp. 37–52.

100. Ramesh, M., Rajeshkumar, L. and Bhuvaneswari, V., 2021. Leaf fibres as reinforcements in green composites: A review on processing, properties and applications. *Emergent Materials. 5*(3), pp. 833–857. https://doi.org/10.1007/s42247-021-00310-6

101. Jojibabu, P., Zhang, Y.X. and Prusty, B.G., 2020. A review of research advances in epoxy-based nanocomposites as adhesive materials. *International Journal of Adhesion and Adhesives, 96*, p. 102454.

102. Goumghar, A., Assarar, M., Zouari, W., Azouaoui, K., El Mahi, A. and Ayad, R., 2022. Study of the fatigue behaviour of hybrid flax-glass/epoxy composites. *Composite Structures, 294*, p. 115790.

103. Deepa, C., Rajeshkumar, L. and Ramesh, M., 2021. Thermal properties of kenaf fiber-based hybrid composites. In: K. Senthilkumar, T.S.M. Kumar, M. Chandrasekar, N. Rajini and S. Siengchin (eds.) *Natural Fiber-Reinforced Composites: Thermal Properties and Applications*. John Wiley & Sons. pp. 167–181. https://doi.org/10.1002/9783527831562.ch10.

104. Abdullah, A.H., Alias, S.K., Jenal, N., Abdan, K. and Ali, A., 2012. Fatigue behavior of kenaf fiber reinforced epoxy composites. *Engineering Journal, 16*(5), pp. 105–114.

105. de Vasconcellos, D.S., Touchard, F. and Chocinski-Arnault, L., 2014. Tension–tension fatigue behaviour of woven hemp fiber reinforced epoxy composite: A multi-instrumented damage analysis. *International Journal of Fatigue, 59*, pp. 159–169.

106. de Vasconcellos, D.S., Sarasini, F., Touchard, F., Chocinski-Arnault, L., Pucci, M., Santulli, C., Tirillò, J., Iannace, S. and Sorrentino, L., 2014. Influence of low velocity impact on fatigue behaviour of woven hemp fiber reinforced epoxy composites. *Composites Part B: Engineering, 66*, pp. 46–57.

107. Liang, S., Gning, P.B. and Guillaumat, L., 2014. Properties evolution of flax/epoxy composites under fatigue loading. *International Journal of Fatigue, 63*, pp. 36–45.

108. Mittal, V., Saini, R. and Sinha, S., 2016. Natural fiber-mediated epoxy composites–a review. *Composites Part B: Engineering, 99*, pp. 425–435.

109. Ramesh, M., Rajeshkumar, L., Balaji, D. and Bhuvaneswari, V., 2021. Keratin-based biofibers and their composites. In: M.R. Sanjay, P. Madhu, J. Parameswaranpillai, S. Siengchin and S. Gorbatyuk (eds.) *Advances in Bio-Based Fiber: Moving towards a Green Society*. Woodhead Publishing, Elsevier. pp. 315–334.

110. Berges, M., Léger, R., Placet, V., Person, V., Corn, S., Gabrion, X., Rousseau, J., Ramasso, E., Ienny, P. and Fontaine, S., 2016. Influence of moisture uptake on the static, cyclic and dynamic behaviour of unidirectional flax fiber-reinforced epoxy laminates. *Composites Part A: Applied Science and Manufacturing, 88*, pp. 165–177.

111. Shahzad, A., 2019. Investigation into fatigue strength of natural/synthetic fiber-based composite materials. In: (M.Jawaid et al., eds.) *Mechanical and Physical Testing of Biocomposites, Fiber-Reinforced Composites and Hybrid Composites* (pp. 215–239). Woodhead Publishing, Duxford, UK.

112. El Sawi, I., Fawaz, Z., Zitoune, R. and Bougherara, H., 2014. An investigation of the damage mechanisms and fatigue life diagrams of flax fiber-reinforced polymer laminates. *Journal of Materials Science*, 49(5), pp. 2338–2346.

113. Shah, D.U., Schubel, P.J., Clifford, M.J. and Licence, P., 2013. Fatigue life evaluation of aligned plant fiber composites through S–N curves and constant-life diagrams. *Composites Science and Technology*, 74, pp. 139–149.

114. Fotouh, A., Wolodko, J.D. and Lipsett, M.G., 2014. Fatigue of natural fiber thermoplastic composites. *Composites Part B: Engineering*, 62, pp.175–182.

115. Bensadoun, F., Vallons, K.A.M., Lessard, L.B., Verpoest, I. and Van Vuure, A.W., 2016. Fatigue behaviour assessment of flax–epoxy composites. *Composites Part A: Applied Science and Manufacturing*, 82, pp. 253–266.

116. Kalam, A., Sahari, B.B., Khalid, Y.A. and Wong, S.V., 2005. Fatigue behaviour of oil palm fruit bunch fiber/epoxy and carbon fiber/epoxy composites. *Composite Structures*, 71(1), pp. 34–44.

117. Shahzad, A., 2012. Effects of alkalization on tensile, impact, and fatigue properties of hemp fiber composites. *Polymer Composites*, 33(7), pp. 1129–1140.

118. Sharba, M.J., Leman, Z., Sultan, M.T., Ishak, M.R. and Hanim, M.A., 2016. Partial replacement of glass fiber by woven kenaf in hybrid composites and its effect on monotonic and fatigue properties. *BioResources*, 11(1), pp. 2665–2683.

119. Asgarinia, S., Viriyasuthee, C., Phillips, S., Dubé, M., Baets, J., Van Vuure, A., Verpoest, I. and Lessard, L., 2015. Tension–tension fatigue behaviour of woven flax/epoxy composites. *Journal of Reinforced Plastics and Composites*, 34(11), pp. 857–867.

120. Bagheri, Z.S., El Sawi, I., Bougherara, H. and Zdero, R., 2014. Biomechanical fatigue analysis of an advanced new carbon fiber/flax/epoxy plate for bone fracture repair using conventional fatigue tests and thermography. *Journal of the Mechanical Behavior of Biomedical Materials*, 35, pp. 27–38.

121. Shahzad, A., 2011. Impact and fatigue properties of hemp–glass fiber hybrid biocomposites. *Journal of Reinforced Plastics and Composites*, 30(16), pp. 1389–1398.

122. Sivakumar, D., Ng, L.F., Lau, S.M. and Lim, K.T., 2018. Fatigue life behaviour of glass/kenaf woven-ply polymer hybrid biocomposites. *Journal of Polymers and the Environment*, 26(2), pp. 499–507.

123. Knoll, J.B., Riecken, B.T., Kosmann, N., Chandrasekaran, S., Schulte, K. and Fiedler, B., 2014. The effect of carbon nanoparticles on the fatigue performance of carbon fiber reinforced epoxy. *Composites Part A: Applied Science and Manufacturing*, 67, pp. 233–240.

124. Jen, Y.M. and Wang, Y.C., 2012. Stress concentration effect on the fatigue properties of carbon nanotube/epoxy composites. *Composites Part B: Engineering*, 43(4), pp. 1687–1694.

125. Yamamoto, I., Higashihara, T. and Kobayashi, T., 2003. Effect of silica-particle characteristics on impact/usual fatigue properties and evaluation of mechanical characteristics of silica-particle epoxy resins. *JSME International Journal Series A Solid Mechanics and Material Engineering*, 46(2), pp. 145–153.

126. Jagannathan, N., Gururaja, S. and Manjunatha, C.M., 2019. Matrix cracking in polymer matrix composites under bi-axial loading. *Procedia Structural Integrity*, 14, pp. 864–871.

127. Manjunatha, C.M., Bojja, R., Jagannathan, N., Kinloch, A.J. and Taylor, A.C., 2013. Enhanced fatigue behavior of a glass fiber reinforced hybrid particles modified epoxy nanocomposite under WISPERX spectrum load sequence. *International Journal of Fatigue*, 54, pp. 25–31.

128. Salih, S.I., Oleiwi, J.K. and Hamad, Q.A., 2015. Numerically and theoretically studying of the upper composite complete prosthetic denture. *Engineering and Technology Journal, Part (A)*, 33(5), pp. 1023–1037.

129. El-Wazery, M.S., 2017. Mechanical characteristics and novel applications of hybrid polymer composites-a review. *Journal of Materials and Environmental Sciences, 8*(2), pp. 666–675.

130. Sprenger, S., 2013. Epoxy resin composites with surface-modified silicon dioxide nanoparticles: A review. *Journal of Applied Polymer Science, 130*(3), pp. 1421–1428.

131. Zou, H., Wu, S. and Shen, J., 2008. Polymer/silica nanocomposites: Preparation, characterization, properties, and applications. *Chemical Reviews, 108*(9), pp. 3893–3957.

132. Saraç, İ., Adin, H. and Temiz, Ş., 2018. Experimental determination of the static and fatigue strength of the adhesive joints bonded by epoxy adhesive including different particles. *Composites Part B: Engineering, 155*, pp. 92–103.

133. Mactabi, R., Rosca, I.D. and Hoa, S.V., 2013. Monitoring the integrity of adhesive joints during fatigue loading using carbon nanotubes. *Composites Science and Technology, 78*, pp. 1–9.

134. Tee, Z.Y., Yeap, S.P., Hassan, C.S. and Kiew, P.L., 2022. Nano and non-nano fillers in enhancing mechanical properties of epoxy resins: A brief review. *Polymer-Plastics Technology and Materials, 61*(7), pp. 709–725.

135. Withers, G.J., Yu, Y., Khabashesku, V.N., Cercone, L., Hadjiev, V.G., Souza, J.M. and Davis, D.C., 2015. Improved mechanical properties of an epoxy glass–fiber composite reinforced with surface organomodified nanoclays. *Composites Part B: Engineering, 72*, pp. 175–182.

136. Pinto, D., Bernardo, L., Amaro, A. and Lopes, S., 2015. Mechanical properties of epoxy nanocomposites using titanium dioxide as reinforcement–a review. *Construction and Building Materials, 95*, pp. 506–524.

137. Wetzel, B., Rosso, P., Haupert, F. and Friedrich, K., 2006. Epoxy nanocomposites–fracture and toughening mechanisms. *Engineering fracture mechanics, 73*(16), pp. 2375–2398.

138. Zhao, S., Schadler, L.S., Duncan, R., Hillborg, H. and Auletta, T., 2008. Mechanisms leading to improved mechanical performance in nanoscale alumina filled epoxy. *Composites Science and Technology, 68*(14), pp. 2965–2975.

139. Zhao, S., Schadler, L.S., Hillborg, H. and Auletta, T., 2008. Improvements and mechanisms of fracture and fatigue properties of well-dispersed alumina/epoxy nanocomposites. *Composites Science and Technology, 68*(14), pp. 2976–2982.

140. Shakil, U.A., Hassan, S.B., Yahya, M.Y. and Nauman, S., 2020. Mechanical properties of electrospun nanofiber reinforced/interleaved epoxy matrix composites—A review. *Polymer Composites, 41*(6), pp. 2288–2315.

141. Brugo, T.M., Minak, G., Zucchelli, A., Saghafi, H. and Fotouhi, M., 2015. An investigation on the fatigue-based delamination of woven carbon-epoxy composite laminates reinforced with polyamide nanofibers. *Procedia Engineering, 109*, pp. 65–72.

142. Daelemans, L., van der Heijden, S., De Baere, I., Rahier, H., Van Paepegem, W. and De Clerck, K., 2017. Improved fatigue delamination behaviour of composite laminates with electrospun thermoplastic nanofibrous interleaves using the Central Cut-Ply method. *Composites Part A: Applied Science and Manufacturing, 94*, pp. 10–20.

143. Brugo, T., Minak, G., Zucchelli, A., Yan, X.T., Belcari, J., Saghafi, H. and Palazzetti, R., 2017. Study on Mode I fatigue behaviour of Nylon 6, 6 nanoreinforced CFRP laminates. *Composite Structures, 164*, pp. 51–57.

144. Phong, N.T., Gabr, M.H., Okubo, K., Chuong, B. and Fujii, T., 2013. Improvement in the mechanical performances of carbon fiber/epoxy composite with addition of nano-(Polyvinyl alcohol) fibers. *Composite Structures, 99*, pp. 380–387.

145. Carvelli, V., Gramellini, G., Lomov, S.V., Bogdanovich, A.E., Mungalov, D.D. and Verpoest, I., 2010. Fatigue behavior of non-crimp 3D orthogonal weave and multilayer plain weave E-glass reinforced composites. *Composites Science and Technology, 70*(14), pp. 2068–2076.

146. Karahan, M., Lomov, S.V., Bogdanovich, A.E. and Verpoest, I., 2011. Fatigue tensile behavior of carbon/epoxy composite reinforced with non-crimp 3D orthogonal woven fabric. *Composites Science and Technology*, 71(16), pp. 1961–1972.

147. Niu, Z., 2011. Experimental investigations and FEM simulations on the three-point bending fatigue of 3D orthogonal woven composite.

148. Van Paepegem, W., 2009. Fatigue damage in structural textile composites: Testing and modelling strategies. In: (M. Miraftab, ed.) *Fatigue Failure of Textile Fibers*. Woodhead Publishing, Cambridge, UK, pp. 201–241.

149. ASTM International, 2012. *Standard Test Methods for Tension-Tension Fatigue of Polymer Matrix Composite Materials*. ASTM International, USA.

150. International Organization for Standardization, 2003. *Fiber-Reinforced Plastics: Determination of Fatigue Properties Under Cyclic Loading Conditions*. International Organization for Standardization, Geneva, Switzerland.

151. GB/T35465.3, 2017. Test method for fatigue properties of polymer matrix composite materials. Chinese national standard category: Fiber reinforced composites.

152. Ma, Z., Zhang, P. and Zhu, J., 2022. Review on the fatigue properties of 3D woven fiber/epoxy composites: Testing and modelling strategies. *Journal of Industrial Textiles*, 51(5_suppl), pp. 7755S–7795S.

13 Epoxy Nanocomposites for Fire-Retardant Applications

Harikrishnan Pulikkalparambil, Sanjay Mavikere Rangappa, Thirawudh Pongprayoon, Suchart Siengchin
King Mongkut's University of Technology

Jithun Lal
Indian Institute of Science

Senthilkumar Krishnasamy
PSG Institute of Technology and Applied Research

Saravana Kumar M
National Institute of Technology

Jyotishkumar Parameswaranpillai
Alliance University

CONTENTS

DOI: 10.1201/9781003271017-13

ABBREVIATIONS

2D	Two-dimensional
CNT	Carbon nanotube
DBS	Dodecyl benzene sulfonate
DOPO	9,10-Dihydro-9-oxa-10-phosphaphenanthrene-10-oxide
EP	Epoxy
FGNS	Functionalized graphite nanoplatelets
FR	Fire retardant
FRGO	Functionalized reduced graphene oxide
GNS	Graphene nanosheets
GNT	Graphitic nanotube
GO	Graphene oxide
LDH	Layered double hydroxide
LOI	Limiting oxygen index
MMT	Montmorillonite
OMMT	Organophilic montmorillonite
PHRR	Peak heat release rate
PPN	Polyphosphazene nanoparticle
PZS	Polyphosphazene microspheres
rGO	Reduced graphene oxide
THR	Total heat release
TSP	Total smoke production
TTI	Time to ignition

13.1 INTRODUCTION

New composite materials are developed and fabricated for many advanced applications such as construction, aerospace, automotive, and electrical and electronics industries. To meet the multifaceted demands in the composite industry, advanced thermosetting polymers with exceptional properties are required. Among the advanced polymers, epoxy (EP) is the most widely accepted polymer due to its versatility in properties. However, the main limitation of EP resin is that many hydrophilic groups are present on the surface of cured composites, which leads to poor hot-wet properties. Additionally, the service temperature of most of the cured EP is limited to <220°C. Thus, EP resin-based composites could not be used in higher-temperature applications [1]. Moreover, the flame resistance of most polymers, including epoxies, is poor. However, flame-retardant properties in EP composites are significant in areas where strict policies are applied, such as public transport, electrical and electronic components, construction, and automotive applications. Thus, such application areas require polymeric materials with modifications to attain fire resistance.

Fire hazard is the potential of destruction caused by fire. A composite's fire hazard is discussed in terms of its burnability, heat release rate, smoke index, weight loss, flame spread, and toxicity [2]. Meanwhile, fire retardants (FRs) are chemicals or additives used with combustible materials to improve fire resistance. Table 13.1 lists traditional FRs used in polymers [3]. Fire retardancy could be introduced into polymers by blending, chemical modifications, and additives [3,4]. Among the different

TABLE 13.1
Traditional Fire Retardants [3]

Type	Class	Examples
Halogenated	-	Tetrabromobisphenol A, polybromodiphenylether, hexabromocyclododecane, and tetrabromophthalic anhydride.
Phosphorous products	-	Phosphates, phosphines, etc.
Nano-flame retardants	Metallic particles	ZnO_2 nanowires, aluminum tri-hydroxide, zinc oxide, etc.
	Clay	Montmorillonite, LDH, MoS_2, etc.
	Biobased fillers	Lignin, starch, proteins, cellulose nanocrystals (treated), etc.
	Carbon family	Carbon nanotube, graphene, fullerene, etc.

methods, the introduction of additives is preferred due to their flexibility in composite processing and manufacturing. These additives must show certain characteristic properties such as (1) thermal stability at polymer processing temperatures, (2) better compatibility with matrix to avoid migration, (3) improved overall performance, and (4) lower toxicity and lower emission of ambiguous/toxic gases [5].

The flammability of a polymer was determined by factors such as ease of ignition, fire durability, rate of heat release, smoke generation, and emission of toxic gas. The flammability in polymeric materials was measured using Underwriters Laboratories 94 (UL-94) flammability test. The UL-94 ratings 5VA, 5VB, V-0, V-1, V-2, and HB are used to characterize the flammability of the composite. Rating 5VA is the most flame resistant, and HB is the least flame resistant.

Depending on the type of additives used as FR in polymers, the mechanism of FR is of three types: gas-phase inhibition mechanism, cooling mechanism, and solid-phase mechanism. In the gas-phase inhibition mechanism, halogenated and phosphorous FRs react with the hydroxyl or oxygen groups present in the polymer to extinguish the flame. In the cooling mechanism, FRs, such as hydrated halogen-free minerals in an exothermic atmosphere, release water molecules to retard the combustion. In the solid-phase mechanism, FR, like cellulose and other carbon-based materials, reacts to combustion to form char that acts as a barrier layer, thus protecting the polymer matrix at elevated temperatures [3]. Many traditional FRs are harmful to the environment and human life; therefore, recently, more attention has been paid to nanoparticles as additives. The EP nanocomposites for FR applications are classified based on the type of fillers used, viz., organic, inorganic, and hybrid nanoparticles. The following sections give an overview of all these materials.

13.2 ORGANIC NANOPARTICLES

Organic nanoparticles are the widely used nanofiller in polymer matrix composites to impart FR properties than conventional additives. Furthermore, one-dimensional nanomaterials such as carbon nanotubes (CNTs) and two-dimensional (2D) nanomaterials such as graphene impart the best flame-retardant properties in the EP nanocomposites system [6,7].

13.2.1 GRAPHENE

Graphene is a layered sheet of sp^2 hybridized carbon atoms with a honeycomb crystal structure [8,9]. As a result of their distinct spatial and bonding through sp^2 hybridization, graphene presents several fascinating characteristics, such as gas impermeability, good flame resistance, thermo-mechanical, and electrical properties. Therefore, adding graphene with very low loadings into a polymer can enhance the flammability, mechanical, thermal, and electrical conductivity properties [10,11]. Graphene oxide (GO), reduced graphene oxide (rGO), graphene nanoparticle, and other functionalized forms of graphene as nanofillers in EP matrix substantially broaden the functionality and enhance the properties of EP /graphene nanocomposites [12–25].

Incorporating GO into the EP matrix improves the ignition time and decreases the heat release rate of EP composites [13]. However, pure graphene has poor dispersion in EP systems; therefore, functionalized or modified graphene is usually used in EP composites [14–17]. Siloxane groups can be used as a good grafting agent on GO. Polydimethylsiloxane (PDMS) chains can effectively graft on GO and can improve the dispersion of GO in the EP matrix. The PDMS-GO endows an improved fire retardation (due to PDMS and GO synergistic effects), which leads to notable toughening and improved moisture absorption resistance. Also, the addition of 0.29 wt.% of PDMS-GO and 0.3 wt.% of phosphorus made the EP to achieve a UL-94 V-0 rating, with a limiting oxygen index (LOI) of 29.2% and a char yield of 21.4 wt.% [14]. Qu et al. [15] functionalized GO (FGO) by octa(propyl glycidyl ether)polyhedral oligomeric silsesquioxane as the grafting agent and γ-aminopropyl triethoxysilane as a chemical bridge for better dispersion in the EP matrix. The researchers observed that incorporating FGO sheets in the EP enhanced smoke-suppressive properties and improved flame-retardant properties. The good barrier properties and tortuous effects of FGO sheets in EP matrix enhanced the flame-retardant properties. The modified EP composite reduced the peak heat release rate (PHRR), total heat release (THR), and total smoke release (TSR) considerably [15]. Another effective graphene modification can be done by cyclophosphazene. The cyclophosphazene FGO suppresses the volatile content by producing a compact char layer during combustion and inhibits heat release. With 1 wt.% FGO, the EP nanocomposites exhibit a distinct decrease in the PHRR, THR, and total smoke production (TSP) by 49.0%, 21.1%, and 51.9%, respectively. The nanocomposite can achieve a V-1 rating in the UL-94 test with a 29.7% LOI value using cyclophosphazene FGO [16]. Li et al. [17] assembled mesoporous MCM-41 (a 2D hexagonal part prepared from an organized low-density mesoporous silicate in the presence of quaternary ammonium halide surfactant) nanosphere over GO to prepare GO@MCM-41 hybrid. The 2 wt.% nanohybrid addition in EP results in a 40% reduction in PHRR compared to pristine GO/EP composite.

rGO has more impact than GO in FR applications. The reduction causes damage to the graphene structure; it reduces the fire effect by removing functional groups. rGO, as such, has only limited dispersion; generally, modified rGO hybrids are exploited to obtain dispersion and maximum FR properties. A hybrid of rGO with 9,10-dihydro-9-oxa-10-phosphaphenanthrene-10-oxide (DOPO)-based flame-retardant additives (FRs-rGO) in EP exfoliates graphene and provides better mixing of rGO. Incorporating 5 wt.% of FRs-rGO into EP achieves a V-0 rating in the UL-94 test and an LOI value of 29.5%, significantly reducing heat release. The improved fire retardance of FRs-rGO in EP was due to the shielding of rGO and the charring of FRs [18]. Functionalized reduced graphene oxide (FRGO) with phosphorous (P) and nitrogen (N) FRs imparts good dispersion and strong interfacial adhesion with EP matrix. The improvement in crosslink density and the confinement of graphene sheet obtained with P-N-wrapped FRGO accelerates the degradation of the EP matrix resulting in the formation of additional char residue. A high thermally stable char layer consisting of graphene sheets reduces the heat release and prevents volatile degradation products. Thus, the synergistic effect reduces PHRR by 37.7% and THR by 30.2% with 4 wt.% P-N-wrapped FRGO-modified EP [19].

As mentioned above, the reduction of GO primarily damages graphene before functionalization; the simultaneous reduction and functionalization of GO can overcome this problem. DOPO-phosphonamidate, piperazine, and zinc hydroxystannate are examples of additives used for simultaneous functionalization and reduction. The functionalization and reduction of GO with DOPO-phosphonamidate develop reactive spaces between GO and the EP matrix resulting in good dispersion of FRGO in the matrix. DOPO-phosphonamidate on the surface of GO promotes flame inhibition and protects the rGO from fire. Also, the adsorption and barrier effect of rGO inhibit the heat and gas release and promote the formation of graphitized carbons. Incorporation of 3 wt.% FRGO in EP reduces PHRR and THR by 31% and 34.3%, respectively, compared to neat EP [20]. Guo et al. [21] reduced and functionalized graphene with piperazine; later the modified graphene was incorporated with piperazine-DOPO-phosphonamidate (PiP-DOPO) to form nanohybrids (PD-rGO). The 4 wt.% PD-rGO5 (5 wt.% rGO in hybrid)-modified EP could pass the UL-94 V0 rating and attain a 28% LOI value. The barrier effect of graphene and flame spread inhibition of PiP-DOPO increased the composite's fire-resistant property. Liu et al. [22] used zinc hydroxystannate to reduce and functionalize graphene to form functionalized graphene oxide hybrid (ZHS/RGO). The researchers reported a significant reduction in the PHRR (−50%) and THR (−39%) compared to pure EP by incorporating 3 wt.% of ZHS/RGO. The physical barrier phenomenon of graphene and the catalytic impact of ZHS suppress smoke and toxic carbon gases and enhance char layer formation [22].

Besides, GO, and RGO, graphene nanosheets (GNS) exhibit good FR properties. The addition of GNS in the EP matrix enhances the compactness of both surface and bottom char residues, which protects the EP more effectively during combustion. Also, the network formation through the GNS–GNS and GNS–EP interaction decreases the melt flow and inhibits the flammable drips of EP. The use of modified GNS and an increase in the amount of GNS can enhance flame retardancy. The 5 wt.% addition of GNS in EP decreases THR by 16.7% and increases LOI value (21.4%) [23]. Allylamine-modified graphene nanoplatelets addition in EP significantly reduced heat release, and smoke production and a reduction of 37% in THR and 32% reduction in TSP were reported. However, the high thermal conductivity of graphene sheets increases PHRR [24]. ZnS-decorated graphene sheets as nanofiller in EP matrix showed an enhanced flame retardancy due to its barrier and catalytic effect. Adding 2% ZNS/GNS in EP decreases PHRR and THR by 47% and 27%, respectively [25]. Table 13.2 shows the different types of graphene and their properties.

TABLE 13.2
Different Graphene Types and Their Fire-Retardant Properties in Epoxy Composites

Graphene Type	Wt.%	Properties	Reference
PDMS-GO	0.29	UL-94 V-0 rating, 29.2% LOI, char yield 21.4 wt.%	[14]
FGO	0.7	49.7%, 34.3%, and 41.5% reduction in PHRR, THR, and TSP, respectively	[15]
Cyclophosphazene FGO	1.0	49.0%, 21.1%, and 51.9% reduction in PHRR, THR, and TSP, respectively UL-94 V-1 rating, 29.7% LOI	[16]
GO@MCM-41	2.0	40% reduction in PHRR compared to pristine GO/EP composite.	[17]
FRs-rGO	5.0	35.0% reduction in PHRR UL-94 V-0 rating, 29.5% LOI	[18]
P-N-wrapped FRGO	4.0	37.7% and 30.2% reduction in PHRR and THR, respectively	[19]
DOPO-rGO	3.0	31.0% and 34.3% reduction in PHRR and THR, respectively	[20]
PD-rGO	4.0	43.0%, 30.2%, and 34% reduction in PHRR, THR, and SPR, respectively UL-94 V-0 rating, 28.0% LOI	[21]
ZHS/RGO	3.0	50.0%, 39.0%, and 31.0% reduction in PHRR, THR, and TSR, respectively	[22]
GNS	5.0	16.7% reduction in THR 21.4% LOI	[23]
FGNS	15.0	37.0 and 32.0% reduction in THR and TSP, respectively	[24]
ZnS/GNS	2.0	47.0%, 27.0%, and 63.0% reduction in PHRR, THR, and TSR, respectively	[25]

13.2.2 CARBON NANOTUBE

There are mainly two different CNTs, namely (1) single-walled CNTs (SWCNTs) and (2) multi-walled CNTs (MWCNTs), which are used for various industrial as well as academic purposes [26]. CNTs-incorporated polymer shows a strong FR effect. During combustion, it forms a network structure on the polymer surface right after ignition. Moreover, as the CNTs burn, they form a thick char layer, preventing combustible materials' diffusion [27].

CNTs wrapped with the phosphorus-nitrogen-containing polymer can protect EP polymer during combustion. Adding 2 wt.% hybrids with 10% polymer content in EP reduced PHRR by 46.9% and THR by 29.3% [28]. Oxygen-doped CNT (COX) in EP shows better flame-resistant results than pristine CNT in EP. A significant reduction in PHRR and THR was obtained with the incorporation of 2 wt.% of the COX nanofiller. The high aspect ratio of COX helps to form a dense percolative char layer that acts as a protective layer [29]. A hybrid of CNT with bismuth selenide (Bi_2Se_3) imparts a barrier effect in EP resin. The 2 wt.% loadings of CNT-Bi_2Se_3 decreased PHRR and THR by 44% and 23%, respectively. The incomplete combustion of modified EP forms excess char, preventing further burning, as it acts as a protective layer. Also, the CNT-Bi_2Se_3 slows down the smoke release and thus decreases the effect of volatile gases [30]. DOPO-functionalized MWCNT (MWCNT-ODOPB) possesses an excellent carbonization effect during combustion. MWCNT-ODOPB into EP along with aluminum diethylphosphinate (AlPi) forms a FR nanocomposite. It achieves UL-94 V-0 rating and a significant reduction in heat release. The synergism between MWCNT-ODOPB and AlPi in the EP matrix reduces smoke production and the release of toxic gases [31]. CNT buckypapers can impart flame retardancy to EP. MWCNT buckypaper in EP remarkably delays the time to ignition of EP nanocomposites. It has high-temperature thermo-oxidation resistance and the ability to form a dense network; this shields the surface from temperature, toxic gases, and smoke generation during fire combustion. However, SWCNT buckypaper easily burns during combustion and does not affect flammability [32].

Transition metals have a remarkable ability to form catalytic char; however, in a nano form, it has limitations in FR application because of nanoparticles' easy agglomeration and oxidization. Fe and Ni can be used to modify CNT to impart better FR properties in EP. The incorporation of 6 wt.% Fe-CNT in EP resin increases fire retardancy, as it reaches 35% LOI value and V-1 rating in the UL-94 vertical test. Fe and CNT nanoparticles' good dispersion and oxidation reaction with EP form a three-dimensional network structure. The formation of carbonaceous ceramic layers during combustion reduced the heat transfer and combustible gas transfer between EP nanocomposite and the flame zone. The overall effect of the nanofiller was a decrease in the PHRR, THR, and TSP by 30.7%, 39.1%, and 48.6%, respectively [33]. Typical CNT may entangle to form bundles in polymer nanocomposites and restricts the polymer chains from entering the narrow channels, thereby blocking the proper dispersion. A graphitic nanotube (GNT) with broad channels can be used to overcome this issue. Boron and nitrogen-doped larger diameter GNT decorated with nickel in EP showed a dramatic improvement in FR properties. A substantial decrease in heat release and smoke release was obtained with the incorporation of 2 wt.% Ni-GNT. The shorter length-to-diameter ratio of the Ni/GNTs strengthens uniform dispersion in the EP, thereby improving the higher contact surface area and better interactions with the matrix. The boron, nitrogen, and nickel increase the char residue, resulting in an enhanced char strength during combustion [34]. Table 13.3 shows the different types of CNTs and their properties.

13.3 INORGANIC NANOPARTICLES

13.3.1 Layered Double Hydroxide (LDH)

LDH is a type of lamellar compound composed of cationic layers with interlayer anions. The general chemical formula is described as $[M^{z+}_{1-x}M^{3+}_x(OH)_2]^{q+}(X^{n-})$ q/n·yH_2O, where M^{z+} represents divalent cation (Zn^{2+}, Ca^{2+} or Mg^{2+}), while M^{3+} is a trivalent cation (Al^{3+} or Fe^{3+}), and X^{n-} is the charge balancing interlayer anion.

TABLE 13.3

Different Types of CNTs and Their Fire-Retardant Properties In Epoxy Composites

CNT Type	Wt.%	Properties	Reference
P-N-wrapped CNT	2.0	46.9% and 29.3% reduction in PHRR and THR, respectively	[28]
Oxygen-doped CNT	2.0	38.7%, and 32.8% reduction in PHRR, and THR respectively and 35.0% LOI	[29]
CNT-Bi$_2$Se$_3$	2.0	44.0% and 23.0% reduction in PHRR and THR, respectively	[30]
MWCNT-ODOPB	1.0	V-0 rating, 39.5% LOI	[31]
MWCNT-Buckypaper		45.0%, 60.0%, and 50.0% reduction in PHRR, THR, and TSP, respectively	[32]
Fe-CNT	6.0	30.7%, 39.1%, and 48.6% reduction in PHRR, THR, and TSP, respectively V-1 rating, 35.0% LOI	[33]
Ni-GNT	2.0	43.5%, 16.3%, and 22.8% reduction in PHRR, THR, and TSP, respectively	[34]

Because of its high water content, non-toxicity, and layer-like structure, pristine and modified forms of LDH can act as a flame-retardant additive [35,36].

LDH can be modified using biobased modifiers. Modifiers from renewable sources like sulfonated cyclodextrin (sCD) and cardanol-BS will reduce nanocomposites' costs, improve filler dispersion, interlayer distance, and char yield during combustion. Wang et al. [37] observed that the cardanol-BS anions intercalated into the interlayer galleries of LDH, resulting in increased interlayer spacing. Therefore, the EP resin can easily impregnate the interlayer spacing and improve dispersion and interaction between filler and matrix. Hence, a compact and continuous residue will be formed during combustion. An increase in the loading of the cardanol-BS modified LDH increases the flame retardancy of EP resin. The 6 wt.% modified LDH loading shows a remarkable decrease in PHRR, THR, and TSP by 62%, 19%, and 45%, respectively. As mentioned above, other studies also reported improved the interlayer space with the modification of LDH with sCD, facilitating the impregnation of EP resin to the interlayer spacing of LDH. The increase in dispersion of LDH in EP increases flame retardancy [38,39]. Li et al. [40] reported the electrostatic assembly of functionalized silica onto LDH-DBS. The addition of 3 wt.% LDH-DBS@silica reported a 29.2% reduction in PHRR compared to 3 wt.% LDH-DBS in EP. The interfacial assembly of silica forms less crystalline-sized char, which reduces the pore size and acts as a barrier. Li et al. [41] modified LDH-DBS with Ni(OH)^2 nanosheet in another study. The researchers reported that the EP nanocomposite with 3 wt.% nanofiller (LDH-DBS@Ni(OH)$_2$) comes under V-0 class in the UL-94 vertical test.

Transition metal compounds exhibit fascinating properties as flame-retardant additives due to their ability to impart synergistic effects. Hence, the modification of LDH with transition metal compounds (Cu-Al, Mg-AL) can significantly improve the FR properties. Ding et al. synthesized LDH with Cu-Al and later modified CuAl-LDH hybrid with sodium phenyl phosphate (SPP). The synthesized CuAl-(SPP)LDH was used to modify the EP and reported significant improvement in FR properties. The composites achieve a 55.6% reduction in PHRR and a 5.8% reduction in THR with a 4 wt.% SPP-modified hybrid [42]. In an interesting study, Li and co-workers [43] synthesized a ZIF@MgAl-LDH hybrid and studied its effect on the flame retardancy of EP composites. The incorporation of ZIF@MgAl-LDH nanofiller in EP reported a 37.5%, 33.9%, and 37.5% decrease in PHRR, THR, and TSP, respectively. The composite comes under a UL-94 V-1 rating with 25.5% LOI. Another transition metal compound Ni-Fe with LDH also provides a shielding effect and good dispersion [44]. The organically modified NiFe-LDH hybrid system reduces the heat and smoke release of nanocomposite during combustion. However, modifying ONiFe-LDH with CNTs significantly improved the composite performance. It was reported that ONiFe-LDH-CNT/EP reduces PHRR and THR by 30% and 20%, respectively, compared to NiFe-LDH/EP. Zhang et al. [45] synthesized a hybrid of two different modified LDH (MgAl-LDH and NiCo-LDH) interconnected with a metal-organic framework. The hybrid nanofiller-modified EP matrix achieved a V-0 rating in UL-94 test. The uniform dispersion of nanofiller in the EP matrix produced strong char layers during combustion. Also, transition metals in the composite capture polymer-free radicals during combustion and increase the crosslinked network's density. This will lead to forming a compact char layer on the surface, thus reducing the release of smoke and volatiles. The EP nanocomposite with only 2.5 wt.% nanofiller significantly reduces the PHRR, THR, and TSP values by 66.6%, 38.7%, and 25%, respectively.

Silica nanoparticles as fillers have a notable effect on FR properties. Flame-retardant additives (DOPO derivatives) modified with silica particles show a synergistic effect of FR and silica nanoparticles. SiO_2 nanoparticles can improve thermal stability and increase char yield [46]. Jiang et al. [47] synthesized mesoporous silica@CoAl LDH (m-SiO_2@Co-Al LDH) spheres by a layer-by-layer process. Incorporating 2 wt.% m-SiO_2@CoAl-LDH reduces PHRR, THR, and TSR by 39.3%, 36.2%, and 23.8%, respectively. Table 13.4 shows the different types of LDH and their properties.

13.3.2 OTHER CLAYS

Other clay materials like montmorillonite (MMT) and bentonite also impart FR properties to EP resin. These clays, without modification, may cause a negative effect. In situ polymerization of MMT modified with alkylammonium ions in EP composite forms intercalated or exfoliated structures during the crosslinking of EP. The synergism between the physical barrier effect of exfoliated nanocomposites and the chemical role of clay acts against fire. The increase in charring during combustion protects the EP surface from the action of oxygen. A significant increase in ignition time and a notable reduction in heat release were observed in the modified EP composites containing in situ polymerized clay composite [48]. Hexadecyltrimethylammonium chloride can change the hydrophilic nature of MMT to organophilic MMT (OMMT). Such OMMT has exfoliated structure within the EP matrix. The barrier effect performed by clay can reduce oxygen and other gases and block the burning of EP. However, the best result can only attain with higher loading of OMMT [49].

TABLE 13.4
Different Types of LDH and Their Fire-Retardant Properties in Epoxy Composites

LDH Type	Wt.%	Properties	Reference
Cardanol-BS-LDH (biobased)	6.0	62.0%, 19.0%, and 45.0% decrease in PHRR, THR, and TSP, respectively UL-94 V-0 rating, 29.2% LOI	[37]
NO₃-LDH	3.0	24.5%, 18.1%, and 14.6% decrease in	[40]
LDH-DBS	3.0	PHRR, THR, and TSP, respectively	
LDH-DBS@silica	3.0	48.1%, 26.5%, and 30.6% decrease in PHRR, THR, and TSP, respectively 24.8% LOI 63.3%, 63.3%, and 64.0% decrease in PHRR, THR, and TSP, respectively Self-extinguishable (near to UL-94 V-1 rating)	
LDH-DBS@Ni(OH)₂	3.0	60.6%, 65.4%, and 66.5% decrease in PHRR, THR, and TSP, respectively UL-94 V-0 rating	[41]
T-LDH/EP	6.0	47.3% decrease in PHRR	[38]
sCD-LDH/EP	6.0	43.6% decrease in PHRR	
sCD-DBS-T-LDH	6.0	65.8% decrease in PHRR, UL-94 V-0 rating, 26.8% LOI	
LDH-CD-Ferr	6.0	36% reduction in PHRR, and UL-94 V-1 rating UL-94 V-1 rating	[39]
CuAl-(SDS)LDH	4.0	55.6% and 5.8% decrease in PHRR and THR, respectively	[42]
MgAl-LDH	2.0	24.5%, 12.5%, and 14.6% decrease in	[43]
ZIF@MgAl-LDH	2.0	PHRR, THR, and TSP, respectively 37.5%, 33.9%, and 37.5% decrease in PHRR, THR, and TSP, respectively UL-94 V-1 rating, 25.5% LOI	
ONiFe-LDH	4.0	33.9% and 12.5% decrease in PHRR and	[44]
ONiFe-LDH-CNT	4.0	THR, respectively 54.0% and 29.8% decrease in PHRR and THR, respectively	
MgAl-LDH/NiCo-LDH hybrid	2.5	66.6%, 38.7%, and 25% decrease in PHRR, THR, and TSP, respectively UL-94 V-0 rating, 26.0% LOI	[45]
CoAl LDH	2.0	19.3%, 4.0%, and 13.4% decrease in	[47]
m-SiO₂@CoAl LDH	2.0	PHRR, THR, and TSR, respectively 39.3%, 36.2%, and 23.7% decrease in PHRR, THR, and TSR, respectively	

Greater separation of MMT nanolayers can be achieved by functionalizing DOPO with MMT. DOPO has greater reactivity and miscibility with EP, which eases the EP chains entering MMT interlayers. Hence, the nanocomposite of EP will have improved nanoscale dispersion. The good distribution of MMT layers can provide adequate insulation and shielding to the EP matrix. This better insulation gave the EP nanocomposite to V-0 rating in the UL-94 vertical burning test [50]. Figure 13.1a and b describes the HRR and THR values of the DOPO-modified EP / MMT composites.

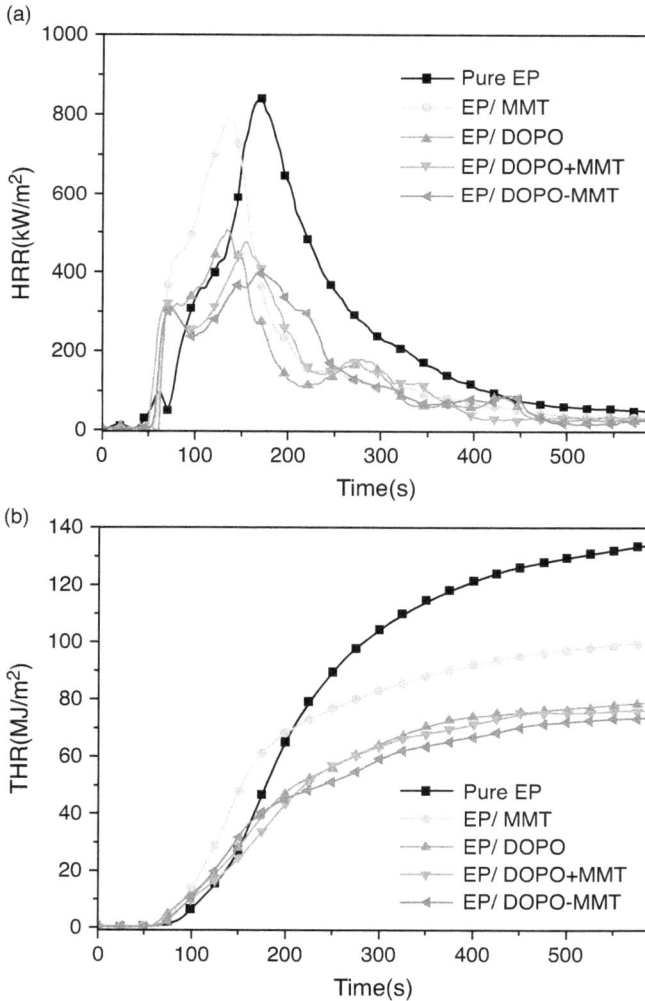

FIGURE 13.1 A graph showing variation of (a) HRR and (b) THR values of DOPO-modified epoxy/MMT with time (Reprinted with permission from Ref. [50]. License Number: 5437100732722.)

Organic modification of bentonites makes them suitable for FR application in EP nanocomposites. Modifying nitrogen-containing groups such as 11-amino-N-(pyridine-2yl) undecanamide (APUA) decreases heat and smoke release during combustion and increases char yield. A 17% decrease in PHRR was obtained with 3 wt.% of bento-APUA [51].

13.3.3 MOLYBDENUM DISULFIDE (MoS₂)

MoS_2 has a unique 2D nanostructure and excellent properties as filler in the polymer matrix. It is widely used for electronic applications, as it has a large direct bandgap and excellent on/off ratio. Since MoS_2 has a similar 2D structure to LDH and graphene, it was used as a FR additive to the polymer matrix.

Several studies on the hybridization/modification of MoS_2 to improve the fire retardancy of EP composites were recently reported [52,53]. Qiu et al. [54] synthesized MoS_2@PPN (polyphosphazene nanoparticle) hybrid using simple ball milling and high-temperature polymerization. A total reduction of 30.7% and 23.6% in PHRR and THR, respectively, was reported with the incorporation of 2 wt.% MoS_2@PPN in EP. Polyphosphazene microspheres (PZS)-modified MoS_2 has shown better flame-retardant properties than PPN-modified MoS_2 [55]. This is because the synergistic effect between PZS and MoS_2 imparts good physical barrier properties, thermal stability, and reinforcement of the microsphere in EP. This will reduce THR by 41.3% and 30.3%, respectively [55]. The application of (4-carboxybutyl)triphenylphosphonium bromide (TPP) has resulted in exfoliated MoS_2 (TPP-MoS₂), hence better dispersion. This helped to restrict the transmission of oxygen and heat to the underlying material and also prevented volatile pyrolysis product diffusion. The EP composite achieved exceptional FR ability with low loading of TPP-MoS₂ hybrid nanofiller [56]. Hou et al. [57] reported that the incorporation of Bi_2Se_3 and N-vinyl pyrrolidone exfoliated MoS_2 hybrid (MB) in EP promotes char formation and suppresses the release of organic decomposition products during combustion. The formation of the char layer can prevent smoke production and release. It was observed that 1 wt.% of MB hybrid in EP decreases the PHRR and THR by 22% and 21.8%, respectively [57]. The modification of MoS_2 using $Mg(OH)_2$ reported an improved physical barrier effect and catalytic charring effect on the EP surface [58]. An efficient char layer's formation restricts oxygen penetration and prevents the spread of flammable gas. This reduces smoke and heat release significantly.

MoS_2 can impart better fire retardancy in combination with another 2D nanomaterial LDH. The hybrid of LDH and MoS_2 in EP resin imparts catalytic carbonization of LDH and barrier effect of MoS_2. Zhou et al. reported that the MoS_2 could form a good synergistic effect with different transition metal compounds-modified LDH. Even though most such LDH have a similar mechanism, a hybrid of NiFe-LDH has more significant FR properties. Homogeneous dispersion of 2 wt.% NiFe-LDH/MoS₂ in EP decreases PHRR and THR by 66% and 34%, respectively [59]. Table 13.5 shows the different types of MoS_2 and their properties.

TABLE 13.5

Different Types of MoS₂ and Their Fire-Retardant Properties in Epoxy Composites

MoS₂ Type	Wt.%	Properties	Reference
PPN-MoS₂	1.0	24.9% and 19.3% decrease in PHRR and THR, respectively	[54]
	2.0	30.7% and 23.6% decrease in PHRR and THR, respectively	
PZS-MoS₂	2.0	37.5% and 27.0% decrease in PHRR and THR, respectively	[55]
	3.0	41.3% and 30.3% decrease in PHRR and THR, respectively	
MoS₂	2.0	18.6%, 18.2%, and 9.7% decrease in PHRR, THR, and TSR, respectively	[56]
TPP-MoS₂	2.0	26.1%, 17.3%, and 21.3% decrease in PHRR, THR, and TSR, respectively	
Bi₂Se₃-MoS₂	1.0	22.0% 21.8%, and 34.0% decrease in PHRR, THR, and TSR, respectively	[57]
Mg(OH)₂-MoS₂	2.0	32.0% and 27.0% decrease in PHRR and SPR, respectively	[58]
CoFe-LDH-MoS₂	2.0	62.0% and 24.8% decrease in PHRR and THR, respectively	[59]
NiFe-LDH-MoS₂		66.0% and 34.0% decrease in PHRR and THR, respectively	

13.4 HYBRID COMPOSITES

13.4.1 GRAPHENE OR CNT-CLAY HYBRIDS

Combining graphene or CNT with clay particles has both organic and inorganic effects of increasing FR properties. LDH reported to achieve good properties with graphene compounds. The incorporation of NiFe-LDH into graphene increases the thermal stability of graphene. The synergistic effect of NiFe-LDH/graphene hybrid enhances the char yield. 2 wt.% NiFe-LDH/graphene in EP reduces PHRR and THR by 60% and 61%, respectively [60]. Modifying rGO/LDH with CuMoO₄ imparts better fire resistance properties in EP composites. The generation of Cu₂O and MoO₃ during combustion enhances the production of dense and hole-free char layers. Thus, the rGO-LDH/CuMoO₄ in EP shows a V-1 rating with a significant reduction in THR and TSP [61]. Zeolitic Imidazolate Framework (ZIF-67) on LDH with rGO results in a synergistic system. Herein, the physical barrier property of rGO and LDH and the catalytic carbonization effect of rGO-LDH/ ZIF-67 retard fire on the composite. In addition, during the combustion process, LDH releases water vapors and protects the EP polymer from burning. Incorporating ZIF-67 enhances the thermal oxidation property of char layers. 2 wt.% addition of this hybrid system is sufficient to significantly reduce heat and smoke production while burning the composite [62].

TABLE 13.6

Different Types of Additives and Their Fire-Retardant Properties in Epoxy Composites

Type of Additive	Wt.%	Properties	Reference
rGO-NiFe-LDH	2.0	60.0% and 61.0% decrease in PHRR and THR, respectively	[60]
rGO-LDH-CuMoO4	2.0	47.6%, 28.5%, and 38.0% decrease in PHRR, THR, and TSP, respectively UL-94 V-1 rating, 24.2% LOI	[61]
MWCNT (0.02)-MMT (2.0)		25.3% LOI, 20.05 mm/min combustion rate	[63]
MWCNT (0.02)-MMT (2.0)-Sb_2O_3-chlorinated paraffins		UL-94 V-0 rating, 33.3% LOI, 10.05 mm/min combustion rate	[64]
MWCNT (0.02)-MMT (2.0) (triethylenetetramine grafting agent)		7.4% and 26.3% decrease in PHRR and SPR, respectively 65.9% increase in TTI	[65]

CNTs show the best fire-resistant properties with MMT clays. Different modified forms of these hybrids can reduce the fire attack on the polymer surface. The MWCNT-MMT combination forms a thin layer on the EP polymer, preventing the oxidizing agents from penetrating the polymer surface. Ash layers were formed on the surface during burning preventing heat transfer and delaying the temperature increase [63]. Adding other flame-retardant additives like chlorinated paraffin and antimony oxide to the CNT-MMT hybrid further increases flame retardancy. Significant reduction in combustion rate and heat release can achieve with small quantity hybrid addition [64]. Grafting of MMT onto CNT by an appropriate grafting agent will result in a more significant dispersion of the hybrid system in EP. Triethylenetetramine grafting agent provides a good interlayer interaction between the two nanoparticles. Thus, a substantial reduction in total smoke and heat production was obtained compared to a typical hybrid system [65]. Table 13.6 shows the different types of additives and their properties.

13.5 CONCLUSIONS

Since thermoplastics and thermosetting polymers are increasingly used in many applications such as electrical and electronics, building and construction, aerospace, and automotive, various organic and inorganic flame-retardant systems have been developed. Besides, non-chemical techniques were developed for flame retardation purposes. In this review work, different inorganic and organic flame retardants were effectively incorporated within the EP matrices and enhanced the flame retardancy property of EP matrix-based composites. Various factors could determine the flame retardancy property of matrices: (1) ease of ignition, (2) fire durability, (3) heat release rate, (4) smoke generation, and (5) emission of toxic gas. Besides, the flammability property of polymeric materials can be measured by UL-94 flammability test.

Among the various flame-retardant materials, graphene, CNT, LDH, other clays (MMT and bentonite), molybdenum disulfide, graphene or CNT-clay hybrids and their different types of organic and inorganic nanofillers were elaborated in this review work. Incorporating graphene with low loadings into the polymer can improve the flammability, mechanical, thermal, and electrical conductivity properties. Likewise, the inclusion of CNT, LDH, etc., can impart a solid FR effect in composites. Thus, the inclusion of flame-retardant materials in composite materials is significant for any composite materials. Also, it is concluded that natural and environmentally friendly flame-retardant materials should be developed in future.

ACKNOWLEDGMENT

The authors would like to thank King Mongkut's University of Technology North Bangkok for the financial support under the foreign postdoctoral scholarship scheme (grant no. KMUTNB-Post-65-01).

REFERENCES

[1] J. Parameswaranpillai, H. Pulikkalparambil, S. M. Rangappa, S. Siengchin (Eds.) *Epoxy Composites: Fabrication, Characterization and Applications*, Wiley-VCH, Weinheim, 2021.

[2] S. V. Levchik, *Fire Retardant Polymer Nanocomposites*, John Wiley & Sons, Hoboken, NJ, pp. 1–30, 2007.

[3] G. Vahidi, D.S. Bajwa, J. Shojaeiarani, N. Stark, A. Darabi, "Advancements in traditional and nanosized flame retardants for polymers—A review", *Journal of Applied Polymer Science*, 138(12), p. 50050, 2021.

[4] L. Costes, F. Laoutid, S. Brohez, P. Dubois, "Bio-based flame retardants: When nature meets fire protection", *Materials Science and Engineering: R: Reports*, 117, pp. 1–25, 2017.

[5] Camino, G. and Costa, L, "Performance and mechanisms of fire retardants in polymers—A review", *Polymer degradation and Stability*, 20(3–4), pp. 271–294, 1988.

[6] X. Yue, C. Li, Y. Ni, Y. Xu, and J. Wang, "Flame retardant nanocomposites based on 2D layered nanomaterials: A review," *Journal of Materials Science*, vol. 54, no. 20, pp. 13070–13105, 2019.

[7] C. Gérard, G. Fontaine, and S. Bourbigot, "New trends in reaction and resistance to fire of fire-retardant epoxies," *Materials*, vol. 3, no. 8, pp. 4476–4499, 2010.

[8] R. Atif, I. Shyha, and F. Inam, "Mechanical, thermal, and electrical properties of graphene-epoxy nanocomposites-A review," *Polymers*, vol. 8, no. 8, p. 281, 2016.

[9] H. Kim, A. A. Abdala, and C. W. MacOsko, "Graphene/polymer nanocomposites," *Macromolecules*, vol. 43, no. 16. pp. 6515–6530, 2010.

[10] X. Fu, C. Yao, and G. Yang, "Recent advances in graphene/polyamide 6 composites: A review," *RSC Advances*, vol. 5, no. 76, pp. 61688–61702, 2015.

[11] J. Jing, Y. Zhang, Z. P. Fang, and D. Y. Wang, "Core-shell flame retardant/graphene oxide hybrid: A self-assembly strategy towards reducing fire hazard and improving toughness of polylactic acid," *Composites Science and Technology*, vol. 165, pp. 161–167, 2018.

[12] M. Ciesielski, B. Burk, C. Heinzmann, and M. Döring, "Fire-retardant high-performance epoxy-based materials," in *Novel Fire Retardant Polymers and Composite Materials*, De-Yi Wang (ed.), Elsevier, Woodhead Publishing, United Kingdom. pp. 3–51, 2017.

[13] Z. Wang, X. Z. Tang, Z. Z. Yu, P. Guo, H. H. Song, and X. S. Duc, "Dispersion of graphene oxide and its flame retardancy effect on epoxy nanocomposites," *Chinese Journal of Polymer Science*, vol. 29, no. 3, pp. 368–376, 2011.

[14] J. Luo, S. Yang, L. Lei, J. Zhao, and Z. Tong, "Toughening, synergistic fire retardation and water resistance of polydimethylsiloxane grafted graphene oxide to epoxy nanocomposites with trace phosphorus," *Composites Part A: Applied Science and Manufacturing*, vol. 100, pp. 275–284, Sep. 2017.

[15] L. Qu et al., "POSS-functionalized graphene oxide hybrids with improved dispersive and smoke-suppressive properties for epoxy flame-retardant application," *European Polymer Journal*, vol. 122, p. 109383, Jan. 2020.

[16] L. Qu, Y. Sui, C. Zhang, P. Li, X. Dai, and B. Xu, "Compatible cyclophosphazene-functionalized graphene hybrids to improve flame retardancy for epoxy nanocomposites," *Reactive and Functional Polymers*, vol. 155, p. 104697, Oct. 2020.

[17] Z. Li, A. J. González, V. B. Heeralal, and D. Y. Wang, "Covalent assembly of MCM-41 nanospheres on graphene oxide for improving fire retardancy and mechanical property of epoxy resin," *Composites Part B: Engineering*, vol. 138, pp. 101–112, Apr. 2018.

[18] X. Qian et al., "Novel organic-inorganic flame retardants containing exfoliated graphene: Preparation and their performance on the flame retardancy of epoxy resins," *Journal of Materials Chemistry A*, vol. 1, no. 23, pp. 6822–6830, Jun. 2013.

[19] B. Yu et al., "Enhanced thermal and flame retardant properties of flame-retardant-wrapped graphene/epoxy resin nanocomposites," *Journal of Materials Chemistry A*, vol. 3, no. 15, pp. 8034–8044, Apr. 2015.

[20] Y. Shi, B. Yu, Y. Zheng, J. Yang, Z. Duan, and Y. Hu, "Design of reduced graphene oxide decorated with DOPO-phosphanomidate for enhanced fire safety of epoxy resin," *Journal of Colloid and Interface Science*, vol. 521, pp. 160–171, Jul. 2018.

[21] W. Guo, B. Yu, Y. Yuan, L. Song, and Y. Hu, "In situ preparation of reduced graphene oxide/DOPO-based phosphonamidate hybrids towards high-performance epoxy nanocomposites," *Composites Part B: Engineering*, vol. 123, pp. 154–164, Aug. 2017.

[22] X. Liu, W. Wu, Y. Qi, H. Qu, and J. Xu, "Synthesis of a hybrid zinc hydroxystannate/reduction graphene oxide as a flame retardant and smoke suppressant of epoxy resin," *Journal of Thermal Analysis and Calorimetry*, vol. 126, no. 2, pp. 553–559, Nov. 2016.

[23] S. Liu, H. Yan, Z. Fang, and H. Wang, "Effect of graphene nanosheets on morphology, thermal stability and flame retardancy of epoxy resin," *Composites Science and Technology*, vol. 90, pp. 40–47, Jan. 2014.

[24] J. Gu et al., "Highly thermally conductive flame-retardant epoxy nanocomposites with reduced ignitability and excellent electrical conductivities," *Composites Science and Technology*, vol. 139, pp. 83–89, Feb. 2017.

[25] S. D. Jiang, Z. M. Bai, G. Tang, Y. Hu, and L. Song, "Synthesis of ZnS decorated graphene sheets for reducing fire Hazards of epoxy composites," *Industrial and Engineering Chemistry Research*, vol. 53, no. 16, pp. 6708–6717, Apr. 2014.

[26] B. T. Marouf, Y. W. Mai, R. Bagheri, and R. A. Pearson, "Toughening of epoxy nanocomposites: Nano and hybrid effects," *Polymer Reviews*, vol. 56, no. 1, pp. 70–112, 2016.

[27] L. Schlagenhauf, Y. Y. Kuo, Y. K. Bahk, F. Nüesch, and J. Wang, "Decomposition and particle release of a carbon nanotube/epoxy nanocomposite at elevated temperatures," *Journal of Nanoparticle Research*, vol. 17, no. 11, Nov. 2015.

[28] S. Wang, F. Xin, Y. Chen, L. Qian, and Y. Chen, "Phosphorus-nitrogen containing polymer wrapped carbon nanotubes and their flame-retardant effect on epoxy resin," *Polymer Degradation and Stability*, vol. 129, pp. 133–141, Jul. 2016.

[29] M. E. Shabestari et al., "Effect of nitrogen and oxygen doped carbon nanotubes on flammability of epoxy nanocomposites," *Carbon*, vol. 121, pp. 193–200, Sep. 2017.

[30] Y. Hou, W. Hu, L. Liu, Z. Gui, and Y. Hu, "In-situ synthesized CNTs/Bi$_2$Se$_3$ nanocomposites by a facile wet chemical method and its application for enhancing fire safety of epoxy resin," *Composites Science and Technology*, vol. 157, pp. 185–194, Mar. 2018.

[31] L. Gu, C. Qiu, J. Qiu, Y. Yao, E. Sakai, and L. Yang, "Preparation and characterization of DOPO-Functionalized MWCNT and Its high flame-retardant performance in epoxy nanocomposites," *Polymers*, vol. 12, no. 3, Mar. 2020.

[32] Q. Wu, W. Zhu, C. Zhang, Z. Liang, and B. Wang, "Study of fire retardant behavior of carbon nanotube membranes and carbon nanofiber paper in carbon fiber reinforced epoxy composites," *Carbon*, vol. 48, no. 6, pp. 1799–1806, May 2010.

[33] J. Zhang, Q. Kong, and D. Y. Wang, "Simultaneously improving the fire safety and mechanical properties of epoxy resin with Fe-CNTs: Via large-scale preparation," *Journal of Materials Chemistry A*, vol. 6, no. 15, pp. 6376–6386, 2018.

[34] Y. Yuan, Y. T. Pan, Z. Zhang, W. Zhang, X. Li, and R. Yang, "Nickle nanocrystals decorated on graphitic nanotubes with broad channels for fire hazard reduction of epoxy resin," *Journal of Hazardous Materials*, vol. 402, p. 123880, Jan. 2021.

[35] M. J. Mochane, S. I. Magagula, J. S. Sefadi, E. R. Sadiku, and T. C. Mokhena, "Morphology, thermal stability, and flammability properties of polymer-layered double hydroxide (LDH) nanocomposites: A review," *Crystals*, vol. 10, no. 7, pp. 1–26, 2020.

[36] Y. Liu, Y. Gao, Q. Wang, and W. Lin, "The synergistic effect of layered double hydroxides with other flame retardant additives for polymer nanocomposites: A critical review," *Dalton Transactions*, vol. 47, no. 42, pp. 14827–14840, 2018.

[37] X. Wang, E. N. Kalali, and D. Y. Wang, "Renewable cardanol-based surfactant modified layered double hydroxide as a flame retardant for epoxy resin," *ACS Sustainable Chemistry and Engineering*, vol. 3, no. 12, pp. 3281–3290, Oct. 2015.

[38] E. N. Kalali, X. Wang, and D. Y. Wang, "Functionalized layered double hydroxide-based epoxy nanocomposites with improved flame retardancy and mechanical properties," *Journal of Materials Chemistry A*, vol. 3, no. 13, pp. 6819–6826, Apr. 2015.

[39] Z. Li, Z. Liu, J. Zhang, C. Fu, U. Wagenknecht, and D. Y. Wang, "Bio-based layered double hydroxide nanocarrier toward fire-retardant epoxy resin with efficiently improved smoke suppression," *Chemical Engineering Journal*, vol. 378, p. 122046, Dec. 2019.

[40] Z. Li, Z. Liu, F. Dufosse, L. Yan, and D. Y. Wang, "Interfacial engineering of layered double hydroxide toward epoxy resin with improved fire safety and mechanical property," *Composites Part B: Engineering*, vol. 152, pp. 336–346, Nov. 2018.

[41] Z. Li, J. Zhang, F. Dufosse, and D. Y. Wang, "Ultrafine nickel nanocatalyst-engineering of an organic layered double hydroxide towards a super-efficient fire-safe epoxy resin: Via interfacial catalysis," *Journal of Materials Chemistry A*, vol. 6, no. 18, pp. 8488–8498, 2018.

[42] J. Ding et al., "Improving the flame-retardant efficiency of layered double hydroxide with disodium phenylphosphate for epoxy resin," *Journal of Thermal Analysis and Calorimetry*, vol. 140, no. 1, pp. 149–156, Apr. 2020.

[43] A. Li, W. Xu, R. Chen, Y. Liu, and W. Li, "Fabrication of zeolitic imidazolate frameworks on layered double hydroxide nanosheets to improve the fire safety of epoxy resin," *Composites Part A: Applied Science and Manufacturing*, vol. 112, pp. 558–571, Sep. 2018.

[44] Q. Kong et al., "Improving thermal and flame retardant properties of epoxy resin with organic NiFe-layered double hydroxide-carbon nanotubes hybrids," *Chinese Journal of Chemistry*, vol. 35, no. 12, pp. 1875–1880, Dec. 2017.

[45] Z. Zhang, J. Qin, W. Zhang, Y. T. Pan, D. Y. Wang, and R. Yang, "Synthesis of a novel dual layered double hydroxide hybrid nanomaterial and its application in epoxy nanocomposites," *Chemical Engineering Journal*, vol. 381, Feb. 2020.

[46] K. Wang et al., "Flame-retardant performance of epoxy resin composites with SiO$_2$ nanoparticles and phenethyl-bridged DOPO derivative," *ACS Omega*, vol. 6, no. 1, pp. 666–674, Jan. 2021.

[47] S. D. Jiang et al., "Synthesis of mesoporous silica@Co-Al layered double hydroxide spheres: Layer-by-layer method and their effects on the flame retardancy of epoxy resins," *ACS Applied Materials and Interfaces*, vol. 6, no. 16, pp. 14076–14086, Aug. 2014.

[48] G. Camino, G. Tartaglione, A. Frache, C. Manferti, and G. Costa, "Thermal and combustion behaviour of layered silicate-epoxy nanocomposites," *Polymer Degradation and Stability*, vol. 90, no. 2, pp. 354–362, 2005.

[49] E. Kaya, M. Tanoğlu, and S. Okur, "Layered clay/epoxy nanocomposites: Thermomechanical, flame retardancy, and optical properties," *Journal of Applied Polymer Science*, vol. 109, no. 2, pp. 834–840, Jul. 2008.

[50] X. He, W. Zhang, and R. Yang, "The characterization of DOPO/MMT nanocompound and its effect on flame retardancy of epoxy resin," *Composites Part A: Applied Science and Manufacturing*, vol. 98, pp. 124–135, Jul. 2017.

[51] T. Benelli et al., "New nitrogen-rich heterocycles for organo-modified bentonites as flame retardant fillers in epoxy resin nanocomposites," *Polymer Engineering and Science*, vol. 57, no. 6, pp. 621–630, Jun. 2017.

[52] X. Wang, W. Xing, X. Feng, L. Song, and Y. Hu, "MoS$_2$/polymer nanocomposites: Preparation, properties, and applications," *Polymer Reviews*, vol. 57, no. 3. pp. 440–466, 2017.

[53] X. Wang, E. N. Kalali, D.-Y. Wang, E. N. Kalali, D. -Y. Wang, and N. Adv, "Two-dimensional inorganic nanomaterials: A solution to flame retardant polymers," *Nano Advances*, vol. 1, p. 155, 2015.

[54] S. Qiu et al., "In situ growth of polyphosphazene particles on molybdenum disulfide nanosheets for flame retardant and friction application," *Composites Part A: Applied Science and Manufacturing*, vol. 114, pp. 407–417, Nov. 2018.

[55] X. Zhou, S. Qiu, W. Xing, C. S. R. Gangireddy, Z. Gui, and Y. Hu, "Hierarchical polyphosphazene@molybdenum disulfide hybrid structure for enhancing the flame retardancy and mechanical property of epoxy resins," *ACS Applied Materials and Interfaces*, vol. 9, no. 34, pp. 29147–29156, Aug. 2017.

[56] S. Wang, B. Yu, K. Zhou, L. Yin, Y. Zhong, and X. Ma, "A novel phosphorus-containing MoS$_2$ hybrid: Towards improving the fire safety of epoxy resin," *Journal of Colloid and Interface Science*, vol. 550, pp. 210–219, Aug. 2019.

[57] Y. Hou, Y. Hu, S. Qiu, L. Liu, W. Xing, and W. Hu, "Bi2Se3 decorated recyclable liquid-exfoliated MoS$_2$ nanosheets: Towards suppress smoke emission and improve mechanical properties of epoxy resin," *Journal of Hazardous Materials*, vol. 364, pp. 720–732, Feb. 2019.

[58] S. Zhao, J. Yin, K. Zhou, Y. Cheng, and B. Yu, "In situ fabrication of molybdenum disulfide based nanohybrids for reducing fire hazards of epoxy," *Composites Part A: Applied Science and Manufacturing*, vol. 122, pp. 77–84, Jul. 2019.

[59] K. Zhou, R. Gao, and X. Qian, "Self-assembly of exfoliated molybdenum disulfide (MoS$_2$) nanosheets and layered double hydroxide (LDH): Towards reducing fire hazards of epoxy," *Journal of Hazardous Materials*, vol. 338, pp. 343–355, 2017.

[60] X. Wang et al., "Self-assembly of Ni-Fe layered double hydroxide/graphene hybrids for reducing fire hazard in epoxy composites," *Journal of Materials Chemistry A*, vol. 1, no. 13, pp. 4383–4390, Apr. 2013.

[61] W. Xu, B. Zhang, X. Wang, G. Wang, and D. Ding, "The flame retardancy and smoke suppression effect of a hybrid containing CuMoO4 modified reduced graphene oxide/layered double hydroxide on epoxy resin," *Journal of Hazardous Materials*, vol. 343, pp. 364–375, 2018.

[62] W. Xu, X. Wang, Y. Liu, W. Li, and R. Chen, "Improving fire safety of epoxy filled with graphene hybrid incorporated with zeolitic imidazolate framework/layered double hydroxide," *Polymer Degradation and Stability*, vol. 154, pp. 27–36, Aug. 2018.

[63] T. A. Nguyen, Q. T. Nguyen, and T. P. Bach, "Mechanical properties and flame retardancy of epoxy resin/nanoclay/multiwalled carbon nanotube nanocomposites," *Journal of Chemistry*, vol. 2019, p. 3105205, 2019.

[64] N. Tuan Anh, N. Quang Tung, B. Trong Phuc, N. Xuan Canh, and T. Liem, "The use of multi-walled carbon nanotubes and nanoclay for simultaneously improving the flame retardancy and mechanical properties of epoxy nanocomposites," *International Journal of Engineering & Technology*, vol. 7, no. 4.36, pp. 1149–1160, 2018.

[65] Y. Xue et al., "A novel strategy for enhancing the flame resistance, dynamic mechanical and the thermal degradation properties of epoxy nanocomposites," *Materials Research Express*, vol. 6, no. 12, p. 125003, Nov. 2019.

14 Suspension Behaviour of Cornstalk Fibre/Fibreglass/ Epoxy Composites Leaf Spring Using Finite Element Analysis

A.S.B. Musamih, C.S. Hassan,
B.B. Sahari, and M.F. Mohamed Nazer
UCSI University

CONTENTS

14.1 INTRODUCTION

A leaf spring is a spring that is long, thin and flat, used for suspension system of wheeled vehicle. The leaf spring, also commonly referred to as a semi-elliptical spring due to its shape[1], has been employed by the automotive industry since the old medieval times. The leaf spring minimizes the vibration and provides a comfortable ride for the passengers[2]. When deflected by force due to irregular road geometry, the leaf spring stores kinetic energy in the form of strain energy and releases the same amount of energy to the environment when freed, without causing any effect [3]. Materials with a high modulus of elasticity are thereby undesirable for leaf spring application as they tend to reduce the strain energy storage. In general, mechanical damping materials can be used to control the vehicle suspension. Damping is the

DOI: 10.1201/9781003271017-14

conversion of vibration energy into energy in the form of heat or other forms. Internal damping, also known as material damping, is the conversion of vibration energy into heat within a material.

Conventional leaf spring primarily uses plain carbon steel material due to its high strength and stiffness. However, it does not have a lifelong span since the steel has high stress and deflection while operating, high deformation value and heavyweight [4,5]. The use of steel also poses a challenge for vehicle manufacturers in reducing fuel consumption while also meeting the emission regulation requirements. It is known that the lighter the vehicle, the less fuel it consumes, resulting in lower CO_2 emissions from the vehicle. As a result, there is growing interest in replacing these metal leaf springs with fibre-reinforced composite leaf springs that are lighter but still stiff and strong.

14.1.1 COMPOSITES FOR LEAF SPRING

Several studies on the use of composite-based materials in automotive suspension systems have been carried out over the years. Nutalapati[1] studied the performance of E-glass/epoxy composites material and steel for a semi-elliptic mono leaf spring application. It was reported that the E-glass/epoxy composites leaf spring had approximately 49% less stress and an approximate 84% weight reduction when compared to the steel leaf spring. The applicability of glass fibre-reinforced epoxy composites for leaf spring material was also investigated by Chaudhari and Barjibhe[6]. Stresses in composite leaf springs were found to be approximately 75% less than those in traditional steel leaf springs under the same loading conditions, while deflection was found to be approximately 10% greater. When compared to conventional steel leaf springs, composite leaf springs are found to be 65% lighter. Noronha et al.[7] compared leaf spring made of steel, aluminium, carbon/glass epoxy composites and Kevlar/epoxy composites. The structural response of these leaf springs was obtained with applied load ranging from 6523.6 to 15,000 N. Composite-based leaf springs have been shown to induce less stress than steel and aluminium leaf springs and to have better strain energy storage, resulting in better ride quality. Due to their superior properties such as greater stiffness and lower mass, the composites-based leaf springs were also reported to have higher natural frequencies and fatigue life. When compared to steel, kevlar/epoxy leaf springs result in 82.14% weight reduction, resulting in lower unsprung mass, which improves ride quality, handling and mechanical efficiency of the vehicle.

While synthetic fibre has gained significant attention as the potential reinforcement material for leaf springs, natural fibre has received very little attention to date. Various studies have been carried out in order to use natural resources efficiently in materials for automotive applications in accordance with sustainable development goals and the desire to reduce environmental impact. In fact, natural fibre-reinforced composites have been reported to be capable of replacing synthetic fibres in interior and exterior parts of automobiles in order to reduce vibration level and improve ride comfort for vehicle occupants[8]. Natural fibre such as kenaf, hemp, jute and other biocomposites have been used in trunk lids, car seats, dashboard covers and roofs by car manufacturers such as Honda, Volkswagen, General Motors and Ford[9]. Several studies [10–16] have shown that natural fibre has a high potential for energy absorption. The material's ability to eliminate or reduce vibration as leaf spring material across a wide range of frequencies, however, has yet to be widely investigated. Genc [17] compared luffa fibre/epoxy composites beam to glass fibre/epoxy composites beam. An impact hammer and an accelerometer were used to obtain the frequency response functions of these free-free beams in order to characterize the material damping. It has been reported that the damping of the luffa composites is approximately twice that of a glass fibre composites. Longana et al. [18] investigated the damping ratio of hybrid reclaimed carbon fibre/flax fibre/epoxy composites. The reclaimed carbon fibre and flax fibre ratios were varied while keeping the total fibre volume ratio of the composites at 35%. The intermingled reclaimed carbon fibre/flax fibre/epoxy composites were characterized in terms of their vibrational response. It has been reported that increasing the flax fibre content leads to an increase in vibration damping properties. The vibration properties of natural fibre-reinforced composites were summarized by Munde et al. [19]. In addition to minimizing vibration, fibrous material also has excellent sound absorption properties, thus ensuring that the weight-saving benefits of biocomposites are not jeopardized by high noise transmission [20].

The main aim of this research is to develop a sustainable and functionalized fibre composite for leaf spring applications by hybridizing corn stalk fibre (CS) and glass fibre. The corn plant is a vital agricultural crop that can be found in many regions of the world. Typically, the corn stalk residue is burned after harvesting, causing air pollution and harming the environment. It is also a waste of renewable biomass resource, which has sparked global concern [21]. The use of the waste in biocomposites applications is thus regarded as one of the innovative ideas for transforming underutilized renewable materials into a value-added product with high economic value. Barriers preventing the automotive industry from further adopting biocomposites are uncertainty, fluctuations and a lack of research into determining the performance of biocomposites in extreme conditions. As a result, the industry remains cautious about investing in uncertain advancement when conventional materials are well established. It is therefore critical for the scientific community to accelerate feasibility research in order to accelerate biocomposite implementation in high-performance and demand automotive components, so that the environmental benefits can be realized.

14.2 MATERIALS AND METHODS

14.2.1 MATERIALS

The materials utilized for this analysis are CS, E-glass fibre (FG) and epoxy based on Liu et al. [21], Gupta & Srivastava [22], Rana & Purohit [23] and Sudheer, Pradyoth & Somayaji [24]. The properties of composites were calculated using the rule of mixture approach. The volume ratio of the fibres was kept at 30% CS and 30% FG with the remainder being epoxy. The high fibre content was intended to improve the damping capability of the composites. The theoretical calculation result of the CS/FG/epoxy composites is shown in Table 14.1. The fibres are assumed to be of random discontinuous type and perfectly distributed throughout the matrix.

TABLE 14.1
Properties of Corn Stalk/E-Glass/Epoxy
Hybrid Composite

Properties	Value
Tensile strength (σ)	1233 MPa
Young's modulus (E)	35.24 GPa
Shear modulus (G)	3.186 GPa
Poisson's ratio (υ)	0.296
Density (ρ)	1572 kg/m^3

14.2.2 Methods

A multi-leaf semi-elliptic leaf spring was created in Solidworks software and imported into ANSYS for simulation and analysis of the spring's response to investigate the behaviour of the proposed composite leaf spring. The leaf spring was designed with two master leaves and eight graduated leaves. The design parameters of the composites-based leaf spring used in this study were as follows:

- Master leave = 1171 mm and 1120 mm
- Leaf thickness = 6 mm
- Width = 50 mm
- Outer eyes diameter = 50 mm
- Inner eyes diameter = 20 mm
- Mid-point bolt diameter = 10 mm

The leaf spring designed with the aforementioned parameters is shown in Figure 14.1. Figure 14.2 depicts the finite element hexagonal meshed composites leaf spring model realized within total of 99,438 nodes and 24,607 elements.

The boundary conditions and the 6855 N load applied were determined by considering the maximum mechanical loads imparted by the vehicle and passenger weight, vehicle acceleration and safety factor of 1.3. The most dominant and critical mechanical load applied to a leaf spring was determined to be vertical load [25]. The load was applied at the spring's midpoint, as shown in Figure 14.3, considering the motion of the wheel when it encounters road bump with the midpoint connected to the axle of the suspension system. The boundary conditions were executed by the front and rear eyes' constraint. The displacement of the front eye was restrained in all directions while rotation degrees of freedom were allowed in z-axis to simulate the pin connection between the front eye and the vehicle frame. The rear eye, on the other hand, was typically attached to swingable free link shackle and thus it was simulated by free translational motion in x-direction and rotational motion along z-axis, while other directions and rotation degrees of freedom were

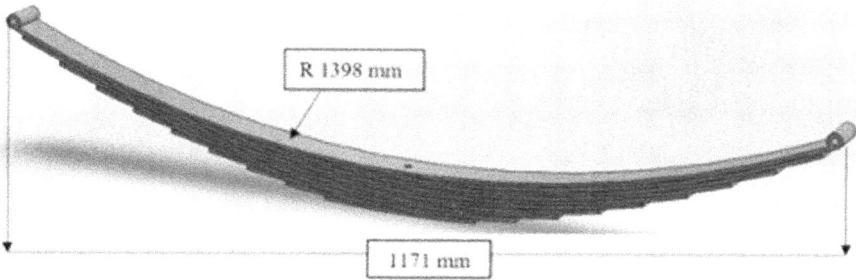

FIGURE 14.1 Solid model of the multi-lead semi-elliptic spring

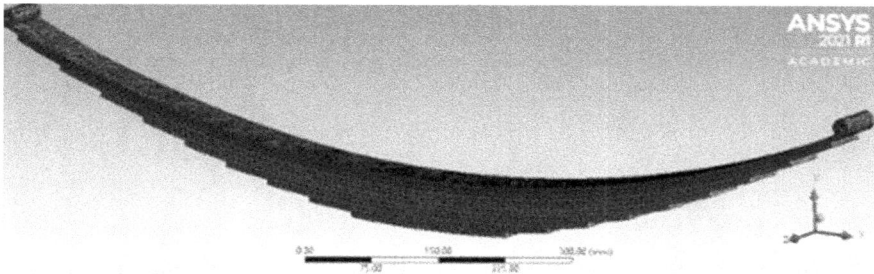

FIGURE 14.2 Meshed multi-leaf semi-elliptic leaf spring

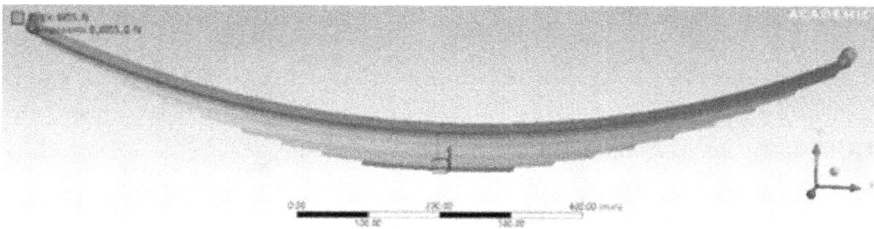

FIGURE 14.3 Applied load

restrained. Figure 14.4 shows the constraints applied for the analysis. The model was then subjected to fatigue and modal analysis.

In fatigue analysis, the model was subjected to periodically varying constant amplitude stress in order to determine the material capacity to withstand load with the number of cycles without failure. The load was varied from initial 6855 to 28,855 N with the assumption that the load was fully reversed, which was represented by zero mean stress.

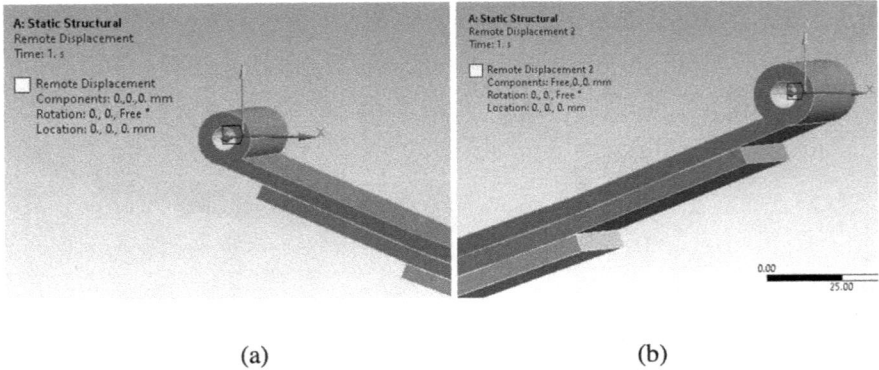

(a)   (b)

FIGURE 14.4   (a) Front, and (b) rear eye constraints view

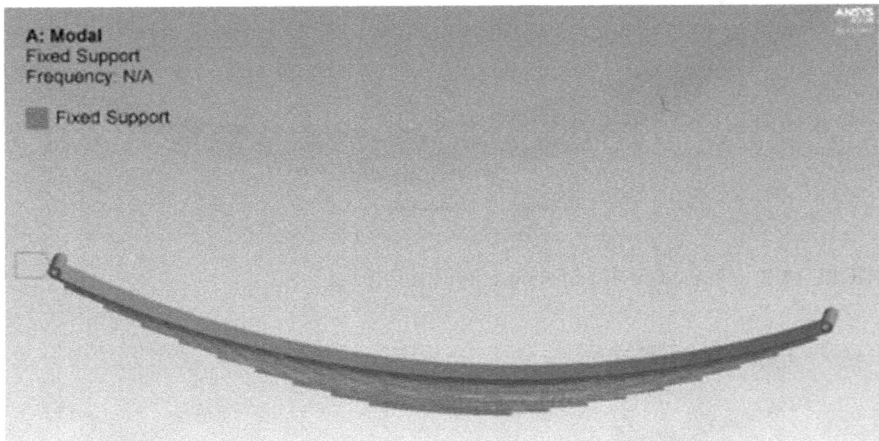

FIGURE 14.5   Boundary condition for modal analysis

The most common method for determining the vibrational characteristics of a leaf spring application is to perform modal analysis, which computes the different mode shape and its natural frequency under free vibration conditions. For modal analysis, the model was remeshed in the shape of tetrahedron with 65,504 nodes and 14,430 elements generated. The modal analysis was performed with no load, with the boundary conditions represented by a fixed front eye and an unrestrained rear eye, as illustrated in Figure 14.5.

14.3   RESULTS AND DISCUSSION

14.3.1   STRUCTURAL RESPONSE OF COMPOSITES LEAF SPRING

The stress, deformation and strain energy contours of the CS/FG/epoxy composites leaf spring are depicted in Figures 14.6–14.8, respectively. The equivalent stress is found to be maximum at the second leaf area near the front eye area and minimum

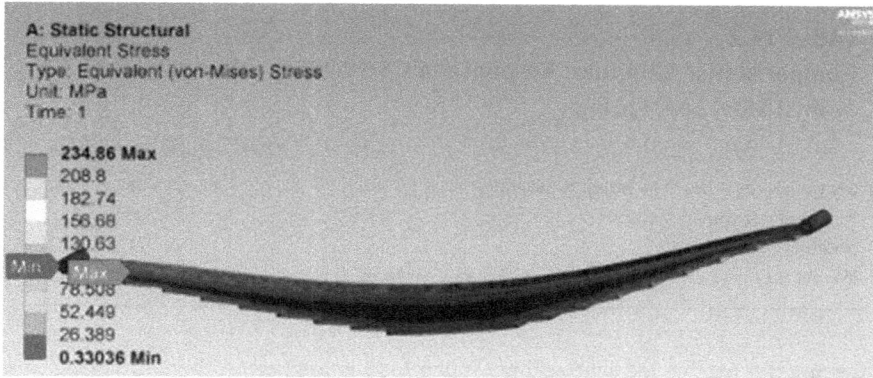

FIGURE 14.6   Equivalent stress contour of CS/FG/epoxy composites leaf spring

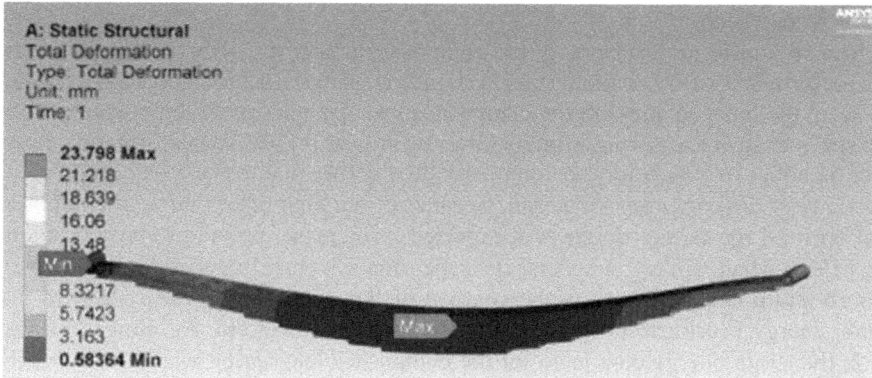

FIGURE 14.7   Deformation contour of CS/FG/epoxy composites leaf spring

FIGURE 14.8   Strain energy contour of CS/FG/epoxy composites leaf spring

TABLE 14.2

Comparison of Structural Response of CS/FG/Epoxy Composites against Steel Leaf Spring

	Steel	CS/FG/Epoxy Composites
Maximum equivalent von Mises stress (MPa)	236.67	234.86
Maximum deformation (mm)	3.9979	23.798
Strain energy (mJ)	7.43	42.971
Weight (kg)	16.88	3.38

at the middle part of the leaf spring. When load is applied to the leaf spring, it is observed that the middle portion of the leaf spring deforms the most. The front eye, on the other hand, had minimum amount of deformation while rear eye displays an average amount of deformation. The second leaf near the rear eye section has the highest strain energy, while the middle part of the leaf spring is observed to have lower strain energy.

The structural performance of the composites leaf spring was evaluated against the performance of conventional steel leaf spring, summarized in Table 14.2. As can be seen, the induced stress in the composites leaf spring is generally comparable to the stress of steel leaf spring, slightly lower by around 0.77%. The deformation, however, is noticeably high for the composites leaf spring due to the lower modulus of elasticity of the composites material. Because of the high deformation of composite leaf springs, the failure duration is expected to increase, providing more cushioning effect, which is very beneficial in suspension systems. In general, the ability to absorb strain energy improves the comfort of the suspension system. As a result, strain energy is critical in the design of the suspension system. As shown in Table 14.2, the strain energy computed for the composites leaf spring was approximately six times greater than the strain energy computed for the steel leaf spring. This indicates that the composites have higher strain energy storage capacity than steel due to lower Young's modulus and lower density. The strain energy capacity of a leaf spring is critical because the vibration caused by surface irregularities in the vehicle is stored in the spring as strain energy, which is then slowly released by the spring. It is also worth noting that, when identical design parameters and load conditions are employed, overall, the composite leaf spring has an approximate weight reduction of 80% when compared to a steel leaf spring.

14.3.2 FATIGUE BEHAVIOUR OF COMPOSITES LEAF SPRING

The fatigue life plots for composites and steel leaf springs are shown in Figures 14.9 and 14.10, respectively. At a load of 6855 N, the minimum fatigue life of the composites leaf spring is 5.4974×10^5 cycles, indicating that it can withstand cyclic loads of up to 5.4974×10^5 cycles, whereas the fatigue life of conventional steel at the same load is 1.4167×10^4 cycles. When compared to conventional steel, the CS/FG/epoxy composites material demonstrated a 97.4% increase in fatigue life. Loganathan,

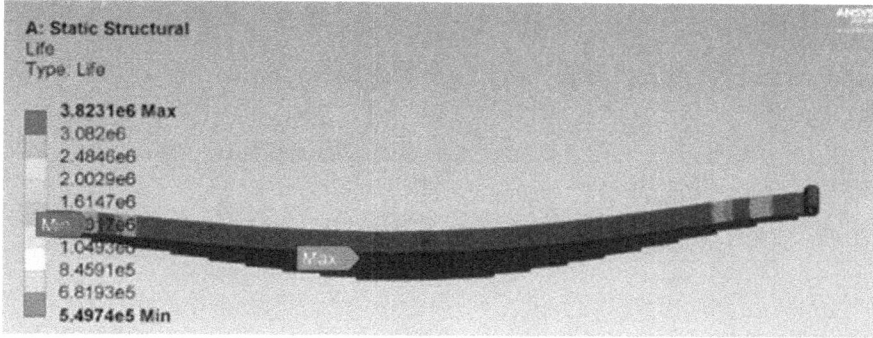

FIGURE 14.9   Fatigue life of CS/FG/epoxy composites leaf spring

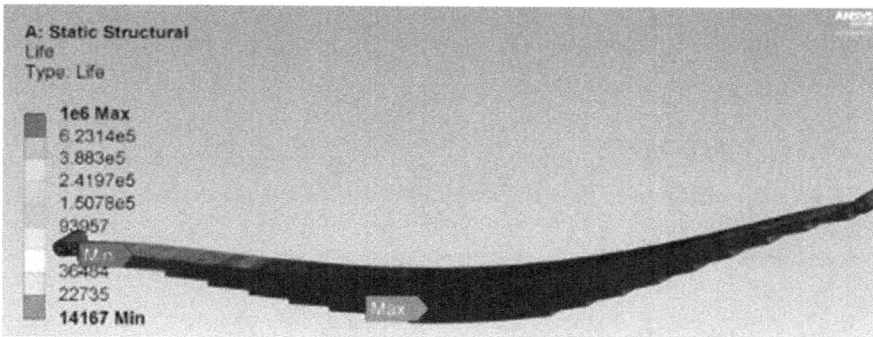

FIGURE 14.10   Steel leaf spring's fatigue life

Kumar and Madhu [26] made a similar observation while researching the fatigue life of Carbon Fibre-Reinforced Polymer (CFRP) composites and SAE 5160 Chromium steel mono leaf springs where it is reported that CFRP has a 97.2% longer fatigue life than SAE 5160 Chromium steel. This is primarily due to the higher alternating stress induced by steel leaf springs as compared to the composites leaf spring, which results in a shorter fatigue life.

14.3.3   Modal Analysis

Leaf springs must be designed to avoid resonant conditions with respect to road frequency in order to provide passenger comfort. Typically, the maximum frequency of road irregularities is 12 Hz. To avoid resonance, leaf springs thus should be designed with a natural frequency that is greater than 12 Hz. Table 14.3 compares the natural frequencies of five mode shapes for CS/FG/epoxy composites leaf spring to the steel leaf spring. The natural frequencies of the CS/FG/epoxy composites leaf spring can be seen to be higher than the 12 Hz natural road frequency, indicating a potentially comfortable ride. The composites leaf spring was also found to exhibit comparable

TABLE 14.3
Comparison of Natural Frequency of CS/FG/Epoxy Composites against Steel Leaf Spring

Mode No.	Steel (Hz)	CS/FG/Epoxy Composites (Hz)	Percentage Difference (%)
1	16.039	14.636	8.75
2	92.197	84.205	8.67
3	278.03	255.37	8.15
4	493.87	453.93	8.09
5	882.19	806.85	8.54

natural frequency to that of the steel leaf spring, lower only by 8%–8.75%. More research is needed, such as considering the fibre orientation in order to increase the natural frequency of the composites leaf spring for better riding comfort for the passengers.

14.4   CONCLUSION

In the present chapter, a finite element analysis was performed using ANSYS software to assess the potential for developing biocomposites leaf springs from CS, fibre glass and epoxy material. Results indicate that

1. The composites leaf spring has higher strain energy capacity, which is an important feature of leaf springs, thereby promising improved suspension system for better ride quality and handling.
2. Weight reduction of around 80% is attainable with the employment of the composites leaf spring instead of the steel leaf spring.
3. When compared to conventional steel, the fatigue life of the CS/FG/epoxy composites material increased by 97.4%.
4. The natural frequency of the composite leaf spring was found to be higher than the road natural frequency and comparable to that of the steel leaf spring.

REFERENCES

[1] Nutalapati, S. Design and analysis of leaf spring by using composite material for light vehicles. *Int. J. Mech. Eng. Technol.* **2015**, *6*(12), 36–59. http://iaeme.com/MasterAdmin/Journal_uploads/IJMET/VOLUME_6_ISSUE_12/IJMET_06_12_005.pdf

[2] Pradeep, S.A.; Iyer, R.K.; Kazan, H.; Pilla, S. Automotive applications of plastics: past, present, and future. In *Applied Plastics Engineering Handbook*, **2017**, Myer Kutz (ed.), (pp. 651–673): Elsevier, Oxford.

[3] Solanki, P.; Kaviti, A.K. Design and computational analysis of semi-elliptical and parabolic leaf spring. *Mater. Today* **2018**, *5*(9), 19441–19455. https://doi.org/10.1016/j.matpr.2018.06.305

[4] Varpe, N.J.; Borkar, B.; Shinde, V.B.; Tajane, R.S. Static & dynamic analysis of EN 47 leaf spring & E-glass fiber with epoxy resin hardner based unidirection laminated composite leaf spring. *Int. J. Adv. Res. Innov. Ideas. Edu.* **2015**. 1(3), 70–75.

[5] Singh, V.; Rastogi, V. Design and static analysis of mono composite leaf spring made of various types of composite materials using finite element method. *IOP Conf. Ser.: Mater. Sci. Eng.* **2021**, 1033. https://doi.org/10.1088/1757-899X/1033/1/012041

[6] Chaudhari, V.; Barjibhe, R.B. Design and optimization of mono composite leaf spring. *Int. Res. J. Eng. Tech.* **2020**, 7(7), 482–487. https://www.irjet.net/archives/V7/i7/IRJET-V7I787.pdf

[7] Noronha, B.; Yesudasan, S.; Chacko, S. Static and dynamic analysis of automotive leaf spring: a comparative study of various materials using ANSYS. *J. Fail. Anal. Prev.* **2020**, 20, 804–818.

[8] Etaati, S.; Abdanan Mehdizadeh, H.; Wang, S.; Pather, S. Vibration damping characteristics of short hemp fibre thermoplastic composites, *J. Reinf. Plast. Compos.* **2014**, 33(4), 330–341. https://doi.org/10.1177/0731684413512228

[9] Pegoretti, T.D.S.; Mathieux, F.; Evrard, D.; Brissaud, D.; Arruda, J.R.D.F. Use of recycled natural fibres in industrial products: a comparative LCA case study on acoustic components in the Brazilian automotive sector, *Resour. Conserv. Recycl.* **2014**, 84 1–14. https://doi.org/10.1016/j.resconrec.2013.12.010

[10] Nasution H; Pandia, S.; Maulida; Sinaga M.S. Impact strength and thermal degradation of waste polypropylene (wPP)/oil palm empty fruit bunch (OPEFB) composites: effect of maleic anhydride-g-polypropylene (MAPP) addition, *Proc. Chem.* **2015**, 16, 432–7. https://doi.org/10.1016/j.proche.2015.12.075

[11] Hassan, C.S.; Durai, V.; Sapuan, S.M.; Abdul Aziz, N.; Mohamed Yusoff, M.Z. Mechanical and crash performance of unidirectional oil palm empty fruit bunch fibre-reinforced polypropylene composite, *BioResources* **2018**, 13, 8310–28.

[12] Hassan, C.S.; Pei, Q.; Sapuan, S.M.; Abdul Aziz, N.; Mohamed Yusoff, M.Z. Crash performance of oil palm empty fruit bunch (OPEFB) fibre reinforced epoxy composites bumper beam using finite element analysis, *Int. J. Automot. Mech. Eng.* **2018**, 15, 5826–36. https://doi.org/10.15282/ijame.15.4.2018.9.0446

[13] Mukhtar, I.; Leman, Z.; Zainuddin, E.S.; Ishak, M.R. Development and performance analysis of hybrid composite side door impact beam: An experimental investigation. In: Sapuan SM and Ilyas RA, editors. *Biocomposite and Synthetic Composites for Automotive Applications*, Woodhead Publishing: Duxford, **2021**, pp. 173–198.

[14] Alkbir, M.F.M.; Januddi, F.; Ariffin, M.A.; Kosnan, M.S.E.; Bakri, A.; Mohamed, S.B. Crashworthiness of circular tube of kenaf fibre composite for automotive applications. In: Sapuan SM and Ilyas RA, editors. *Biocomposite and Synthetic Composites for Automotive Applications*, Woodhead Publishing: Duxford, **2021**, pp. 217–232.

[15] Hassan, C.S.; Qiang, P.; Sapuan, S.M.; Nuraini, A.A.; Zuhri, M.Y.M.; Ilyas, R.A. Unidirectional oil palm empty fruit bunch (OPEFB) fibre reinforced epoxy composite car bumper beam—effects of different fibre orientations on its crash performance. In: Sapuan SM and Ilyas RA, editors. *Biocomposite and Synthetic Composites for Automotive Applications*, Woodhead Publishing: Duxford, **2021**, pp. 233–254.

[16] Wong, K.P.; Hassan, C.S.; Qiang, P.; Sapuan, S.M.; Abdul Aziz, N.; Mohamed Yusoff, M.Z. Crash behaviour of unidirectional oil palm empty fruit bunch fibre-reinforced polypropylene composite car bumper fascias using finite element analysis. *Funct. Compos. Struct.* **2021**, 3, 044001. https://doi.org/10.1088/2631-6331/ac33b7

[17] Genc, G. Dynamic properties of luffa cylindrica fiber reinforced bio-composite beam, *J. Vibroengineering* **2015**, 17, 1615–1622.

[18] Longana, M.L.; Ondra, V.; Yu, H.; Potter, K.D.; Hamerton, I. Reclaimed carbon and flax fibre composites: manufacturing and mechanical properties, *Recycling* **2018**, 3(4), 52–64. https://doi.org/10.3390/recycling3040052

[19] Munde, Y.S.; Ingle, R.B.; Siva, I. A comprehensive review on the vibration and damping characteristics of vegetable fiber-reinforced composites, *J. Reinf. Plast. Compos.* **2019**, 38, 822–832.https://doi.org/10.1177/0731684419838340

[20] Arenas, J.P.; Li, Z. Composites and biocomposites for noise and vibration control in automotive structures. In: Sapuan SM and Ilyas RA, editors. *Biocomposite and Synthetic Composites for Automotive Applications*, Woodhead Publishing **2021**, pp. 305–346.

[21] Liu, Y.; Xie, J.; Wu, N.; Ma, Y.; Menon, C.; Tong, J. Characterization of natural cellulose fiber from corn stalk waste subjected to different surface treatments. *Cellulose* **2019**, 26(8), pp. 4707–4719.

[22] Gupta, M.K.; and Srivastava, R.K. Mechanical properties of hybrid fibers-reinforced polymer composite: A review. *Polym-plast. Tech. Mat.* **2016**, 55(6), pp. 626–642. https://doi.org/10.1080/03602559.2015.1098694

[23] Kumre, A.; Rana, R.S.; Purohit, R. J. A Review on mechanical property of sisal glass fiber reinforced polymer composites. *Mater. Today* **2017**, 4(2), pp. 3466–3476. https://doi.org/10.1016/j.matpr.2017.02.236

[24] Sudheer, M.; Pradyoth, K.R.; Somayaji, S. Analytical and numerical validation of epoxy/glass structural composites for elastic models, *Am. J. Mater. Sci.* **2015**, 5(3C), pp. 162–168. https://doi.org/10.5923/c.materials.201502.32

[25] Shokrieh, M.M.; Rezaei, D. Analysis and optimization of a composite leaf spring. *Compos. Struct.* **2003**, 60, pp. 317–325. https://doi.org/10.1016/S0263-8223(02)00349-5

[26] Loganathan, T.G.; Kumar, K.V.; Madhu, S. Flexural and fatigue of a composite leaf spring using finite element analysis. *Mater. Today* **2020**, 22, pp. 1014–1019. https://doi.org/10.1016/j.matpr.2019.11.265

15 Natural Fibre-Reinforced Epoxy Composites for Marine Applications

Tarkan Akderya
Izmir Bakırçay University

Uğur Özmen
Manisa Celal Bayar University

CONTENTS

15.1 INTRODUCTION

The use of fibre-reinforced polymer (FRP) composite materials has increased steadily over the past few decades as new application markets are found, and traditional metal-based materials are replaced by composite materials (Ramasamy et al., 2020). Composite materials have found widespread applications as metal, ceramic, and polymer composites in the aviation, automobile, energy, construction, biomedical, and marine industries due to their outstanding features such as ultra-lightness, high mechanical properties, and easy formability (Ashik & Sharma, 2015; Gupta & Srivastava, 2016; Sivakumar et al., 2018). Although most of the composite forms consist of a matrix, it is reinforced with fibres to increase the matrix's mechanical strength (Senthilkumar et al., 2015, 2016, 2021, 2022). The properties of the newly formed composite material depend on the individual properties of the matrix and fibres and their compatibility with each other (Gonzalez-Murillo & Ansell, 2009; Sivasaravanan et al., 2019; Torres-Arellano et al., 2020).

DOI: 10.1201/9781003271017-15

The fact that FRP is generally composed of petroleum-based non-renewable matrix and fibre components causes their recyclability to be difficult and, therefore, unsustainable at the end of their shelf life (Mohanty et al. 2018). The most commonly preferred reinforcement materials such as carbon and glass for obtaining FRP composite materials are not renewable, and current methods for recycling composite materials reinforced with these fibres are not feasible (Fitzgerald et al., 2021; Kandemir et al., 2021; Sukanto et al., 2021). The limitations that arise when they are desired to be reused and recycled at the end of their shelf-life and the infeasibility of these processes cause them to be incinerated or directly thrown away. Thereby, the degradation behaviour of composite materials with synthetic components raises environmental concerns, considering their widespread use. When focusing on the subject with these sensitivities with a modern engineering approach, it is obvious that materials with easy biodegradability, disposable and shelf-life controllable materials should be preferred (Akderya et al., 2020, 2021).

With the development of sustainable environmental awareness, natural fibres have replaced synthetic fibres due to engineering studies and research. Low density, low cost, recyclability, and high flexibility are just a few of the prominent features of natural fibres (Nabi Saheb & Jog, 1999; Akil et al., 2011; Mukherjee & Kao, 2011). Natural fibres can be produced sustainably and can be easily processed for recycling at the end of their shelf life (Dhakal, 2015). In addition, synthetic fibres are obtained from limited sources, whereas natural fibres can be obtained in various ways from all over the world (Nabi Saheb & Jog, 1999). Besides, they also have some features such as poor wettability, hydrophilic characteristic, high moisture absorption ability, and low thermal stability, which make them difficult to use in a composite material (Asumani et al., 2012).

Organic fibres obtained from natural sources are most generally classified as fibres of vegetable and animal origin (Saba et al., 2014; Ramamoorthy et al., 2015). Plant-derived fibres are categorised as fruit, grass, leaf, root, stalk, stem, seed, and wood fibres (Mohanty et al., 2000; Reddy & Yang, 2005; John & Thomas, 2008; Ramamoorthy et al., 2015; Madhu et al., 2021). This categorisation is schematised in Figure 15.1a. Fibres of animal origin are subject to the main classification as animal wool/hair, avian feathers, and silk fibres (Nabi Saheb & Jog, 1999; Liu et al., 2004; Hunter, 2020; Kumar et al., 2020). This classification is schematised in Figure 15.1b. In addition to that, examples of natural organic fibres classified as plant-based and animal-based fibres, and the studies conducted on these examples are given in Table 15.1.

Biocomposite materials are classified as green composites if both the fibres and the matrix are sustainable and partly eco-friendly composites if a component is sustainable. Green composite material refers to innovative engineering materials with zero environmental impact, with the ability to fully decompose, and have renewable components (Mohanty et al., 2005; Mitra, 2014; Paulo et al., 2018). Biocomposite materials reinforced with biofibres, which can be obtained with high diversity by utilising the richness of geographical regions, contribute to sustainability and ecological efficiency while supporting and leading the green industry formation (Alothman et al., 2020; Chandrasekar et al., 2020). It supports the green industrial revolution by obtaining from local and renewable resources, leading the production of

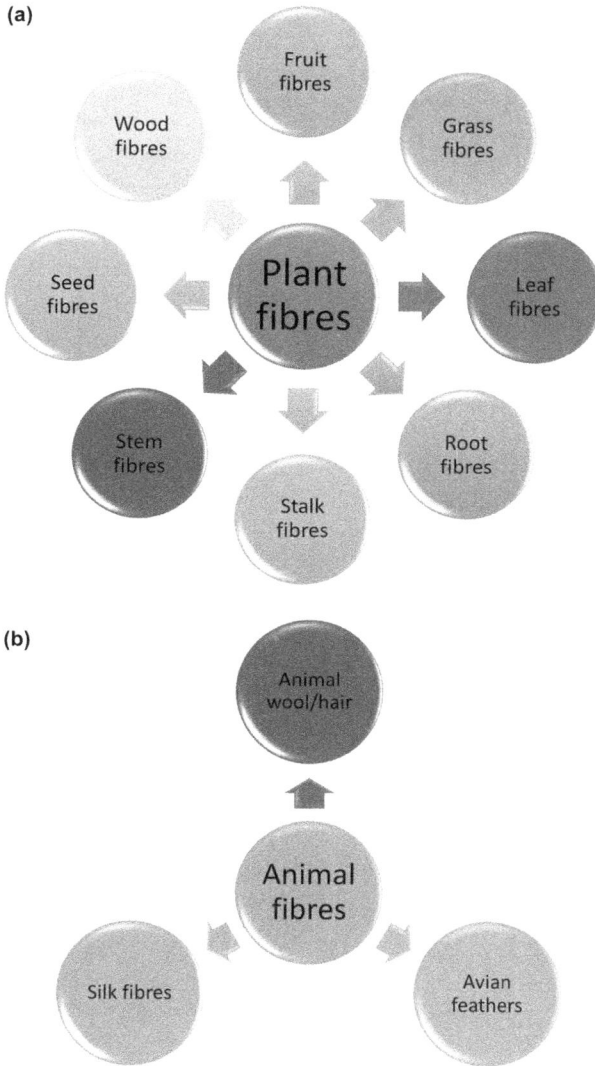

FIGURE 15.1 (A)   Plant-based and (b) animal-based fibres classification.

new-generation engineering materials and products, increasing ecological efficiency, and reducing the carbon footprint caused by the industry (Bharath & Basavarajappa, 2016; Akderya et al., 2022).

It is possible to evaluate the impact of a process or product on the environment sensitively and comprehensively with the life cycle approach. Accordingly, the environmental impact of a product or activity is evaluated by considering its use and interactions throughout its lifecycle, from its design to its disposal (Chandrasekar & Senthilkumar, 2021; Senthil Muthu Kumar et al., 2021). Since maritime activities take place in coastal or inland waters of the marine ecosystem in direct and

TABLE 15.1

Classification of Natural Organic Fibres

Classification				Examples
Natural fibres			Fruit fibres	• Betelnut (Jayamani et al., 2014) • Coir (Walte et al., 2020) • Coconut (Bujang et al., 2007) • Oil palm (Ahmad et al., 2021)
			Grass/reeds fibres	• Bagasse (Siddique et al., 2021) • Bamboo (Qiu et al., 2021) • Cogon grass (Jumaidin et al., 2019) • Corn (Luo et al., 2014) • Elephant grass (Rao et al., 2007) • Esparto (Maghchiche, 2013) • Rape (Kısmet & Dogan, 2022) • Switchgrass (Van Den Oever et al., 2003)
			Leaf fibres	• America century plant (Geethika & Rao, 2017) • Abaca (Bledzki et al., 2007) • Areca leaf stalk (Shanmugasundaram et al., 2018) • Banana (Rao et al., 2019) • Cantala fibre (Raharjo et al., 2020) • Carnauba leaves (Melo et al. (2012) • Date palm (Al-Sulaiman, 2002) • Ethiopian palm leaf (Atalie & Gideon, 2018) • Henequen fibre (Serra-Parareda et al., 2021) • Pineapple fibre (Motaleb et al., 2018) • Sisal (Liang et al., 2021)
	Organic fibres	Plant fibres	Root fibres	• Broom (Al Amin et al., 2021)
			Stalk fibres	• Barley (Serra-Parareda et al. (2019) • Maise (Saravana Bavan & Mohan Kumar, 2012) • Oat (Reddy et al., 2013) • Rice (Mohamed et al., 2020) • Rye (Mamun et al., 2015) • Wheat (Mittal & Sinha, 2017)
			Stem fibres	• Flax (Moudood et al., 2019) • Hemp (Sair et al., 2018) • Jute (Karim et al., 2021) • Kenaf (Mahjoub et al., 2014) • Ramie (Djafar et al., 2021) • Roselle (Kazi et al., 2021)
			Seed fibres	• Cotton (Parikh & Gohil, 2017) • Kapok (Lyu et al., 2020) • Loofah (Li et al., 2022)
			Wood fibres	• Hardwood (Tawfik et al., 2017) • Softwood (Prasad et al., 2022)

(Continued)

TABLE 15.1 (*Continued*)
Classification of Natural Organic Fibres

Classification		Examples
Animal fibres	Animal wool/hair	• Alpaca (Vinoth et al., 2021)
		• Angora (Narendiranath Babu et al., 2018)
		• Bison (McGregor, 2012)
		• Cashmere wool (Selli et al., 2018)
		• Human hair (Rao et al., 2018)
		• Horse hair (Wutabachew & Alemu, 2019)
		• Goat hair (Rao et al., 2020)
		• Sheep wool (Straznicky et al., 2020)
		• Qiviut wool (Rowell et al., 2001)
	Avian feathers	• Chicken feather (Kurien et al., 2022)
		• Emu feather (Sekhar et al., 2014)
		• Goose feather (Büyükkaya, 2017)
		• Turkey feather (Soykan, 2022)
	Silk fibres	• Silkworm silk (Faizah et al., 2018)
		• Spider silk (Mayank et al., 2022)

prolonged contact, evaluating it with a life cycle approach is vital in terms of sustainability (Moreau, 2009; Routray et al., 2020; Boljat et al., 2021; Kappenthuler & Seeger, 2021). This study aims to examine the epoxy composite materials reinforced with natural fibres and their properties to reveal sustainable products' use in the maritime industry.

15.2 NATURAL FIBRE-REINFORCED EPOXY COMPOSITES FOR THE MARINE APPLICATIONS

Considering the expected lifespan of a boat that has been used in the maritime industry for 30–40 years, it seems that this boat has reached the end of its useful life, and this is the inevitable end for numerous boats around the world, circulatory for the maritime industry (Moreau, 2009; Fragassa, 2017). Although these abandoned boats do not have any environmental damage, they occupy space in marinas and shipyards and cause visual pollution. The recreational maritime sector, which foresees such risks, has recently been conducting research for the use of recyclable production processes and products and has been conducting a few pilot projects for the use of biofibre-reinforced epoxy composite materials in the production of both structural and non-structural elements (Fragassa, 2017; Routray et al., 2020; Boljat et al., 2021; Kappenthuler & Seeger, 2021).

The resins used in FRP structures in boat construction are mainly thermoset polymers such as polyester, epoxy, and vinyl ester. Thermoplastic resins such as polyamide and polypropylene have only begun to be used in boat construction or equipment in the maritime industry. Although the existence of low-emission and low-styrene-containing resins has been mentioned recently in the marine industry, it is difficult to say that bio-based resins have yet found an area of use. In addition, it is possible to talk about using natural fibre-reinforced epoxy composite materials in the maritime industry (Shahroze et al., 2021; Thomas et al., 2021). Increasing attention and sensitivity towards sustainability may encourage using composite materials and natural fibres and preferring ecological resins (Moreau, 2009; Qin et al., 2020; Boljat et al., 2021).

Biocomposites with new aesthetic properties and different functionalities could develop a new market for the marine industry. For example, basalt is dark black in colour, more homogeneous, less expensive than carbon, and has better physicomechanical properties than glass fibre. Basalt, an inert, naturally occurring volcanic rock with worldwide availability, is environmentally friendly, non-hazardous, non-toxic, highly chemically stable, non-combustible, and resistant to high temperatures (Wei et al., 2010a, b; Fiore et al., 2011; Li et al., 2018). It is used in the manufacture of compressed natural gas cylinders, which must have good mechanical properties, lightweight, and temperature resistance (Pavlovski et al., 2007), and is also used as a reinforcement material for concretes due to its high physical and mechanical properties (Zhou et al., 2020; Lu et al., 2022). Basalt, in terms of its use in the maritime industry, can replace asbestos, which is used as an insulation and sealing material, due to its better thermal insulation properties (Jamshaid & Mishra, 2016). Thanks to fibre diameter control, it can eliminate contact with ultrafine fibres that are harmful to human health (Kogan & Nikitina, 1994). In addition, the shipyards want to use basalt as an alternative material to glass fibres in the manufacture of boats because it can be found naturally, is economical, and does not pose a health risk to the workers since its dimensions are too large to be taken into the body through the respiratory tract (Shi et al., 2017; Davies & Verbouwe, 2018; Li et al., 2018). Flax has a warm colour scale ranging from walnut and oak tones and can be used in developing products for human use. In addition, there is no need to perform the surface stabilisation processes applied to eliminate the problems of spreading to the environment or stinging in contact with human skin, as is the case with glass fibre.

15.3 MECHANICAL PROPERTIES OF NATURAL FIBRE-REINFORCED EPOXY COMPOSITES FOR MARINE APPLICATIONS

It is essential to determine the characteristics of natural fibre-reinforced epoxy composites used in structural, semi-structural, or non-structural applications, especially in the maritime industry, in terms of usage and interaction. Harish et al. (2009) investigated the mechanical properties such as tensile, flexural, and impact strengths of natural fibre coir-reinforced composites and compared results with glass fibre-reinforced plastics (GFRPs). They found that the strengths of GFRPs were higher than those of coir/epoxy composites. Ashok Kumar et al. (2010) conducted a study to obtain properties of sisal-glass fibre epoxy hybrid composites and presented frictional coefficient, hardness, impact, and chemical resistance results. They also analysed the effect of treatment on properties. It was observed that the treated composites showed higher hardness and impact strength values compared to untreated composites. The mechanical properties of jute and banana fibre-reinforced epoxy hybrid composites were investigated by Boopalan et al. (2013). Tensile, flexural, and impact tests were carried out, and results were evaluated for various weight ratios. The maximum values of tensile strength, flexural strength, and impact strength are obtained for 50/50 weight ratios of jute and banana reinforcements. Codispoti et al. (2015) analysed tensile properties (elongation at break, modulus of elasticity, and tensile strength) of natural fibres (jute, sisal, coir, flax, and hemp)-reinforced epoxy composites. Results showed that flax had higher mechanical properties, followed by jute and hemp. Sisal-reinforced epoxy composites had lower mechanical properties compared to other fibres. Also, warp direction showed higher strength compared to weft direction. Anand et al. (2018) presented tensile, flexural, and impact properties of jute/kenaf hybrid epoxy composites. While tensile strengths were higher for all treated composites than untreated composites, flexural, compressive, and impact strengths were higher for some treated composites. Chaudhary et al. (2018) investigated the mechanical characterisation of jute/hemps/flax-reinforced epoxy composites. Results showed that jute/hemp/flax/epoxy composites achieved the highest tensile strength, tensile modulus, and impact strength, while jute/hemp/epoxy composites had the highest flexural strength. The mechanical properties of jute/epoxy composites were investigated by Mishra et al. (2000). It was shown that untreated (control) 50% jute fibre-reinforced epoxy-hot curing (JEH-50(C)) had more tensile strength values compared to JEH-40(C) and JEH-57(C). Bleached JEH-50 (JEH-50(B)) showed higher flexural properties and impact strength than JEH-50(C). The mechanical properties of natural fibre-reinforced epoxy composites for marine applications are shown in Table 15.2.

15.4 THERMAL PROPERTIES OF NATURAL FIBRE-REINFORCED EPOXY COMPOSITES FOR MARINE APPLICATIONS

The selection of thermally stable and durable materials for the marine industry is crucial because of its harsh conditions of use. For this reason, the thermal properties of natural fibre-reinforced epoxy composite materials used in the maritime industry have been compiled and are given in Table 15.3. Azwa and Yousif (2013) studied the thermal properties of kenaf fibre/epoxy composites. Decomposition temperature, final weight after decomposition, increment in thermal stability, and increment in char production from epoxy values were given for pure epoxy, glass fibre/epoxy, treated kenaf-fibre/epoxy, untreated kenaf-fibre/epoxy, treated kenaf-fibre, and untreated kenaf-fibre. The thermal properties of jute and banana-reinforced epoxy composites were investigated by Boopalan et al. (2013). Results showed that 50% jute and 50% banana fibre-reinforced epoxy composites sustain their character at a higher temperature compared to other reinforced composites.

TABLE 15.2

Mechanical Properties of Natural Fibre-Reinforced Epoxy Composites

Study Done By	Material	Application Areas	Mechanical Properties		
			Tensile Strength (MPa)	Flexural Strength (MPa)	Impact Strength (kJ/m²)
Harish et al. (2009)	Coir-reinforced epoxy	Marine structures such as boat hull structures	17.86 ± 2.32	31.08 ± 6.01	11.49 ± 0.99
	Glass fibre-reinforced plastics		85.35 ± 4.32	132.39 ± 11.85	52.66 ± 3.13

Study Done By	Material	Application Areas	Mechanical Properties			
			Rockwell Hardness (Untreated)	Rockwell Hardness (Treated)	Impact Strength (J/m) (Untreated)	Impact Strength (J/m) (Treated)
Ashok Kumar et al. (2010)	Sisal-glass reinforced epoxy (1 cm fibre length)	Marine applications for making water and chemical storage tanks	110	114	10.887	11.190
	Sisal-glass-reinforced epoxy (2 cm fibre length)		113	119	11.455	12.871
	Sisal-glass-reinforced epoxy (3 cm fibre length)		105	111	10.030	10.809

Study Done By	Material	Application Areas	Mechanical Properties				
			Tensile Strength (MPa)	Tensile Modulus (MPa)	Flexural Strength (MPa)	Flexural Modulus (MPa)	Impact Strength (kJ/m²)
Boopalan et al. (2013)	Jute and banana-reinforced epoxy composites	Marine applications for ceilings, floorings, storage filo, and liquid containers					
	Weight ratios of jute/jute banana 100/0		16.62	664	57.22	8956	13.44
	Weight ratios of jute/jute banana 75/25		17.89	682	58.60	9065	15.81
	Weight ratios of jute/jute banana 50/50		18.96	724	59.84	9170	18.23
	Weight ratios of jute/jute banana 25/75		18.25	720	59.30	9056	17.89
	Weight ratios of jute/jute banana 0/100		17.92	718	58.06	9048	16.92

(Continued)

TABLE 15.2 (Continued)
Mechanical Properties of Natural Fibre-Reinforced Epoxy Composites

Study Done By	Material	Application Areas	Mechanical Properties		
			Elongation at break (%)	Tensile Strength (MPa)	Elasticity Module (MPa)
Codispoti et al. (2015)	Jute (255 g/m²) warp (90°)	Marine applications for structure, ceilings, floorings, storage filo, and liquid containers	4.8	32.91	691.49
	Jute (255 g/m²) weft (0°)		3.3	22.83	699.48
	Jute (398 g/m²) warp (90°)		8.6	31.41	364.74
	Jute (398 g/m²) weft (0°)		6.2	34.50	554.29
	Jute (1099 g/m²) warp (90°)		6.0	55.20	863.33
	Jute (1099 g/m²) weft (0°)		1.7	30.68	178.87
	Sisal (1768 g/m²) warp (90°)		4.4	17.73	402.00
	Sisal (1768 g/m²) weft (0°)		7.2	7.43	103.60
	Sisal (1375 g/m²) warp (90°)		4.0	9.94	239.23
	Sisal (1375 g/m²) weft (0°)		5.4	15.34	28.25
	Hemp (454 g/m²)		7.5	46.68	618.65
	Flax (388 g/m²)		4	68.81	1746.91

Study Done By	Material	Application Areas	Tensile Strength (MPa)	Flexural Strength (MPa)	Compressive Strength (MPa)	Impact Strength
Anand et al. (2018)	Jute/kenaf-reinforced epoxy composites	Marine applications such as structure, furniture, and absorbent				
	UX KHHK (untreated)		32.56	780.5295	539.0112	0.62
	KHHK		53.64	866.286	603.564	0.79
	KHKH		42.43	813.4755	446.94	0.55
	HKKH		45.5	440.895	556.14	0.45

(Continued)

TABLE 15.2 (Continued)

Mechanical Properties of Natural Fibre-Reinforced Epoxy Composites

Study Done By	Material	Application Areas	Mechanical Properties					
			Tensile Strength (MPa)	Tensile Modulus (GPa)	Elongation at Break (%)	Flexural Strength (MPa)	Flexural Modulus (GPa)	Impact Strength (kJ/m²)
Chaudhary et al. (2018)	Jute, hemp, flax-reinforced epoxy composites	Marine applications for structure, flooring, and fuel storage						
	Neat epoxy		30	1.1	3.2	34.69	0.6	4
	Jute epoxy		43.32	1.64	3.7	60	1.42	7.68
	Hemp epoxy		36.48	1.43	3.5	85.59	1.78	5.18
	Flax epoxy		46.21	1.58	4.5	80	0.69	4.71
	Jute/hemp epoxy		42.19	1.49	4.4	86.6	0.78	6.93
	Hemp/flax epoxy		44.17	1.56	4.7	44.6	0.74	4.18
	Jute/hemp/flax epoxy		58.59	1.88	5.9	65	1.22	10.19

Study Done By	Material	Application Areas	Mechanical Properties					
			Tensile Strength (MPa)	Tensile Modulus (GPa)	Elongation at Break (%)	Flexural Strength (MPa)	Flexural Modulus (GPa)	Impact Strength (kJ/m)
Mishra et al. (2000)	Jute-reinforced epoxy composites	Marine applications for storage filo, flooring, furniture, absorbent						
	JEH-40		139.8	2826	7.316	155.816	14.232	94.46
	JEH-50 (C)		148.3	3184	6.293	196.109	20.445	107.94
	JEH-50 (B)		131.085	2348	7.412			
	JEH-57		143.355	3060	5.978			

TABLE 15.3

Thermal Properties of Natural Fibre-Reinforced Epoxy Composites

Study Done By	Material	Application Areas	Thermal Properties			
			Decomposition temperature (°C)	Final Weight after Decomposition (%)	Increment in Thermal Stability from Epoxy (%)	Increment in Char Production from Epoxy (%)
Azwa and Yousif (2013)	Kenaf fibre-reinforced composites	Marine applications for containers, structures, flooring, ceilings, absorbent, and hull structure				
	Epoxy		372	8.29	-	-
	Glass fibre/epoxy		380	39.83	2.19	380.46
	Treated kenaf-fibre/epoxy		373	12.46	0.40	50.30
	Untreated kenaf-fibre/epoxy		378	13.67	1.79	64.90
	Treated kenaf-fibre		350	18.96	-	-
	Untreated kenaf-fibre		346	16.21	-	-
			Heat Deflection Temperature (°C)	% Weight Loss		
Boopalan et al. (2013)	Jute and banana-reinforced epoxy composites	Marine applications for ceilings, floorings, storage filo, and liquid containers				
	Weight ratios of jute/jute banana 100/0		64.5	0.81		
	Weight ratios of jute/jute banana 75/25		78	0.83		
	Weight ratios of jute/jute banana 50/50		90	0.98		
	Weight ratios of jute/jute banana 25/75		89	1.03		
	Weight ratios of jute/jute banana 0/100		82	1.24		

15.5 CONCLUSION

The fact that natural fibre-reinforced epoxy composites have a wide range of uses in the maritime industry, including sailing applications, means a proper step towards a sustainable maritime perspective. Maritime activities' long-term future and economic sustainability depend on a sensitive and meticulously protected environment. This review proves that the maritime industry can make a difference in activities and products with the selected material. The maritime industry's transformation into a sustainable sector comes to life when it prefers environmentally friendly products and processes from design to production, and the products can be recycled at the end of their useful life. It is also essential to bring the awareness level of end-users to a level where they prefer green products and technologies.

REFERENCES

Ahmad MN, Ishak MR, Taha MM, Mustapha F, Leman Z. 2021. Rheological and morphological properties of oil palm fiber-reinforced thermoplastic composites for fused deposition modeling (FDM). *Polymers*. 13(21): 3739.

Akderya T, Özmen U, Baba BO. 2022. Effects of natural weathering on aesthetics, thermal and mechanical properties of the bio-composites. *In: Aging Effects on Natural Fiber-Reinforced Polymer Composites* (pp. 137–157). Eds. Muthukumar C, Krishnasamy S, Thiagamani SMK, Siengchin S. Springer, Singapore.

Akderya T, Özmen U, Baba BO. 2020. Investigation of long-term ageing effect on the thermal properties of chicken feather fibre/poly(lactic acid) biocomposites. *J Polym Res*. 27(6): 162.

Akderya T, Özmen U, Baba BO. 2021. Revealing the long-term ageing effect on the mechanical properties of chicken feather fibre/poly(lactic acid) biocomposites. *Fibers Polym*. 22(9): 2602–2611.

Akil HM, Omar MF, Mazuki AAM, Safiee S, Ishak ZAM, Abu Bakar A. 2011. Kenaf fiber reinforced composites: A review. *Mater Des*. 32(8–9): 4107–4121.

Al Amin MA, Mahjabin T, Hasan M. 2021. Effect of fibre hybridization on mechanical properties of nylon-broom grass/root-broom grass fibre reinforced hybrid polypropylene composites. *J Appl Sci Process Eng*. 8(2): 965–976.

Alothman OY, Jawaid M, Senthilkumar K, Chandrasekar M, Alshammari BA, Fouad H, Hashem M, Siengchin S 2020. Thermal characterization of date palm/epoxy composites with fillers from different parts of the tree. *J Mater Res Technol*. 9: 15537–15546.

Al-Sulaiman FA. 2002. Mechanical properties of date palm fiber reinforced composites. *Appl Compos Mater*. 9(6): 369–377.

Anand P, Rajesh D, Senthil Kumar M, Saran Raj I. 2018. Investigations on the performances of treated jute/Kenaf hybrid natural fiber reinforced epoxy composite. *J Polym Res*. 25(4): 1–9.

Ashik KP, Sharma RS. 2015. A review on mechanical properties of natural fiber reinforced hybrid polymer composites. *J Miner Mater Charact Eng*. 3(5): 420.

Ashok Kumar M, Ramachandra Reddy G, Siva Bharathi Y, Venkata Naidu S, Naga Prasad Naidu V. 2010. Frictional coefficient, hardness, impact strength, and chemical resistance of reinforced sisal-glass fiber epoxy hybrid composites. *J Compos Mater*. 44(26): 3195–3202.

Asumani OML, Reid RG, Paskaramoorthy R. 2012. The effects of alkali-silane treatment on the tensile and flexural properties of short fibre non-woven kenaf reinforced polypropylene composites. *Compos Part A Appl Sci Manuf*. 43(9): 1431–1440.

Atalie D, Gideon RK. 2018. Extraction and characterization of Ethiopian palm leaf fibers. *Res J Text Appar*. 22(1): 15–25.

Azwa ZN, Yousif BF. 2013. Characteristics of kenaf fibre/epoxy composites subjected to thermal degradation. *Polym Degrad Stab*. 98(12): 2752–2759.

Bharath KN and Basavarajappa S. 2016. Applications of biocomposite materials based on natural fibers from renewable resources: A review. *Sci Eng Compos Mater*. 23(2): 123–133.

Bledzki AK, Mamun AA, Faruk O. 2007. Abaca fibre reinforced PP composites and comparison with jute and flax fibre PP composites. *Express Polym Lett*. 1(11): 755–762.

Boljat HU, Grubišić N, Slišković M. 2021. The impact of nautical activities on the environment —A systematic review of research. *Sustainability*. 13(19): 10552.

Boopalan M, Niranjanaa M, Umapathy MJ. 2013. Study on the mechanical properties and thermal properties of jute and banana fiber reinforced epoxy hybrid composites. *Compos Part B Eng*. 51: 54–57.

Bujang IZ, Awang MK, Ismail AE. 2007. Study on the dynamic characteristic of coconut fibre reinforced composites. In *Regional Conference on Engineering Mathematics, Mechanics, Manufacturing & Architecture* (pp. 185–202).

Büyükkaya K. 2017. Effects of the fiber diameter on mechanic properties in polymethyl-methacrylate composites reinforced with goose feather fiber. *Mater Sci Appl*. 8(11): 811–827.

Chandrasekar M, Senthilkumar K 2021. Effect of adding sisal fiber on the sliding wear behavior of the coconut sheath fiber-reinforced composite. In *Tribology of Polymer Composites* (pp. 115–125). Elsevier.

Chandrasekar M, Siva I, Kumar TSM, Senthilkumar K, Siengchin S, Rajini N 2020. Influence of fibre inter-ply orientation on the mechanical and free vibration properties of banana fibre reinforced polyester composite laminates. *J Polym Environ* 28: 2789–2800.

Chaudhary V, Bajpai PK, Maheshwari S. 2018. Studies on mechanical and morphological characterization of developed jute/hemp/flax reinforced hybrid composites for structural applications. *J Nat Fibers*. 15(1): 80–97.

Codispoti R, Oliveira D V., Olivito RS, Lourenço PB, Fangueiro R. 2015. Mechanical performance of natural fiber-reinforced composites for the strengthening of masonry. *Compos Part B Eng*. 77: 74–83.

Davies P, Verbouwe W. 2018. Evaluation of basalt fibre composites for marine applications. *Appl Compos Mater*. 25(2): 299–308.

Dhakal HN. 2015. Mechanical performance of PC-based biocomposites. In: *Biocomposites. Design and Mechanical Performance* (pp. 303–317). Eds Misra M, Pandey JK and Amar Mohanty K.. Woodhead Publishing. Sawston, UK.

Djafar Z, Renreng I, Jannah M. 2021. Tensile and bending strength analysis of ramie fiber and woven ramie reinforced epoxy composite. *J Nat Fibers*. 18(12): 2315–2326.

Faizah A, Murdiyanto D, Widyawati YN, Dewi NL. 2018. Effects of silkworm fiber position on flexural and compressive properties of silk fiber-reinforced composites. *Majalah Kedokt Gigi*. 51(2): 57–61.

Fiore V, Di Bella G, Valenza A. 2011. Glass-basalt/epoxy hybrid composites for marine applications. *Mater Des*. 32(4): 2091–2099.

Fitzgerald A, Proud W, Kandemir A, Murphy RJ, Jesson DA, Trask RS, Hamerton I, Longana ML. 2021. A life cycle engineering perspective on biocomposites as a solution for a sustainable recovery. *Sustainability* 13(3): 1160.

Fragassa C. 2017. Marine applications of natural fibre-reinforced composites: A manufacturing case study. In: *Advances in Applications of Industrial Biomaterials* (pp. 21–47). Eds. Pellicer E, Nikolic D, Sort J, Baró M, Zivic F, Grujovic N, Grujic R, Pelemis S. Springer, Cham.

Geethika VN, Rao VDP. 2017. Study of tensile strength of Agave americana fibre reinforced hybrid composites. *Mater Today Proc*. 4(8): 7760–7769.

Gonzalez-Murillo C, Ansell MP. 2009. Mechanical properties of henequen fibre/epoxy resin composites. *Mech Compos Mater*. 45(4): 435–442.

Gupta MK, Srivastava RK. 2016. Mechanical properties of hybrid fibers-reinforced polymer composite: A review. *Polym - Plast Technol Eng*. 55(6): 626–642.

Harish S, Michael DP, Bensely A, Lal DM, Rajadurai A. 2009. Mechanical property evaluation of natural fiber coir composite. *Mater Charact*. 60(1): 44–49.

Hunter L. 2020. Mohair, cashmere and other animal hair fibres. In *Handbook of Natural Fibres* (pp. 279–383). Eds. Kozłowski RM and Mackiewicz-Talarczyk M. Woodhead Publishing. Sawston, UK.

Jamshaid H, Mishra R. 2016. A green material from rock: Basalt fiber – A review. *J Text Inst*. 107(7): 923–937.

Jayamani E, Hamdan S, Rahman MR, Bakri MKB. 2014. Investigation of fiber surface treatment on mechanical, acoustical and thermal properties of betelnut fiber polyester composites. *Procedia Eng*. 97: 545–554.

John MJ, Thomas S. 2008. Biofibres and biocomposites. *Carbohydr Polym*. 71(3):343–364.

Jumaidin R, Saidi ZAS, Ilyas RA, Ahmad MN, Wahid MK, Yaakob MY, Maidin NA, Ab Rahman MH, Osman MH. 2019. Characteristics of cogon grass fibre reinforced thermoplastic cassava starch biocomposite: Water absorption and physical properties. *J Adv Res Fluid Mech Therm Sci*. 62(1): 43–52.

Kandemir A, Longana ML, Panzera TH, del Pino GG, Hamerton I, Eichhorn SJ. 2021. Natural fibres as a sustainable reinforcement constituent in aligned discontinuous polymer composites produced by the HiPerDiF method. *Materials*. 14(8): 1885.

Kappenthuler S, Seeger S. 2021. Assessing the long-term potential of fiber reinforced polymer composites for sustainable marine construction. *J Ocean Eng Mar Energy*. 7(2): 129–144.

Karim N, Sarker F, Afroj S, Zhang M, Potluri P, Novoselov KS. 2021. Sustainable and multifunctional composites of graphene-based natural jute fibers. *Adv Sustain Syst*. 5(3): 2000228.

Kazi AM, Ramasastry DVA, Waddar S, Shaikh TM, Tamboli AA, Shaikh SAM. 2021. Mechanical characterization of stacked roselle fibre composites: Effect of fibre orientation. *Int J Veh Struct Syst*. 13(5): 678–682.

Kısmet Y, Dogan A. 2022. Characterization of the mechanical and thermal properties of rape short natural-fiber reinforced thermoplastic composites. *Iran Polym J*. 31(2): 143–151.

Kogan FM, Nikitina O V. 1994. Solubility of chrysotile asbestos and basalt fibers in relation to their fibrogenic and carcinogenic action. *Environ Health Perspect*. 102: 205–206.

Kumar N, Singh A, Debnath K, Ranjan R. 2020. Mechanical characterization of animal fibre-based composites. *Indian J Fibre Text Res*. 45(3): 293–297.

Kurien RA, Biju A, Raj KA, Chacko A, Joseph B, Koshy CP. 2022. Chicken feather fiber reinforced composites for sustainable applications. *Mater Today Proc*. 58: 862–866.

Li L, Song J, Wang Y, Du M, Wei Q, Cai Y. 2022. Fabrication and performance of shape-stable phase change composites supported by environment-friendly and economical loofah sponge fibers for thermal energy storage. *Energy Fuels*. 36(7): 3938–3946.

Li Z, Ma J, Ma H, Xu X. 2018. Properties and applications of basalt fiber and its composites. *IOP Conf Ser Earth Environ Sci*. 86: 012052.

Liang Z, Wu H, Liu R, Wu C. 2021. Preparation of long sisal fiber-reinforced polylactic acid biocomposites with highly improved mechanical performance. *Polymers*. 13(7): 1124.

Liu X, Wang L, Wang X. 2004. Evaluating the softness of animal fibers. *Text Res J*. 74(6):535–538.

Lu L, Han F, Wu S, Qin Y, Yuan G, Doh JH. 2022. Experimental study on durability of basalt fiber concrete after elevated temperature. *Struct Concr*. 23(2): 682–693.

Luo H, Xiong G, Ma C, Chang P, Yao F, Zhu Y, Wan Y. 2014. Mechanical and thermo-mechanical behaviors of sizing-treated corn fiber/polylactide composites. Polym. Test. 39: 45–52.

Lyu L, Tian Y, Lu J, Xiong X, Guo J. 2020. Flame-retardant and sound-absorption properties of composites based on kapok fiber. *Materials* 13(12): 2845.

Madhu P, Praveenkumara J, Sanjay MR, Siengchin S, Gorbatyuk S. 2021. Introduction to bio-based fibers and their composites. In *Advances in Bio-Based Fiber* (pp. 1–20). Eds. Rangappa SM, Puttegowda M, Parameswaranpillai J, Siengchin S, and Gorbatyuk S. Woodhead Publishing, Sawston.

Maghchiche A. 2013. Characterisation of esparto grass fibers reinforced biodegradable polymer composites. *Biosci Biotechnol Res Asia*. 10(2): 665–673.

Mahjoub R, Yatim JM, Mohd Sam AR, Raftari M. 2014. Characteristics of continuous unidirectional kenaf fiber reinforced epoxy composites. *Mater Des*. 64: 640–649.

Mamun AA, Heim HP, Bledzki AK. 2015. The use of maize, oat, barley and rye fibres as reinforcements in composites. In *Biofiber Reinforcements in Composite Materials* (pp. 454–487). Eds. Faruk O and Sain M. Woodhead Publishing. Sawston, UK.

Mayank BA, Sethi V, Gudwani H. 2022. Spider-silk composite material for aerospace application. *Acta Astronaut*. 193: 704–709.McGregor BA. 2012. Production, properties and processing of American bison (Bison bison) wool grown in southern Australia. *Anim Prod Sci*. 52(7): 431–435.

Melo JDD, Carvalho LFM, Medeiros AM, Souto CRO, Paskocimas CA. 2012. A biodegradable composite material based on polyhydroxybutyrate (PHB) and carnauba fibers. *Compos Part B Eng*. 43(7): 2827–2835.

Mishra HK, Dash BN, Tripathy SS, Padhi BN. 2000. A study on mechanical performance of jute-epoxy composites. *Polym Plast Technol Eng*. 39(1): 187–198.

Mitra B. 2014. Environment friendly composite materials: Biocomposites and green composites. *Def Sci J*. 64(3): 244–261.

Mittal V, Sinha S. 2017. Effect of alkali treatment on the thermal properties of wheat straw fiber reinforced epoxy composites. *J Compos Mater*. 51(3): 323–331.

Mohamed SAN, Zainudin ES, Sapuan SM, Azaman MD, Arifin AMT. 2020. Energy behavior assessment of rice husk fibres reinforced polymer composite. *J Mater Res Technol*. 9(1): 383–393.

Mohanty AK, Misra M, Hinrichsen G. 2000. Biofibres, biodegradable polymers and biocomposites: An overview. *Macromol Mater Eng*. 276–277: 1–24.

Mohanty AK, Misra M, Drzal L. 2005. *Natural Fibers, Biopolymers, and Biocomposites*. CRC Press, Florida.

Mohanty AK, Vivekanandhan S, Pin JM, Misra M. 2018. Composites from renewable and sustainable resources: Challenges and innovations. *Science* 362(6414): 536–542.

Moreau R. 2009. Nautical activities: What impact on the environment? A life cycle approach for clear blue. Commissioned by the European Confederation of Nautical Industries-ECNI.

Motaleb KZMA, Islam MS, Hoque MB. 2018. Improvement of physicomechanical properties of pineapple leaf fiber reinforced composite. *Int J Biomater*. 2018: 7384360.

Moudood A, Rahman A, Öchsner A, Islam M, Francucci G. 2019. Flax fiber and its composites: An overview of water and moisture absorption impact on their performance. *J Reinf Plast Compos*. 38(7): 323–339.

Mukherjee T, Kao N. 2011. PLA Based biopolymer reinforced with natural fibre: A review. *J Polym Environ*. 19(3):714–725.

Nabi Saheb D, Jog JP. 1999. Natural fiber polymer composites: A review. *Adv Polym Technol*. 18(4):351–363.

Narendiranath Babu T, Aravind SS, Naveen Kumar KS, Sumanth Rao MS. 2018. Study on mechanical and tribological behaviour of Angora, kenaf and ramie hybrid reinforced epoxy composites. *Int J Mech Eng Technol*. 9(4): 11–20.

Parikh HH, Gohil PP. 2017. Experimental investigation and prediction of wear behavior of cotton fiber polyester composites. *Friction*. 5(2): 183–193.

Pavlovski D, Mislavsky B, Antonov A. 2007. CNG cylinder manufacturers test basalt fibre. *Reinf Plast.* 51(4): 36–39.

Paulo P, Hugon C, Hafiz S, Marco L. 2018. Natural fibre composites and their applications: A review. *J Compos Sci.* 2(4): 66.

Prasad K, Nikzad M, Sbarski I. 2022. Mechanical, viscoelastic and gas transport behaviour of rotationally molded polyethylene composites with hard- and soft-wood natural fibres. *J Polym Res.* 29(4): 131.

Qin Y, Summerscales J, Graham-Jones J, Meng M, Pemberton R. 2020. Monomer selection for in situ polymerization infusion manufacture of natural-fiber reinforced thermoplastic-matrix marine composites. *Polymers.* 12(12): 2928.

Qiu Z, Wang J, Fan H. 2021. Low velocity flexural impact behaviors of bamboo fiber reinforced composite beams. *Polym Test.* 94: 107047.

Raharjo WW, Kusharjanto B, Triyono T. 2020. Alkali treatment effect on the tensile and impact properties of recycled high-density polyethylene composites reinforced with short cantala fiber. *J Southwest Jiaotong Univ.* 55(3):1–10.

Ramamoorthy SK, Skrifvars M, Persson A. 2015. A review of natural fibers used in biocomposites: Plant, animal and regenerated cellulose fibers. *Polym Rev.* 55(1):107–162.

Ramasamy M, Daniel AA, Nithya M, Kumar SS, Pugazhenthi R. 2020. Characterization of natural - Synthetic fiber reinforced epoxy based composite - hybridization of kenaf fiber and kevlar fiber. *Mater Today Proc.* 37: 1699–1705.

Rao DN, Mukesh G, Ramesh A, Anjaneyulu T. 2020. Investigations on the mechanical properties of hybrid goat hair and banana fiber reinforced polymer composites. *Mater Today Proc.* 27: 1703–1707.

Rao KMM, Prasad AVR, Babu MNVR, Rao KM, Gupta AVSSKS. 2007. Tensile properties of elephant grass fiber reinforced polyester composites. *J Mater Sci.* 42(9): 3266–3272.

Rao PD, Kiran CU, Prasad KE. 2018. Tensile studies on random oriented human hair fiber reinforced polyester composites. *J Mech Eng.* 47(1): 37–44.

Rao RH, P V DKP, Raju L, Rajendran RL, Kotabagi VK. 2019. Study on mechanical behaviour of banana fiber reinforced epoxy composites. *Asian J Multidiscip Res.* 5(1): 1–4.

Reddy JP, Misra M, Mohanty A. 2013. Injection moulded biocomposites from oat hull and polypropylene/polylactide blend: Fabrication and performance evaluation. *Adv Mech Eng.* 5: 761840.

Reddy N, Yang Y. 2005. Biofibers from agricultural byproducts for industrial applications. *Trends Biotechnol.* 23(1):22–27.

Routray S, Sundaray A, Pati D, Jagadeb AK. 2020. Preparation and assessment of natural fiber composites for marine application. *J Inst Eng Ser D.* 101(2): 215–221.

Rowell JE, Lupton CJ, Robertson MA, Pfeiffer FA, Nagy JA, Whit RG. 2001. Fiber characteristics of qiviut and guard hair from wild muskoxen (*Ovibos moschatus*). *J Anim Sci.* 79(7): 1670–1674.

Saba N, Tahir PM, Jawaid M. 2014. A review on potentiality of nano filler/natural fiber filled polymer hybrid composites. *Polymers* 6(8): 2247–2273.

Sair S, Oushabi A, Kammouni A, Tanane O, Abboud Y, El Bouari A. 2018. Mechanical and thermal conductivity properties of hemp fiber reinforced polyurethane composites. *Case Stud Constr Mater.* 8: 203–212.

Saravana Bavan D, Mohan Kumar GC. 2012. Morphological and thermal properties of maize fiber composites. *Fibers Polym.* 13(7): 887–893.

Sekhar VC, Pandurangadu V, Subba Rao T. 2014. Prediction of mechanical properties of polymer composites reinforced with feather fibers of "emu" bird. In *Applied Mechanics and Materials* (Vol. 592, pp. 694–699). Eds. Balasubramanian KR, Sivapirakasam SP and Anand R. Trans Tech Publications Ltd., Baach.

Selli F, Seki Y, Erdoğan ÜH. 2018. The effect of surface treatments on properties of various animal fibers as reinforcement material in composites. *Tekst ve Muhendis* 25(112): 292–302.

Senthil Muthu Kumar T, Senthilkumar K, Chandrasekar M, Karthikeyan S, Ayrilmis N, Rajini N, Siengchin S 2021. Mechanical, thermal, tribological, and dielectric properties of biobased composites. In *Biobased Composites: Processing, Characterization, Properties, and Applications* (pp. 53–73). Eds. Khan A, Rangappa SM, Siengchin S and Asiri AM. John Wiley & Sons, Inc., New Jersey.

Senthilkumar K, Saba N, Chandrasekar M, Jawaid M, Rajini N, Siengchin S, Ayrilmis N, Mohammad F, Al-Lohedan HA 2021. Compressive, dynamic and thermo-mechanical properties of cellulosic pineapple leaf fibre/polyester composites: Influence of alkali treatment on adhesion. *Int J Adhes Adhes* 106: 102823.

Senthilkumar K, Siva I, Jappes JTW, Amico SC, Cardona F, Sultan MTH 2016. Effect of inter-laminar fibre orientation on the tensile properties of sisal fibre reinforced polyester composites. *IOP Conf Ser Mater Sci Eng* 152: 12055.

Senthilkumar K, Siva I, Rajini N, Jeyaraj P 2015. Effect of fibre length and weight percentage on mechanical properties of short sisal/polyester composite. *Int J Comput Aided Eng Technol* 7: 60.

Senthilkumar K, Subramaniam S, Ungtrakul T, Kumar TSM, Chandrasekar M, Rajini N, Siengchin S, Parameswaranpillai J 2022. Dual cantilever creep and recovery behavior of sisal/hemp fibre reinforced hybrid biocomposites: Effects of layering sequence, accelerated weathering and temperature. *J Ind Text* 51: 2372S–2390S.

Serra-Parareda F, Tarrés Q, Delgado-Aguilar M, Espinach FX, Mutjé P, Vilaseca F. 2019. Biobased composites from biobased-polyethylene and barley thermomechanical fibers: Micromechanics of composites. *Materials* 12(24): 4182.

Serra-Parareda F, Vilaseca F, Aguado R, Espinach FX, Tarrés Q, Delgado-aguilar M. 2021. Effective Young's modulus estimation of natural fibers through micromechanical models: The case of henequen fibers reinforced-PP composites. *Polymers* 13(22): 3947.

Shahroze RM, Chandrasekar M, Senthilkumar K, Senthil Muthu Kumar T, Ishak MR, Rajini N, Siengchin S, Ismail SO 2021. Mechanical, interfacial and thermal properties of silica aerogel-infused flax/epoxy composites. *Int Polym Proc* 36: 53–59.

Shanmugasundaram N, Rajendran I, Jayaraj M, Aravindhaguru I. 2018. Mechanical and water absorption behavior of continuous untreated and Alkali treated areca Palm leaf stalk fiber reinforced polymer composites. *Ecol Environ Conserv*. 24: 63–66.

Shi J, Wang X, Wu Z, Zhu Z. 2017. Fatigue behavior of basalt fiber-reinforced polymer tendons under a marine environment. *Constr Build Mater*. 137: 46–54.

Siddique SH, Faisal S, Ali M, Gong RH. 2021. Optimization of process variables for tensile properties of bagasse fiber-reinforced composites using response surface methodology. *Polym Polym Compos*. 29(8): 1304–1312.

Sivakumar D, Ng LF, Zalani NFM, Selamat MZ, Ab Ghani AF, Fadzullah SHSM. 2018. Influence of kenaf fabric on the tensile performance of environmentally sustainable fibre metal laminates. *Alexandria Eng J*. 57(4): 4003–4008.

Sivasaravanan S, Bupesh Raja VK, Avinash Babu K, Chandra Mouli B. 2019. Mechanical characterization of GFRP/CFRP/natural fiber laminated in epoxy resin composite. *Mater Today Proc*. 16: 934–938.

Soykan U. 2022. Development of turkey feather fiber-filled thermoplastic polyurethane composites: Thermal, mechanical, water-uptake, and morphological characterizations. *J Compos Mater*. 56(2): 339–355.

Straznicky P, Rusnakova S, Zaludek M, Bosak O, Kubliha M, Gross P. 2020. The technological properties of polymer composites containing waste sheep wool filler. *Mater Sci Forum*. 994: 170–178.

Sukanto H, Raharjo WW, Ariawan D, Triyono J. 2021. Carbon fibers recovery from CFRP recycling process and their usage: A review. *IOP Conf Ser Mater Sci Eng.* 1034(1): 012087.

Tawfik ME, Eskander SB, Nawwar GAM. 2017. Hard wood-composites made of rice straw and recycled polystyrene foam wastes. *J Appl Polym Sci.* 134(18): 1–9.

Thomas SK, Parameswaranpillai J, Krishnasamy S, Begum PMS, Nandi D, Siengchin S, George JJ, Hameed N, Salim Nisa V, Sienkiewicz N 2021. A comprehensive review on cellulose, chitin, and starch as fillers in natural rubber biocomposites. *Carbohydr Polym Technol Appl* 2: 100095.

Torres-Arellano M, Renteria-Rodríguez V, Franco-Urquiza E. 2020. Mechanical properties of natural-fiber-reinforced biobased epoxy resins manufactured by resin infusion process. *Polymers* 12(12): 2841.

Van Den Oever MJA, Elbersen HW, Keijsers ERP, Gosselink RJA, De Klerk-Engels B. 2003. Switchgrass (*Panicum virgatum* L.) as a reinforcing fibre in polypropylene composites. *J Mater Sci.* 38(18): 3697–3707.

Vinoth N, Rajkumar K, Kumar RS, Mohanavel V, Ravichandran M, Sathish T, Subbiah R. 2021. Tensile and impact strength of alpaca fiber epoxy matrix hybrid composites prepared by injection moulding process. *J Phys Conf Ser.* 2027: 012011.

Walte AB, Bhole K, Gholave J. 2020. Mechanical characterization of coir fiber reinforced composite. *Mater Today Proc.* 24: 557–566.

Wei B, Cao H, Song S. 2010a. Tensile behavior contrast of basalt and glass fibers after chemical treatment. *Mater Des.* 31(9): 4244–4250.

Wei B, Cao H, Song S. 2010b. RETRACTED: Environmental resistance and mechanical performance of basalt and glass fibers. *Mater Sci Eng A.* 527: 4708–4715.

Wutabachew M, Alemu DN. 2019. Investigation of mechanical properties of horse hair and glass fiber rein forced hybrid polymer composite. *Int J Mech Eng.* 6(5): 32–40.

Zhou H, Jia B, Huang H, Mou Y. 2020. Experimental study on basic mechanical properties of basalt fiber reinforced concrete. *Materials.* 13(6): 1362.

Index

For Product Safety Concerns and Information please contact our EU
representative GPSR@taylorandfrancis.com
Taylor & Francis Verlag GmbH, Kaufingerstraße 24, 80331 München, Germany

www.ingramcontent.com/pod-product-compliance
Lightning Source LLC
Chambersburg PA
CBHW060808220326
41598CB00022B/2569